Lecture Notes in Networks and Systems

Volume 603

The series "Lecture Notes in Networks and Systems" publishes the latest developments in Networks and Systems—quickly, informally and with high quality. Original research reported in proceedings and post-proceedings represents the core of LNNS.

Volumes published in LNNS embrace all aspects and subfields of, as well as new challenges in, Networks and Systems.

The series contains proceedings and edited volumes in systems and networks, spanning the areas of Cyber-Physical Systems, Autonomous Systems, Sensor Networks, Control Systems, Energy Systems, Automotive Systems, Biological Systems, Vehicular Networking and Connected Vehicles, Aerospace Systems, Automation, Manufacturing, Smart Grids, Nonlinear Systems, Power Systems, Robotics, Social Systems, Economic Systems and other. Of particular value to both the contributors and the readership are the short publication timeframe and the world-wide distribution and exposure which enable both a wide and rapid dissemination of research output.

The series covers the theory, applications, and perspectives on the state of the art and future developments relevant to systems and networks, decision making, control, complex processes and related areas, as embedded in the fields of interdisciplinary and applied sciences, engineering, computer science, physics, economics, social, and life sciences, as well as the paradigms and methodologies behind them.

Indexed by SCOPUS, INSPEC, WTI Frankfurt eG, zbMATH, SCImago.

All books published in the series are submitted for consideration in Web of Science.

For proposals from Asia please contact Aninda Bose (aninda.bose@springer.com).

Vicente Julián · João Carneiro · Ricardo S. Alonso · Pablo Chamoso · Paulo Novais
Editors

Ambient Intelligence—Software and Applications—13th International Symposium on Ambient Intelligence

Springer

Editors
Vicente Julián
Department of Computer Systems
and Computation, Institut Valencià
d'Intel·ligència Artificial
Universitat Politècnica de València
Valencia, Spain

Ricardo S. Alonso
Deep Tech Lab
IoT Digital Innovation Hub
AIR Institute
Valladolid, Spain

Paulo Novais
Departamento de Informática
ALGORITMI Centre
University of Minho
Braga, Portugal

João Carneiro
Research Group on Intelligent Engineering
and Computing for Advanced Innovation
and Development (GECAD)
ISEP/IPP—Engineering Institute
Polytechnic of Porto
Porto, Portugal

Pablo Chamoso
Bioinformatics Intelligent Systems
and Educational Technology (BISITE)
Research Group, IoT Digital Innovation
Hub, Department of Computer Science
Faculty of Science
University of Salamanca
Salamanca, Spain

ISSN 2367-3370 ISSN 2367-3389 (electronic)
Lecture Notes in Networks and Systems
ISBN 978-3-031-22355-6 ISBN 978-3-031-22356-3 (eBook)
https://doi.org/10.1007/978-3-031-22356-3

This Springer imprint is published by the registered company Springer Nature Switzerland AG
The registered company address is: Gewerbestrasse 11, 6330 Cham, Switzerland

Organization

General Chair

Paulo Novais	Universidade do Minho, Portugal

Program Chair

Vicente Julián	Technical University of Valencia, Spain

Organizing Committee Chairs

Ricardo S. Alonso	AIR Institute, Spain
João Carneiro	ISEP/GECAD, Portugal
Pablo Chamoso	University of Salamanca, Spain

Program Committee

Héctor Alaiz Moretón	University of León, Spain
Ana Almeida	ISEP, Portugal
Cesar Analide	University of Minho, Portugal
Francisco Bellas	Universidade da Coruña, Spain
Carlos Bento	University of Coimbra, Portugal
José Luis Calvo Rolle	Universidade da Coruña, Spain
Valerie Camps	University of Toulouse III—IRIT, France
Javier Carbo	Universidad Carlos III de Madrid, Spain

Davide Carneiro	CIICESI/ESTG, Polytechnic Institute of Porto, Portugal
Fabio Cassano	Università degli Studi di Bari Aldo Moro, Italy
Álvaro Castro-González	Universidad Carlos III de Madrid, Spain
Silvio Cesar Cazella	UFCSPA, Brazil
Pablo Chamoso	University of Salamanca, Spain
Stefano Chessa	Department of Computer Science, University of Pisa, Italy
Stephanie Combettes	University of Toulouse III—IRIT, France
Luis de la Fuente Valentín	Universidad Internacional de La Rioja, Spain
Patricio Domingues	ESTG—Leiria, Portugal
John Dowell	University College London, UK
Luiz Faria	Institute of Engineering—Polytechnic of Porto, Portugal
Florentino Fdez-Riverola	University of Vigo, Spain
Marta Fernandes	GECAD—Polytechnic of Porto, Portugal
Antonio Fernández-Caballero	Universidad de Castilla-La Mancha, Spain
Adina Magda Florea	University Politehnica of Bucharest, AI-MAS Laboratory, Romania
Daniela Fogli	Università di Brescia, Italy
Óscar García	Universidad Internacional de La Rioja, Spain
Jose García-Alonso	University of Extremadura, Spain
Marco Gomes	University of Minho, Portugal
Celestino Goncalves	Instituto Politecnico da Guarda, Portugal
Sérgio Gonçalves	University of Minho, Portugal
Alfonso González Briones	University of Salamanca, Spain
Junzhong Gu	East China Normal University, China
Esteban Guerrero	School of Technology and Innovations, Computer Science, University of Vaasa, Finland
Hans W. Guesgen	Massey University, New Zealand
Guillermo Hernández	University of Salamanca, Spain
Miguel Hornos	University of Granada, Spain
Carlos A. Iglesias	Universidad Politécnica de Madrid, Spain
Esteban Jove Pérez	UDC, Spain
Alexander Kocian	University of Pisa, Italy
Igor Kotenko	St. Petersburg Federal Research Center of the Russian Academy of Sciences, Russia
Afonso Lima	ABLN, Brazil
Guillaume Lopez	Aoyama Gakuin University, College of Science and Technology, Japan
José Machado	University of Minho, Portugal
Gonçalo Marques	Polytechnic of Coimbra, Portugal
Constantino Martins	Knowledge Engineering and Decision Support Research (GECAD)—Institute of Engineering—Polytechnic of Porto, Portugal

Rene Meier	Lucerne University of Applied Sciences, Switzerland
José Pascual Molina	Universidad de Castilla-La Mancha, Spain
Andrés Muñoz Ortega	Universidad de Cádiz, Spain
Tatsuo Nakajima	Waseda University, Japan
Elena Navarro	University of Castilla-La Mancha, Spain
Paulo Novais	University of Minho, Portugal
Hugo Peixoto	University of Minho, Portugal
Antonio Pereira	ESTG, Polytechnic of Porto, Portugal
Francisco Pinto	University of Salamanca, Spain
Tiago Pinto	GECAD Research Group, School of Engineering, Polytechnic of Porto, Portugal
António Pinto	ESTG, Polytechnic of Porto, Portugal
Isabel Praça	GECAD/ISEP, Portugal
Héctor Quintián	Universidade da Coruña, Spain
Joao Ramos	School of Management and Technology—Polytechnic Institute of Porto (IPP), Portugal
Luís Paulo Reis	APPIA, University of Porto/LIACC, Portugal
Jaime A. Rincón	Universitat Politècnica de València, Spain
Iván Rodríguez Conde	University of Arkansas at Little Rock, USA
Altino Sampaio	Instituto Politécnico do Porto, Escola Superior de Tecnologia e Gestão de Felgueiras, Portugal
Fernando Silva	School of Technology and Management, Computer Science and Communications Research Centre, Polytechnic of Leiria, Portugal
Radu-Daniel Vatavu	University Stefan cel Mare of Suceava, Romania
Gianni Vercelli	DIBRIS—University of Genoa, Italy
Raúl López-Blanco	University of Salamanca, Spain

Local Organizing Committee

Pierpaolo Vittorini (Co-chair)	University of L'aquila, Italy
Tania Di Mascio (Co-chair)	University of L'aquila, Italy
Federica Caruso	University of L'aquila, Italy
Anna Maria Angelone	University of L'aquila, Italy

Organizing Committee

Juan M. Corchado Rodríguez	University of Salamanca, Spain AIR Institute, Spain
Fernando de la Prieta	University of Salamanca, Spain

Sara Rodríguez-González	University of Salamanca, Spain
Javier Prieto Tejedor	University of Salamanca, Spain AIR Institute, Spain
Pablo Chamoso Santos	University of Salamanca, Spain
Liliana Durón	University of Salamanca, Spain
Belén Pérez Lancho	University of Salamanca, Spain
Ana Belén Gil González	University of Salamanca, Spain
Ana De Luis Reboredo	University of Salamanca, Spain
Angélica González Arrieta	University of Salamanca, Spain
Emilio S. Corchado Rodríguez	University of Salamanca, Spain
Alfonso González Briones	University of Salamanca, Spain
Yeray Mezquita Martín	University of Salamanca, Spain
Beatriz Bellido	University of Salamanca, Spain
María Alonso	University of Salamanca, Spain
Sergio Márquez Sánchez	University of Salamanca, Spain
Marta Plaza-Hernández	University of Salamanca, Spain
Guillermo Hernández González	University of Salamanca, Spain
Ricardo S. Alonso	University of Salamanca, Spain AIR Institute, Spain
Raúl López-Blanco	University of Salamanca, Spain
Sergio Alonso	University of Salamanca, Spain
Andrea Gil	University of Salamanca, Spain
Javier Parra-Domínguez	University of Salamanca, Spain

Preface

Ambient Intelligence (AmI) represents a vision of a future in which users will be surrounded by environments endowed with devices with specific intelligent capabilities, sensitive to people and reacting to them. AmI technologies combine concepts of ubiquitous computing and intelligent systems, placing people at the center of all developments, not only in technology but also in the different disciplines related to people's existence (health, socio-economics, psychology, well-being, pedagogy, etc.).

The main goal of AmI is to facilitate and improve people's lives by collecting vast amounts of information, analyzing it, and providing an exclusive personalized environment better suited to its occupants. Thanks to AmI, it is possible to achieve the people's well-being and a new, more friendly, rational, sustainable, and secure relationship between the individual and their environment.

In this sense, ISAmI is the International Symposium on Ambient Intelligence, which seeks to bring together scientists working in all areas related to AmI, as seen in the papers included in the conference. This year, ISAmI is in its 13th edition, including several papers focused on the different key aspects of the AmI. Among others, the reader can find papers on security, quality of life, smart homes, and affective computing, which are some examples of the most ambitious expressions of AmI.

In total, 24 articles have been selected from a total of 37 submissions from many different countries (Ecuador, India, Indonesia, Italy, Japan, Nigeria, Portugal, Romania, Spain, and Vietnam).

As in previous editions, there are special issues in JCR-ranked journals such as the *Journal of Interactive Multimedia and Artificial Intelligence*, *Electronics*, or *Smart Cities Journal*.

The symposium is organized by the Universidad de Salamanca, the Universitat Politècnica de València, the Instituto Superior de Engenharia do Porto, and the Universidade do Minho. The present edition was held at the University of L'Aquila, Italy, from July 13 to 15, 2022.

We would like to thank all colleagues who submitted their work, all the members of the Program Committee for their work in promoting the ISAmI, who reviewed papers, and those who helped us find reviewers. Without their efforts, it would have been impossible to put together such an interesting and actual program.

Finally, we appreciate the efforts of local organizers, for hosting and organizing these events, without which ISAmI could not have taken place.

Vicente Julián
João Carneiro
Ricardo S. Alonso
Pablo Chamoso
Paulo Novais

Contents

Deep Learning Based Automated Chest X-ray Abnormalities Detection

Vraj Parikh[1], Jainil Shah[1], Chintan Bhatt[1(✉)], Juan M Corchado[2], and Dac-Nhuong Le[3]

[1] U & P U. Patel Department of Computer Engineering, Charotar University of Science and Technology (CHARUSAT), Gujarat, India
chintanbhatt.ce@charusat.ac.in

[2] BISITE Research Group, University of Salamanca, 37007 Salamanca, Spain
corchado@usal.es

[3] Faculty of Information Technology Haiphong University, Haiphong, Vietnam
Nhuongld@dhhp.edu.vn

Abstract. Abnormalities related to the chest are a fairly common occurrence in infants as well as adults. The process of identifying these abnormalities is relatively easy but the task of actually classifying them into specific labels pertaining to specific diseases is a much harder endeavour. COVID-19 sufferers are multiplying at an exponential rate, putting pressure on healthcare systems all around the world. Because of the limited number of testing kits available, it is impractical to test every patient with a respiratory ailment using traditional methods. Thus in such dire circumstances, we propose the use of modern deep learning techniques to help in the detection and classification of a number of different thoracic abnormalities from a chest radiograph. The goal is to be able to automatically identify and localize multiple points of interest in a provided chest X-ray and act as a second level of certainty after the radiologists. On our publically available chest radiograph dataset, our methods resulted in a mean average precision of 0.246 for the detection of 14 different thoracic abnormalities.

Keywords: Deep learning · Chest x-rays · Pulmonary diseases · Object detection · Chest abnormalities

1 Introduction

A chest X-ray is an imaging test that utilises low doses of radiation in short blasts to create images of the inside of a patient's chest. They're the most commonly recommended medical images. Any improvement in the reading and reporting of these pictures has the potential to have a significant influence on the radiology

V. Julián et al. (Eds.): ISAmI 2022, LNNS 603, pp. 1–12, 2023.
https://doi.org/10.1007/978-3-031-22356-3_1

workflow. The availability of a variety of open-source datasets for disease identification in chest X-rays has sparked an increase in academic research interest in AI model building for disease detection in chest X-rays. Researchers have succeeded in labelling X-ray pictures with global labels (such as "Consolidation"), but not in mapping them to a specific lung location. As a result, developing interpretable AI models for detecting disease on chest X-rays remains a tough and unsolved research problem [14].

ChestX-ray8, ChestX-ray1410, Padchest11, CheXpert3, and MIMIC-CXR12 are some of the most well-known public CXR datasets. The US National Institutes of Health (NIH) released ChestX-ray14 [16], an expanded version of ChestX-ray8, comprising over 112,000 CXR scans from over 30,000 individuals. This dataset has serious difficulties with the quality of its labels because it has not been manually annotated. CheXNet [13] is one such simple model architecture trained on ChestX-ray14 giving a much better performance. 4. Significant progress has been achieved in medical science in recent years as a result of AI and deep learning, which now allows doctors to diagnose diseases early and readily, which was previously tedious and time-consuming. Hence in our study with the help of the chest radiographs dataset, we provide a method for classifying and localizing various different thoracic abnormalities using various convolutional neural network model architectures.

It can work as a pair of second eyes to assist and validate the base finding that a radiologist might make. The model will be able to automatically identify and localize multiple points of interest in the data provided to it and may be integrated in conjunction with other similar operations to increase the overall efficiency in medical diagnosis and procedures. This project also provides understanding of the social, gender and environmental influences.

2 Related Works

Since the research was a part of a Kaggle Competition there were various other teams that also collaborated on the dataset and produced commendable results on the dataset. The team having notebook name "VinBigData complete pipeline" scored 0.286 and used a 2-class classifier complete pipeline. Different things done by them includes, using "timm" (pytorch-image-models, providing a lot of popular SoTA CNN models with pretrained weights). They also used "Pytorch Ignite" (Training/Evaluation abstraction framework on top of pytorch) and "Pytorch pfn extras" which is used to add more feature-rich functionality on Ignite Useful for logging, printing, evaluating, saving the model, scheduling the learning rate during training.

This has been a subject of study ever since the 1960s, when the first papers describing an automated abnormality detection system using CXR images were published [10]. The method of choice for image processing tasks in recent years has been deep learning, which has had a significant impact on the area of medical imaging. The publishing of multiple sizable labelled databases in recent years has benefited the CXR research community, primarily made possible by the development of labels by computerised parsing of radiological reports. With the publication of 112,000 photos from the NIH clinical centre, this trend got underway in 2017. More than 755,000 pictures were made available in 3 labelled datasets in 2019 alone (CheXpert, MIMIC-CXR, PadChest) [3, 9, 10].

Among the most well-known public CXR datasets are MIMIC-CXR12, Padchest11, CheXpert3, ChestX-ray1410, ChestX-ray8, ChestX-ray8, and ChestX-ray1410. The US National Institutes of Health (NIH) created ChestX-ray14, an updated version of ChestX-ray8, which includes more than 112,000 CXR scans from more than 30,000 people. This dataset has serious problems with the accuracy of its labels without being manually annotated. CheXNet is one such basic model architecture that performs significantly better after being trained on ChestX-ray14.

Deep learning for medical image processing and computer-aided diagnosis for CXR have both been the subject of previous reviews [2]. Contrarily, recent analyses of deep learning in chest radiography are by no means thorough in terms of the literature and research methodologies, the descriptions of the publicly accessible datasets, or the evaluation of the field's potential and advancements.

Significant progress has been achieved in medical science in recent years as a result of AI and deep learning, which now allows doctors to diagnose diseases early and readily, which was previously tedious and time-consuming. Hence, in our study with the help of the chest radiographs dataset, we provide a method for classifying and localizing various different thoracic abnormalities using various convolutional neural network model architectures.

3 Dataset Specifics

The main goal was to automatically identify and categorise 14 different types of thoracic anomalies from chest radiographs. We used a dataset with 18,000 images that was labelled by radiologists with experience. Our model can be trained on 15,000 images that have been independently labelled, and it will be assessed on a test set of 3,000 images. Through VinLab, a web-based tool from VinBigData, these annotations were gathered. You can read more about how they created the dataset in their most recent work [12]. We have used a public dataset hosted on Kaggle in the form of a competition. The dataset is available in DICOM format with a total size of 195 GB. DICOM files are implementations of a predetermined medical file standard that packages the original radiograph images along with the relevant metadata might be useful for visualizing and classifying. Any DICOM medical image consists of two parts-a header and the actual image itself. The header consists of data that describes the image, the most important being patient data. The above specified original dataset can be found here [8] and the resized dataset wherein all the original images have been scaled down to a 256×256 format can be found here [18]. All images were labeled for the presence of 14 critical radiographic findings as listed below:

1. Aortic enlargement
2. Atelectasis
3. Calcification
4. Cardiomegaly
5. Consolidation
6. ILD
7. Infiltration
8. Lung Opacity
9. Nodule/Mass
10. Other lesion
11. Pleural effusion
12. Pleural thickening
13. Pneumothorax
14. Pulmonary fibrosis
15. "No finding" observation was intended to capture the absence of all findings above

4 Approach

4.1 Exploratory Data Analysis

We started by performing exploratory data analysis on the available dataset to gather some important insights to the data and to summarize and visualize the main characteristics gathered from the dataset in Fig. 1.

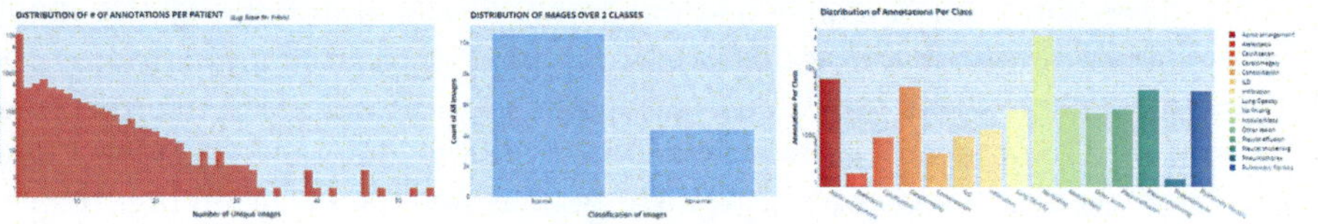

Fig. 1. The distribution of annotations per patient

The **image_id** column contains a Unique IDentifier (UID) that indicates which patient the respective row (object) relates to. As there can be up to three radiologists annotating the same image and potentially multiple objects/bboxes per image, it is possible for a single image UID to occur many times. However, it is stated in the competition data details that there exists only one image for one patient. This means that if a specific **image_id** appears 12 times, that there are 4 objects in the image, and each object was annotated by all three radiologists. The following is a visualization of the total number of objects/annotations per images as per the dataset. We use a log y-scale to essentially normalize the final values for better visualization.

Gained Insights: Each image has at least 3 annotations including the No Finding class i.e. one each from a radiologist. The majority of images (10000 out of 15000) have 3 annotations associated with them. The maximum possible annotations on an image from the dataset are 57. The data columns specify the relevant class ids and class names corresponding to the annotations. There is a possibility to detect 15 classes including the "Normal" class which we just saw is in abundance. To identify the distribution of the different annotated classes over the complete group of training images, we can employ a bar graph. We again use a logarithmic scale on the y-axis.

Gained Insights: Pneumothorax and Atelectatsis are the target abnormality classes with the least number of annotations in the training subset. Aortic Enlargement and Cardiomegaly are the target abnormality classes with the most number of annotations in the training subset. The `rad_id` column of the dataframe is determinant of the particular radiologist that made that particular ground truth annotations. We have already derived that each image has been annotated by 3 distinct radiologists each identified by a unique id. Since there are 17 radiologists in total, each has been denoted by an id between R1 and R17. However one great issue with having a large number of radiologists annotate these radiographs independently of each other, the labels and bounding-boxes can greatly differ from one radiologist to the next even for a single image. Furthermore we can deduce the distribution of different radiologists with respect to amount of annotations they made pertaining to a specific target class. We can figure out how the annotations of different classes were divided among the available radiologists using the following visualizations.

Gained Insights: The three radiologists with the highest number of annotations made have also annotated the majority of the "Abnormal" class or the classes from 1 to 14 denoting a specific target abnormality class. Among the other 14 radiologists, a majority of them have only ever annotated images as "No finding" or "Normal" class. This might allow us to estimate that radiologists other than R8, R9, and R10, are much more likely to annotate images as No finding.

4.2 Data Pre-processing

We started our data pre-processing by creating helper functions to read a DICOM file and to convert the raw DICOM data into a human readable format by applying a VOI LUT transformation on the numpy array. The images were also converted into a 3-channel format with normalization of the values between 0 and 1 to assist in the scaling down of the images prior to providing them as inputs to a deep learning neural network model during the training or evaluation of the model.

We also created a few visualizing modules to help in drawing the annotated boxes on the original image. Here we came across our first issue with the overlapping of various different annotations by different radiologists creating an erratic ground truth boxes compilation which would cause an overflow of false information during the training phase thus bringing down the mean average precision.

Overlapping Annotations : One of the major problems as discussed above is the variation in annotations made by different radiologists on the same image thus leading to contradicting and overlapping bounding boxes which will essentially pass wrong ground truth values to the model thus gaining wrong inference (see Fig. 2).

One solution to this problem is to employ the IoU function i.e. ratio of intersection over union, over the overlapping bounding boxes (see Fig. 3).

Several important parameters for performance evaluation in target detection models are IoU, precision and recall. The IoU serves as a very important function of the performance mean average precision(mAP) calculation of the target

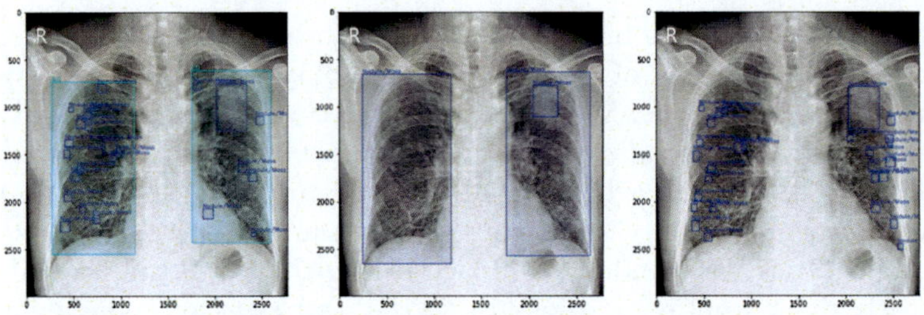

Fig. 2. Bounding boxes grouped by radiologists

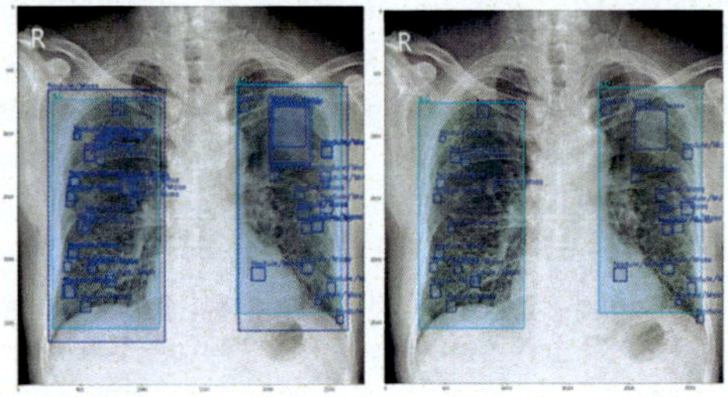

Fig. 3. Original boxes versus reduced boxes

detection algorithm. Mean Average Precision is essentially a metric to gauge the performance of object detection tasks, by taking the mean of precision-recall slope calculated over different IOUs between the ground and predicted bounding boxes. We further elaborate on this technique while discussing the results.

The overview of IOU is that when the IoU is calculated for a pair of overlapping bounding boxes and it is greater than a certain threshold, a new bounding box i.e. annotation is created from the common area between the two original boxes. This not only reduces the amount of excess bounding boxes in the image but also simplifies the annotations by essentially narrowing down the overall feature detection to a smaller area. The target class of the resulting box is assigned as the target class of the larger box out of the original boxes.

Data Augmentations is a strategy for artificially increasing the quantity and complexity of existing data [4]. We know that training a deep neural network needs a large amount of data to fine-tune the parameters (see Fig. 4).

We applied data augmentation techniques on our training dataset by adding modifications to our images by making minor changes, such as flipping, rescaling, rotation, blurring and brightness and contrast. It will increase our training data size and our model will consider each of these small changes as a distinct image, and it will enable our model to learn better and perform well on unseen data.

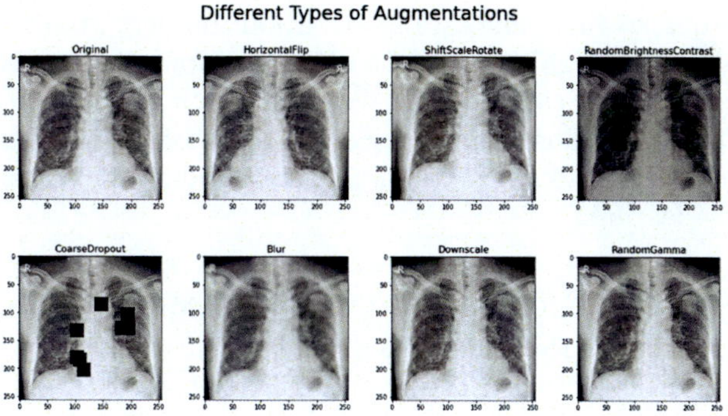

Fig. 4. Data augmentations over dataset

4.3 Proposed CNN Architectures

ResNet50/ResNet101: ResNet, short for Residual Network is a specific type of neural network that was introduced in [6]. It was built on the concept of residual functions.

ResNet50 is a 50-layer Residual Network with 26M parameters. The network can take the input image having height, width as multiples of 32 and 3 as channel width. Every ResNet architecture performs the initial convolution and max-pooling using 7×7 and 3×3 kernel sizes respectively. Afterward, Stage 1 of the network starts and it has 3 Residual blocks containing 3 layers each. The size of kernels used to perform the convolution operation in all 3 layers of the block of stage 1 are 64, 64 and 128 respectively. For deeper networks like ResNet50, ResNet152, etc., bottleneck design is used. For each residual function F, 3 layers are stacked one over the other. The three layers are 1×1, 3×3, 1×1 convolutions. The 1×1 convolution layers are responsible for reducing and then restoring the dimensions. The 3×3 layer is left as a bottleneck with smaller input/output dimensions. Finally, the network has an Average Pooling layer followed by a fully connected layer having 1000 neurons (ImageNet class output) (see Fig. 5).

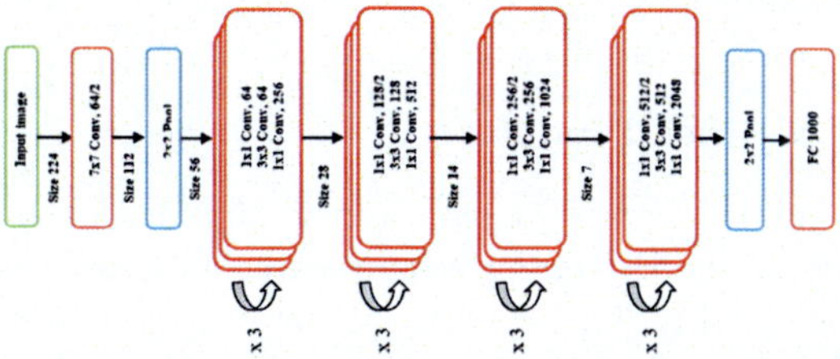

Fig. 5. ResNet50 model architecture

We employed the Detectron2 framework [17] to use a pre-trained ResNet50 model in our experiment and fine-tuned it. The Detectron2 framework contains pytorch high-quality implementations of various modern architectures and their variants. It also has a highly function API which allows both abstraction and customization to create efficient models. We format out augmented data in dataset dictionaries and start the training process. The hyperparameters tuned included the number of ROIs detected in an image, the number of iterations made by the model and the learning rate.

YoloV5 : Unlike the models in the R-CNN family, YoloV5 is a single stage detector model architecture [15]. What this actually means is that instead of proposing a bunch of regions of interest, the model directly runs detection directly over a dense sampling of possible locations. It reframes object detection as a single regression problem, straight from image pixels to bounding box coordinates and class probabilities. This enables the model to decrease its training time and perform inference faster thus becoming more simpler with a slight tradeoff over accuracy and performance. The base model is similar to GoogLeNet with inception module replaced by 1×1 and 3×3 conv layers. The final prediction of shape $S \times S \times (5B + K)$ is produced by two fully connected layers over the whole conv feature map (see Fig. 6).

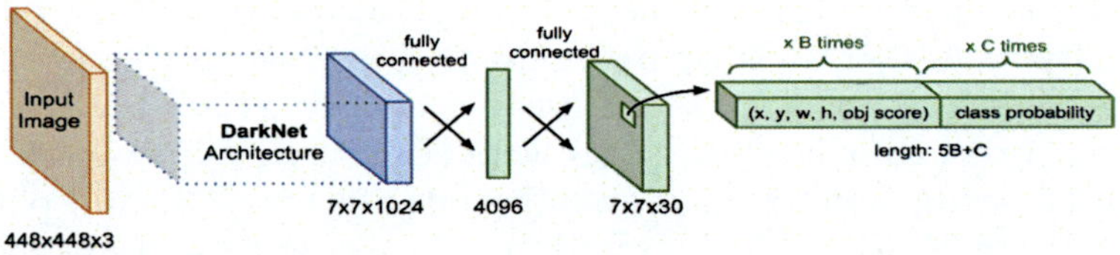

Fig. 6. YoloV5 model architecture

We used the large YoloV5 model architecture after pre-processing the data to obtain annotations in COCO format and dividing the original dataset in specific directories of images and their specific labels in text files with the same name. We hyper-tuned the parameters in the config file and ran the model for 40 epochs with a batch-size of 4.

4.4 2-Class Classifier Pipeline

In order to increase the overall accuracy of the pipeline, we add a 2-class classifier on top of the original model. This classifier can classify an image into two classes i.e. Normal or the "No Finding" target class or Abnormal which is all of the other target classes. Two thresholds are chosen arbitrarily each representing a lower and a higher limit. The final predictions are modified as follows:

- If the prediction on Normal is greater than the maximum threshold, the final image is marked with a "No Finding" prediction and all the previous detected predictions are dropped.

- If the prediction is between the limits, an extra prediction of "xNo Finding" class is appended to the existing detected predictions.
- If the prediction is less than the minimum threshold, no changes are made to the existing predictions.

EfficientNet is developed as a new baseline network by performing a neural architecture search using the AutoML MNAS framework, which optimizes both accuracy and efficiency (FLOPS) [5]. The resulting architecture uses mobile inverted bottleneck convolution (MBConv), similar to MobileNetV2 and MnasNet, but is slightly larger due to an increased FLOP budget (see Fig. 7).

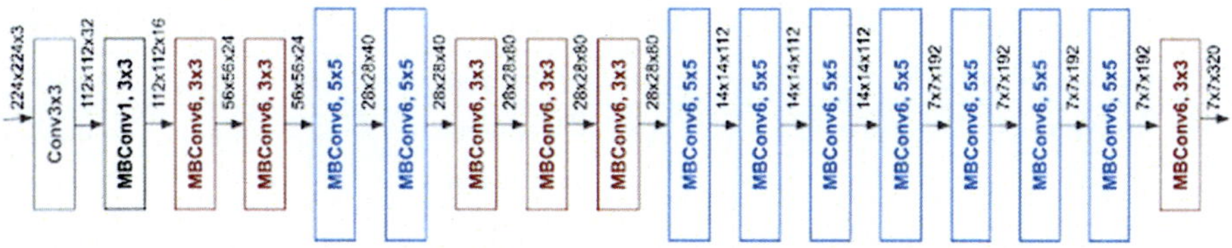

Fig. 7. EfficientNet-B0 model architecture

We developed a custom model with a pre-trained Tensorflow and Keras implementation of EfficientNet as the base model with customized dense layers with ReLU and Sigmoid as the activation methods along with a dropout.

ResNet-18 A variation of the Resnet model family includes 4 convolutional layers in each module (excluding the 1×1 convolutional layer). Together with the first 7×7 convolutional layer and the final fully-connected layer, there are 18 layers in total. Therefore, this model is commonly known as ResNet-18 [1,7,11]. We use a pytorch implementation of the ResNet-18 model and fine-tuned the model according to the target data to avoid overfitting.

5 Results

The paper uses mean average precision (mAP) as the evaluation metric. It acquires a higher preference since this is an object detection task primarily utilizing the PASCAL Visual Object Classes (VOC) as bounding box annotations. To obtain the mAP for a set of object predictions, we calculate the ratio of the precision and recall as average precision i.e. the area under the precision-recall curve, and then take a mean of the various iterations of AP thus producing the name mAP.

We experimented with various combinations of model architectures with various data pipelines inclusive of data augmentations and reduced bounding boxes to find the best suitable model for the current task. Our hardware configurations includes 32GB RAM DDR4 2400 MHz, CPU Intel i7 9750 H 2.6 GHz and

Nvidia GeForce RTX 2070 with Max-Q design 8 Gb. During the training, a validation set was used to validate the model predictions before gathering actual predictions on the unlabeled test dataset. After the various combinations, the final submissions on the test data as validated by Kaggle itself can be used to compare the different architectures as follows:

Table 1. Proposed training pipeline implementations

Model	Data Augmentation	Annotations	2-Class Classifier	mAP
ResNet50 - FPN	No	Original	–	0.136
ResNet50 - FPN	Yes	Original	–	0.044
ResNet50 - FPN	No	Reduced	–	0.092
ResNet101 - FPN	Yes	Original	–	0.067
ResNet50 - FPN	Yes	Reduced	–	0.040
YoloV5	No	Original	–	0.110
YoloV5	Yes	Original	–	0.110
YoloV5	No	Original	EfficientNet-B0	0.198
YoloV5	No	Original	Resnet-18	0.079
ResNet50 - FPN	No	Original	EfficientNet-B0	0.246
ResNet50 - FPN	No	Original	Resnet-18	0.198

(HorizontalFlip: "p": 0.4; ShiftScaleRotate: "scale_limit": 0.20, "rotate_limit": 20, "p": 0.4).

As seen in the Table 1, the ResNet-50 model architecture performs the best on the original dataset without implementing any data augmentations and accompanied by the EfficientNet-B0 as a 2-classifier pipeline. Having trained for a considerably high number of iterations, the training time is comparatively greater than other architectures and will only increase with the increase in iterations over the dataset or epochs. After a certain threshold, overfitting might become a considerable problem that will have to be dealt with manually to preserve the accuracy offered by the model.

6 Conclusion

Deep learning based on neural networks offers a new way to process and analyze medical imaging results. Radiographs, as an important health tool for diagnosis, is an essential component of clinical assessment and treatment planning. On specific tasks, computer-aided systems based on deep learning have shown good success in particular activities, such as assisting clinicians in reducing diagnosis time, improving accurate diagnosis rate, and designing treatment plans. We have presented some initial results on detecting various chest abnormalities from chest radiographs using a custom deep-learning model. As stated earlier, due the high ambiguity of inter-applicable labels pertaining to chest abnormalities, the results still look quite promising despite the large size of the publicly available dataset. We plan to try out various other approaches to smooth out the errors in our current ways and to effectively increase the overall accuracy of our predictions. With the assistance of the continued advancement of deep learning algorithms, the growth of high performance parallel computing technologies, the growing sample set of medical image labels and the improvement of medical image quality, we believe that deep learning based medical image processing will be very promising in the future.

References

1. Al-Waisy, A., Mohammed, M.A., Al-Fahdawi, S., Maashi, M., Garcia-Zapirain, B., Abdulkareem, K.H., Mostafa, S., Kumar, N.M., Le, D.N.: Covid-deepnet: hybrid multimodal deep learning system for improving covid-19 pneumonia detection in chest x-ray images. Comput. Mater. Continua **67**(2) (2021)
2. Anis, S., Lai, K.W., Chuah, J.H., Ali, S.M., Mohafez, H., Hadizadeh, M., Yan, D., Ong, Z.C.: An overview of deep learning approaches in chest radiograph. IEEE Access **8**, 182347–182354 (2020)
3. Bustos, A., Pertusa, A., Salinas, J.M., de la Iglesia-Vayá, M.: Padchest: a large chest x-ray image dataset with multi-label annotated reports. Med. Image Anal. **66**, 101797 (2020)
4. Shorten, C., Khoshgoftaar, T.M.: A survey on image data augmentation for deep learning. J. Big Data **6**(1), 1–48 (2019). https://doi.org/10.1186/s40537-019-0197-0
5. Freeman, I., Roese-Koerner, L., Kummert, A.: Effnet: An Efficient Structure for Convolutional Neural Networks (2018)
6. He, K., Zhang, X., Ren, S., Sun, J.: Deep Residual Learning for Image Recognition (2015)
7. Hore, S., Chakraborty, S., Chatterjee, S., Dey, N., Ashour, A.S., Van Chung, L., Le, D.N.: An integrated interactive technique for image segmentation using stack based seeded region growing and thresholding. Int. J. Electrical Comput. Eng. (2088–8708) **6**(6) (2016)
8. Institute, V.B.D.: Vinbigdata Chest X-ray Abnormalities Detection (2020). https://www.kaggle.com/c/vinbigdata-chest-xray-abnormalitiesdetection
9. Johnson, A., Pollard, T., Mark, R., Berkowitz, S., Horng, S.: Mimic-cxr database. PhysioNet10 13026, C2JT1Q (2019)
10. Litjens, G., Ciompi, F., Wolterink, J.M., de Vos, B.D., Leiner, T., Teuwen, J., Išgum, I.: State-of-the-art deep learning in cardiovascular image analysis. JACC: Cardiovasc. Imaging **12**(8 Part 1), 1549–1565 (2019)
11. Mohammed, M.A., Abdulkareem, K.H., Garcia-Zapirain, B., Mostafa, S.A., Maashi, M.S., Al-Waisy, A.S., Subhi, M.A., Mutlag, A.A., Le, D.N.: A comprehensive investigation of machine learning feature extraction and classification methods for automated diagnosis of covid-19 based on x-ray images. Comput. Mater. Continua **66**(3) (2020)
12. Nguyen, H.Q., Lam, K., Le, L.T., Pham, H.H., Tran, D.Q., Nguyen, D.B., Le, D.D., Pham, C.M., Tong, H.T., Dinh, D.H., et al.: Vindr-cxr: An Open Dataset of Chest X-rays with Radiologist's Annotations. arXiv preprint arXiv:2012.15029 (2020)

13. Rajpurkar, P., Irvin, J., Zhu, K., Yang, B., Mehta, H., Duan, T., Ding, D., Bagul, A., Langlotz, C., Shpanskaya, K., Lungren, M.P., Ng, A.Y.: Chexnet: Radiologist-Level Pneumonia Detection on Chest X-rays with Deep Learning (2017)
14. Razzak, M.I., Naz, S., Zaib, A.: Deep Learning for Medical Image Processing: Overview, Challenges and the Future, pp. 323–350. Springer International Publishing, Cham (2018). https://doi.org/10.1007/978-3-319-65981-7_12
15. Redmon, J., Divvala, S., Girshick, R., Farhadi, A.: You Only Look Once: Unified, Real-time Object Detection (2016)
16. Wang, X., Peng, Y., Lu, L., Lu, Z., Bagheri, M., Summers, R.: Chestx-ray14: Hospital-Scale Chest X-ray Database and Benchmarks on Weakly-Supervised Classification and Localization of Common Thorax Diseases (2017)
17. Wu, Y., Kirillov, A., Massa, F., Lo, W.Y., Girshick, R.: Detectron2. https://github.com/facebookresearch/detectron2 (2019)
18. xhlulu: Vinbigdata Chest X-ray Resized png (256×256) (2020). https://www.kaggle.com/xhlulu/vinbigdata-chest-xray-resized-png-256x256

Acquisition and Synchronisation of Multi-source Physiological Data Using Microservices and Event-Driven Architecture

Roberto Sánchez-Reolid[1,2], Daniel Sánchez-Reolid[2], António Pereira[3,4], and Antonio Fernández-Caballero[1,2,5](✉)

[1] Departamento de Sistemas Informáticos, Universidad de Castilla-La Mancha, 02071 Albacete, Spain
Roberto.Sanchez@uclm.es

[2] Unidad Multidisciplinar en Investigación de la Neurocognición y Emoción en Entornos Reales y Virtuales, Instituto de Investigación en Informática de Albacete, 02071 Albacete, Spain
Daniel.Sanchez@uclm.es

[3] Polytechnic Institute of Leiria, School of Technology and Management,Leiria, Portugal
apereira@ipleiria.pt

[4] INOV INESC INOVAÇÃO, Institute of New Technologies-Leiria Office, 2411-901 Leiria, Portugal

[5] CIBERSAM-ISCIII (Biomedical Research Networking Center in Mental Health, Instituto de Salud Carlos III), 28016 Madrid, Spain
Antonio.Fdez@uclm.es

Abstract. This paper introduces an architecture based on microservices and events for the acquisition and synchronisation of physiological signals. The proposed architecture incorporates the biosensors and a gateway that synchronises, secures and connects the traffic with the back-end infrastructure. This infrastructure is in turn composed of numerous services that are responsible for acquiring, storing and classifying the synchronised data. It has been chosen to integrate two non-invasive devices in a first version of the architecture. These are an electrocardiogram sensor and a wearable device that monitors electrodermal activity, blood volume pressure, time between heartbeats, skin surface temperature and acceleration. Tests have shown that the proposed approach improves synchronisation times using the edge compared to direct data delivery to the cloud.

Keywords: Physiological signals · Multi-source data acquisition · Synchronisation · Microservices · Event-driven architecture

V. Julián et al. (Eds.): ISAmI 2022, LNNS 603, pp. 13–23, 2023.
https://doi.org/10.1007/978-3-031-22356-3_2

1 Introduction

Multiple sensor technologies, cloud computing and mobile technology are enabling the Internet of Things (IoT) to be deployed in systems based on Ambient Intelligence (AmI) [6,7]. The increasing availability of connected systems, the performance of the sensors they incorporate and their low price make their integration into everyday environments attractive. IoT is also increasingly being deployed as a solution for automated monitoring and diagnosis of people's condition and well-being [2]. Monitoring of physical and emotional conditions are mainly based on the acquisition of physiological signals [3,8–10,19].

The IoT technologies linking humans with digital devices and/or services need to provide platforms for the integration of sensors, data processing and analytics dashboards. Consequently, three main components are involved in an IoT-based architecture aimed at assisting people. The first is a sensor gateway to configure, collect and synchronise data. Secondly, there has to be a platform for storing and manipulating the data collected from the different physiological sensors. Finally, a dashboard is required for the monitoring of users [13]. Two concepts can be used to deploy these, namely microservices- and event-driven-based architectures. A microservice is a software component that can perform a specific function very quickly and efficiently on the servers side [18]. On the other hand, an event-driven architecture is based on an internal or external event that triggers workflows in the infrastructure [17]. In this case, event producers and consumers are available and can be decoupled using such an architecture. An important aspect of microservice communication then emerges, as message routing, topic maintenance, and message validation may be facilitated by a message broker [16]. In this sense, asynchronous processing takes place in an event-driven architecture. Direct connection between producer and consumer is not feasible, which leads to the reduction of dependency between services and to the optimisation of the throughput capacity of each service.

This work introduces a microservices- and event-driven-based architecture for the acquisition, synchronisation and processing of physiological signals. The signals, coming from different sources and different devices, need to be available and analysed throughout the whole process, monitoring the user on a continuous basis and allowing the generation of reports. The emphasis is put on the need to improve synchronisation times between different biosignal sources.

2 Related Works

On the foreseeable horizon, IoT-based well-being applications will have a major impact on both the healthcare setting and everyday life. These applications based on a set of heterogeneous wireless sensor networks are used for real-time remote monitoring of health-related biosignals. Some of these physiological signals, together with their characteristics, are shown in Table 1.

In addition, requirements such as reliability, connectivity, user interaction and moderate costs are often considered. Moreover, as the number of participants

Table 1. Biosignals and their main properties.

Signal		Channels	SF[1] (Hz)	RF[2] (Hz)	AL[3]
Electrocardiogram	ECG	1–12	0.05–150	250–1 K	0.1–5
Electromyography	EMG	1–32	25–5 K	512–10 K	0.1–100
Blood Volume Pressure	BVP	1	0.25–40	5–500	−10–10
Electrooculography	EOG	2	0–100	1–100	50–3.5 K
Pupillography	PUP	2	120	240	–
Electroencephalography	EEG	1–128	128–2 K	128–2 K	1–150 mV
Near-Infrared Spectroscopy	NIRS	1–256	0–5	2.5–240	−20–20 μM/L
Inter-Breath	IBR	1	1–20	1–20	−0.05–0.05
Acceleration	ACC	3	20-2 K	20-2 K	−1–1
Skin Temperature	SKT	1	1–200	2–50 K	−50–50
Electrodermal Activity	EDA	1	1–16	16–128	0–100 μS

[1] significance frequency; [2] record frequency; [3] amplitude

continues to increase over the years, AmI applications require scalability to serve this need. At the moment, a lot of research based on wireless technologies (WiFi), Bluetooth, ZigBee and RFID has been carried out to partially implement architectures that meet these requirements. Wireless (WiFi) and Bluetooth-based architectures have presented sensor-based implementations and cloud architectures for wearable devices [14]. In most cases, they describe the entire environment from signal acquisition, processing and classification, but leave out the part of fault tolerance, high availability, fault tolerance and scalability [5]. ZigBee, on the other hand, has been used in applications where environmental monitoring plays an important role. The focus has typically been on establishing boundary conditions [4], both indoors and outdoors, in conjunction with technologies such as 5G [1]. Finally, in the case of heterogeneous sensor architectures, research has investigated how to integrate these devices using low-speed wireless personal area networks (LR-WPAN), such as wireless body area networks (WBAN) [15].

The main motivation of this work is to focus on providing a microservices- and event-driven-based architecture that improves aspects such as fault tolerance, high availability and scalability with high performance using minimal resources on the server side. Therefore, our research differs from previous studies in that we aim to perform all these improvements in the same architecture. The use of event-driven architectures and horizontal scaling enabled by the use of microservices simplify and help solve these problems. For example, to achieve scalability and increase performance, services simply need to be replicated with a technology like Docker and Kubernetes that allows us to scale the architecture [11]. This also adds an extra layer that minimises errors and improves consistency by increasing the fault tolerance of the system, thereby raising the system quality.

3 Design of the Architecture

This section describes the design and implementation of the proposed architecture. Emphasis will be put on the acquisition devices used so far, and the diagram of the proposed architecture will be shown, as well as a detailed explanation of the elements that compose it. Finally, the synchronisation option used for the different signals will be addressed.

3.1 Acquisition Devices

An important aspect to consider for the correct development of the architecture is the signal acquisition devices. In this case, we have decided to use two devices, the first one a commercial Empatica E4 wristband (https://www.empatica.com/en-eu/research/e4/) and the second one an electrocardiogram sensor based on the AD8232 sensor and the ESP32 microcontroller. The devices, the variables measured by each one and their main features are shown in more detail below.

Empatica E4 Wristband The Empatica E4 device (see Fig. 1a) has been used to acquire the following biosignals: electrodermal activity, blood volume pressure, inter-beat interval, acceleration and skin temperature. This commercial device is a wearable that allows us to capture physiological signals in real time via Bluetooth protocol. The wearable must be securely attached to the wrist to avoid signal artefacts.

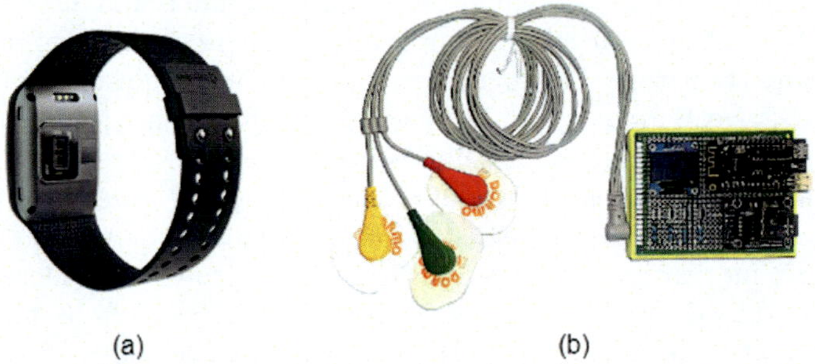

Fig. 1. (a) Empatica E4 wearable device. (b) ECG on AD8232 and ESP32 devices.

Electrodermal activity (EDA) has been used to determine the level of arousal of subjects. This phenomenon is due to the fact that the skin responds to stimuli by causing a variation in skin sweating. Blood volume pressure (BVP) is the pressure of circulating blood against the walls of blood vessels. Most of this pressure is the result of the heart pumping blood through the circulatory system. Inter-beat interval (IB) is defined as the time between one complete heartbeat and the next. Skin temperature (SKT) is the temperature of the external surface of the body. The normal human skin temperature at the torso is between 33.5 and 36.9 °C, while at the extremities it is usually slightly lower. Acceleration (ACC) is the ratio of body motion measured in the three Cartesian axes (XYZ).

Wireless Electrocardiogram AD8232 Device Electrocardiogram signals have been used to record the electrical activity of the heart muscle over a period of time. In this case an analogue AD8232 device has been chosen (see Fig. 1b). This sensor has a sampling rate 360 Hz. Due to this low power consumption it allows us to monitor continuously over a long period of time. To give it the necessary connectivity, this sensor has been mounted on an ESP32 microcontroller, which has 2.4 GHz WiFi and Bluetooth connectivity. It also has numerous analogue and digital inputs to connect additional different sensors. It also includes SPI and I2C ports that allow us to add a data-logger and a 0.96-inch Oled display.

3.2 Microservices and Event-Driven-Based Architecture

The architecture of the IoT-based monitoring system consists of WiFi and Bluetooth star nodes, a gateway (or edge) and a back-end, as shown in Fig. 2. On the left side of the figure are the sensors responsible for the acquisition of physiological signals. These sensors are connected wirelessly (WiFi and Bluetooth) to a hub that will act as a gateway. This hardware/software element is a router with a Linux-based distribution called *Openwrt*. The router is responsible for hosting a number of services that will secure connections to the rest of the infrastructure through a virtual private network (VPN) connection. It also synchronises and tags the images, as well as caches them, and then sends a copy of the signals. Connected to this element, we find three services which are API, WEB and BACK-OFFICE.

The *Application Programming Interface* (API) service is characterised by its adaptability and ability to listen to the different events that take place in the architecture. It is responsible for controlling the configuration, acquisition and control of the different sensors. In addition, it is also in charge of configuring the services of the entire domain, taking care of the auto-discovery and assembly of the different services and agents. In addition, it also establishes the processing and classification route of the different data to the respective classifiers. The *Control and Reporting Web* (WEB) service provides a web-based control and reporting interface. In this sense, it achieves a multi-platform system that can always be accessed regardless of the hardware used (PC, tablet, mobile phone, etc.). The *Back-office Control System* (BACK-OFFICE) service is the control interface for all the components that make up the architecture. In other words, it is the control panel used by the administrator to activate the different components and manage the databases of the architecture.

The three elements described above lead to the *Reverse Proxy*, which is actually the entry point to the infrastructure. It is used to secure internal and external traffic by encrypting it via an SSL/TLS security protocol. It is also responsible

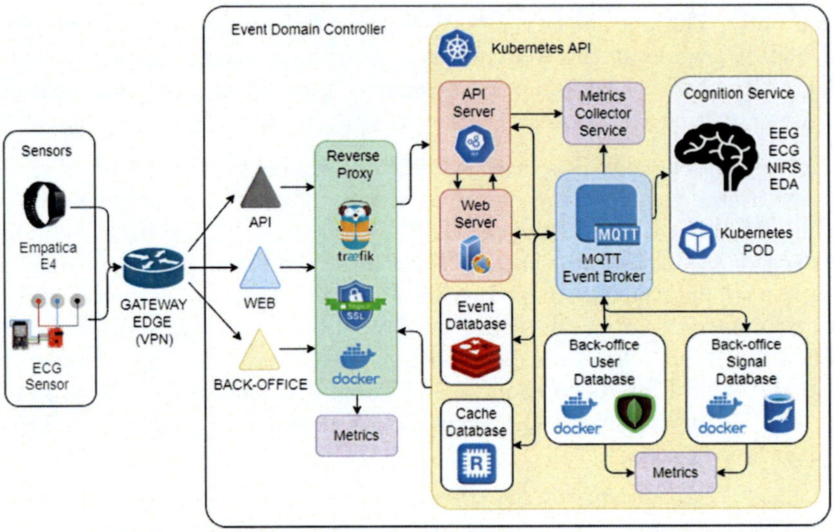

Fig. 2. Overall architecture.

for enabling traffic to services through a domain name server (DNS), balancing the load of requests (load-balancer) between the multiple requests and services hosted on the different servers that make up the architecture. The *Kubernetes API* software component is responsible for scaling the system based on the metrics collected by a metrics collection algorithm. It is one layer above the event handler and the domain controller. Strictly speaking, it manages the number of services and is in charge of replicating them on different servers, thereby achieving the most optimal configuration of resources at each moment.

Within the *Kubernetes API* we find a series of services and databases. The *Metrics Collection Service*, which collects and centralises all the metrics (service and server metrics), is a software component widely used in this type of architecture. The metrics extracted from the services are used to determine whether they are running optimally or inefficiently in order to provide more resources if necessary or replicating the services on the available servers. On the other hand, the metrics extracted from the server allow us to establish a relationship between the workload and the available resources. Setting in an effective way the number of machines (servers) for each case. The *Cognition Service* performs various processing and sorting operations on the different signals acquired. Different stages and configurations have to be set up for the signals. Depending on their origin, they need to go through different processing stages. *Databases* are software elements that are used to store different data. In our case, 4 different types have been considered, namely a *Cache Database* used to quickly store sensor data, an *Event Database* used to store events and maintain the list of different events in the system and two databases to store signals and user data separately (*Back-*

office User Database and *Back-office Signal Database*). Of course, all of them have to be in a shared database configuration.

Finally, the *MQTT Broker* is a building block to interconnect, communicate and set up the different relationships between events and services is the message broker based on the MQTT protocol. This service will listen to the different topics and allow the routing of data extracted from devices, event statuses and others. This is the most fault-tolerant element, as all other services are based on it. In our case, we have opted for fragmentation (3 replicas) of different services across several servers, establishing quality of service (QoS) values between them, so that if one is degraded, another is capable of routing and balancing the message traffic [12].

3.3 Synchronisation Strategy

As this is a microservices and event-driven-based architecture, data synchronisation is a crucial issue that must be solved. As can be seen in Fig. 3, two environments have been established to face synchronisation, namely local and cloud. The first corresponds to the local environment that stores local microservices for adding, configuring and merging sensors and sensor data, while the second responds to cloud services that host different agents and microservices designed to store, process and classify the various signals.

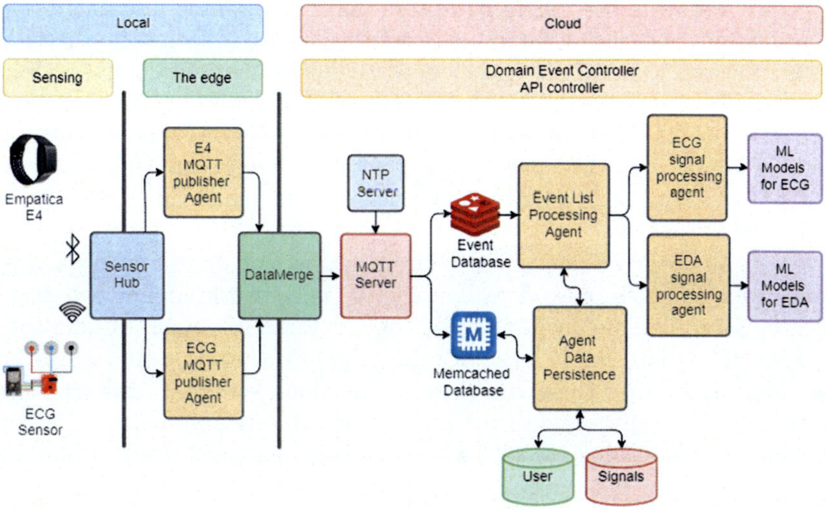

Fig. 3. Diagram for acquisition and synchronisation of physiological signals based on edge computing.

Furthermore, the local environment has been divided into two zones: *Sensing* and *The edge*. In the *Sensing* zone, the connection to the infrastructure has been established via the sensor hub. This was necessary because the Empatica E4 wearable uses the Bluetooth protocol while the AD8232 ECG device uses the 2.4G wireless protocol. Within *The edge*, a number of agents are responsible for preparing the data (signals) by tagging them with the local time and then storing or sending them to the rest of the infrastructure via the MQTT protocol.

After the data were sent via the MQTT protocol, they were re-tagged with the clock time of the NTP time server. This ultimately served to set a latency level for the data transmission. The signals were temporarily stored in a cache database because the process of writing to disk is too slow. Then, an agent observing this event stored the associated user and signal data on disk. Furthermore, at the same time as the process occurred, an event was stored in the event database. This event allows, among other things, to know the state of the signal (processing and storage system), the work that remains to be done. Finally, following the guidelines of the event database, a series of microservices (agents) for signal processing indicate the path to follow for each of the signals to be processed and classified.

4 Results

A primary consideration when evaluating a multi-source architecture is its latency in synchronising and transmitting data. In order to test whether a real-time data stream can be produced, synchronism and latency tests have been performed between the different sensors and the architecture. Figure 4 shows the results obtained using direct data delivery and synchronisation in the cloud versus local synchronisation at the edge and subsequent data delivery. Variables such electrocardiogram (ECG), captured from the AD8232 ECG device, and temperature (TMP), acquired from the Empatica E4 wristband, have a higher latency than others. This may be due to some delay in the Bluetooth communication layer with the system sending the data to the cloud storage versus the wireless ECG device.

It can be observed that in the direct connection to the cloud the system takes longer to synchronise data, largely due to network latency, whereas if the synchronisation is done locally at the edge, it takes less time. In addition, data storage will be more stable when a local (edge) synchronisation is performed, as there is no latency in the connection to the cloud services. This approach has allowed obtaining, for most of the sensors tested, latencies between 50 and 100 ms, which are promising values for the real-time implementation of any system.

5 Conclusions

Largely due to the rise of IoT devices, within Ambient Intelligence it is imperative to create self-scalable, fault-tolerant and highly available platforms that allow the management and aggregation of numerous devices or participants to be monitored. This work has presented an event-driven- and microservices-based

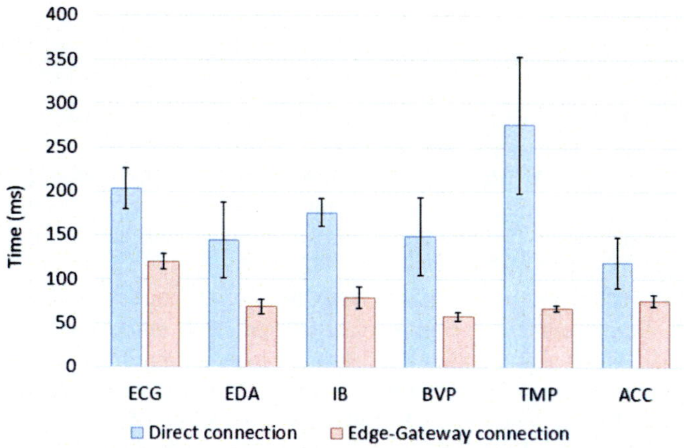

Fig. 4. Time delay from acquisition to insertion into the system using direct dispatch versus edge synchronisation via the MQTT broker.

architecture for the acquisition, monitoring and processing of physiological signals. It is intended to supplement and/or complement current IoT-based architectures for the monitoring of numerous participants. In addition, its design allows for the aggregation and configuration of devices from many different locations, enabling simultaneous signal acquisition from multiple sources. The main limitations of other architectures, such as high availability, scalability and fault tolerance in the architecture, can be largely overcome thanks to our proposal.

Preliminary results show that the use of the edge improves transfer times and signal delay times, providing a real-time performance. This means that the proposed acquisition and synchronisation approach is an improvement over existing architectures. By leaving such a complex issue as the synchronism between the biosignals well addressed, a firm basis for extending the architecture with a larger number of sensors has been established.

As future work remains to be carried out different tasks such as the inclusion of different algorithms for the classification of the physiological signals collected, the addition of more capture devices such as EEG, NIRS and a camera device to analyse emotions, body positions, etc. Our goal is to create a generic platform aimed at improving the quality of life of the users to introduce these devices in the most automatic way possible, facilitating the work of quality of life experts as well as researchers in the field.

Acknowledgements. Grants PID2020-115220RB-C21 and EQC2019-006063-P funded by MCIN/AEI/10.13039/501100011033 and by "ERDF A way to make Europe". Grants BES-2017-081958 and PTA2019-016876-I funded by MCIN/AEI/10.13039/501100011 033 and by "ESF Investing in your future". This work was also partially supported by Portuguese Fundação para a Ciência e a Tecnologia - FCT, I.P. under project UIDB/04524/2020 and by Portuguese National funds through FITEC - Programa Interface, with reference CIT "INOV - INESC Inovação - Financiamento Base". This work was also partially supported by CIBERSAM of the Instituto de Salud Carlos III.

References

1. Al-Asadi, A.M.Q.K., Muttair, K.S., Wadday, A.G., Mosleh, M.F.: Wireless body-area network monitoring with ZigBee, 5G and 5G with MIMO for outdoor environments. Bull. Electrical Eng. Inf. **11**(2), 893–900 (2022)
2. Cai, Y., Genovese, A., Piuri, V., Scotti, F., Siegel, M.: IoT-based architectures for sensing and local data processing in ambient intelligence: research and industrial trends. In: 2019 IEEE International Instrumentation and Measurement Technology Conference, pp. 1–6. IEEE (2019)
3. Castillo, J.C., Fernández-Caballero, A., Castro-González, Á., Salichs, M.A., López, M.T.: A framework for recognizing and regulating emotions in the elderly. In: Pecchia, L., Chen, L.L., Nugent, C., Bravo, J. (eds.) Ambient Assisted Living and Daily Activities, pp. 320–327. Springer International Publishing, Cham (2014)
4. Del-Valle-Soto, C., Valdivia, L.J., Velazquez, R., Visconti, P.: Remotely vital signs capturer for older adults applied in residential zones. In: IEEE Consumer Electronics Magazine (2022)
5. Dilmaghani, R.S., Bobarshad, H., Ghavami, M., Choobkar, S., Wolfe, C.: Wireless sensor networks for monitoring physiological signals of multiple patients. IEEE Trans. Biomed. Circuits Syst. **5**(4), 347–356 (2011)
6. Dohr, A., Modre-Opsrian, R., Drobics, M., Hayn, D., Schreier, G.: The internet of things for ambient assisted living. In: 2010 Seventh International Conference on Information Technology, pp. 804–809. IEEE (2010)
7. Guerrero, G., da Silva, F.J.M., Fernández-Caballero, A., Pereira, A.: Augmented humanity: a systematic mapping review. Sensors **22**(2), 514 (2022)
8. Lozano-Monasor, E., López, M.T., Vigo-Bustos, F., Fernández-Caballero, A.: Facial expression recognition in ageing adults: from lab to ambient assisted living. J. Ambient Intell. Humanized Comput. **8**(4), 567–578 (2017). https://doi.org/10.1007/s12652-017-0464-x
9. Lozano-Monasor, E., López, M.T., Fernández-Caballero, A., Vigo-Bustos, F.: Facial expression recognition from webcam based on active shape models and support vector machines. In: Pecchia, L., Chen, L.L., Nugent, C., Bravo, J. (eds.) Ambient Assisted Living and Daily Activities, pp. 147–154. Springer International Publishing, Cham (2014)
10. Martínez-Rodrigo, A., García-Martínez, B., Alcaraz, R., González, P., Fernández-Caballero, A.: Multiscale entropy analysis for recognition of visually elicited negative stress from eeg recordings. Int. J. Neural Syst. **29**(2), 1850038 (2019)
11. Merkel, D.: Docker: lightweight linux containers for consistent development and deployment. Linux J. **2014**(239), 2 (2014)
12. Mokhtar, S.B., Liu, J., Georgantas, N., Issarny, V.: QoS-aware dynamic service composition in ambient intelligence environments. In: Proceedings of the 20th IEEE/ACM International Conference on Automated Software Engineering, pp. 317–320 (2005)
13. Nakashima, H., Aghajan, H., Augusto, J.C.: Handbook of Ambient Intelligence and Smart Environments. Springer Science & Business Media (2009)
14. Phan, D.T., Nguyen, C.H., Nguyen, T.D.P., Tran, L.H., Park, S., Choi, J., Lee, B.i., Oh, J.: A flexible, wearable, and wireless biosensor patch with internet of medical things applications. Biosensors **12**(3), 139 (2022)
15. Soni, G., Selvaradjou, K.: Optimal GTS distribution to heterogeneous sensors in IEEE 802.15. 4 network for healthcare monitoring applications. Personal and Ubiquitous Computing **26**(1), 131–153 (2022)

16. Surantha, N., Utomo, O.K., Lionel, E.M., Gozali, I.D., Isa, S.M.: Intelligent sleep monitoring system based on microservices and event-driven architecture. IEEE Access (2022)
17. Taylor, H., Yochem, A., Phillips, L., Martinez, F.: Event-Driven Architecture: How SOA Enables the Real-time Enterprise. Pearson Education (2009)
18. Thönes, J.: Microservices. IEEE Softw. **32**(1), 116–116 (2015)
19. Zangróniz, R., Martínez-Rodrigo, A., López, M.T., Pastor, J.M., Fernández-Caballero, A.: Estimation of mental distress from photoplethysmography. Appl. Sci. **8**(1), 69 (2018)

Percutaneous Electrolysis (EPI®), a Promising Technology in the Treatment of Insertional Patellar Tendinopathy in Soccer Players

Laura Calderón-Díez[1](✉), José Luis Sánchez-Sánchez[1], José Manuel Sánchez-Ibáñez[2], and Pedro Belón-Pérez[3]

[1] Department of Physiotherapy, University of Salamanca, Salamanca, Spain
lauca@usal.es
[2] CEREDE Sports Recovery Center, Barcelona, Spain
[3] Real Madrid Football Club S.A.D, Madrid, Spain

Abstract. Percutaneous electrolysis (EPI®) is an ultrasound-guided minimally invasive new technology consisting of the application of a galvanic electrolytic current through an acupuncture. The objective of this therapy is to generate a local inflammatory response and start regeneration mechanisms. Different studies report promising clinical results with the EPI® technique in the treatment of musculoskeletal pathologies, mainly tendinopathies, such as the patellar tendinopathy. This study provides an analysis of patellar tendinopathy treated with the combination of EPI and eccentric exercises. The working hypothesis is that intratissue percutaneous electrolysis obtains a significant improvement in terms of pain and function in the patellar tendinopathies. A prospective study was performed involving 23 soccer players diagnosed with patellar tendinopathy. All patients underwent EPI® technique that was applied with an intensity of 3 mA during a period of 10 s. In all cases, the EPI® procedure was performed guided by high-resolution Doppler ultrasound. All patients received a weekly session of EPI combined with two weekly sessions of eccentric exercise consisting of three sets of ten repetitions at low velocity and without pain. Functional improvement was assessed in 91.7% of the total sample (n = 22), measured by the Victorian Institute of Sport Assessment-Patellar Tendon (VISA-P) questionnaire. The present study has shown promising results obtained with EPI® combined with eccentric exercise in the treatment of proximal patellar enthesopathy in a population of soccer players.

Keywords: Percutaneous electrolysis · Patellar tendinopathy · Football soccer

1 Introduction

Percutaneous electrolysis (EPI) is an ultrasound-guided minimally invasive technology consisting of the application of a galvanic electrolytic current through an acupuncture needle placed in the affected soft tissue [1–3]. The objective of this therapy is to generate an analgesic effect and reactivate the local inflammatory response that allows phagocytosis and repair in the region of treatment [1, 4, 5].

V. Julián et al. (Eds.): ISAmI 2022, LNNS 603, pp. 24–31, 2023.
https://doi.org/10.1007/978-3-031-22356-3_3

Regarding the underlying mechanisms of this technique, they are not completely defined, and both mechanical and biochemical effects are currently being researched [5–9]. Firstly, it is proposed that the mechanical stimulation produced by the needle would induce remodeling of the extracellular matrix following fibrocyte activation [2]. Secondly, the electrical stimulation provided by the galvanic current in a saline solution would generate a non-thermal electrolytic reaction that promotes a controlled inflammatory response and wound healing in damaged and/or degenerated tissue [4, 8–10]. Although these hypotheses have been proposed on the basis of clinical results, and mechanistic research contrasting them is scarce, pioneering works have examined the histologic and molecular response produced by the EPI application to tendon or muscle in rats. These authors have reported an increase in the expression of angiogenic factors, and some genes related to collagen regeneration [11, 12].

Different studies report promising clinical results with the EPI technique in the treatment of musculoskeletal pathologies, mainly tendinopathies [3, 10, 13]. Patellar tendinopathy or "jumper's knee" is considered more as a degenerative than a non-inflammatory process. Recent histopathological studies have shown the absence of inflammatory cells in this type of pathology and demonstrated the presence of myxoid degeneration, disruption of the collagen fibers and signs of hypoxia in tenocytes and resident macrophages. Percutaneous electrolysis (EPI®) is a non-thermal electrochemical ablation through a cathodic flow in the region or clinical focus of the degenerated tendon. EPI® produces the dissociation of water, salts and amino acids of the extracellular matrix creating new molecules through ionic instability. The organic reaction that occurs in the cathodic needle leads to an immediate and brief inflammatory response that occurs, only and exclusively, in the region of treatment, and increases the expression of some genes related to collagen regeneration and remodeling of extracellular matrix, particularly the *Cox2, Mmp9 and Vegf* genes [12].

This study provides an analysis of patellar tendinopathy treated with the combination of EPI and eccentric exercises. The working hypothesis is that intratissue percutaneous electrolysis obtains a significant improvement in terms of pain and function in the patellar tendinopathies.

2 Material and Methods

A prospective study was performed between January and December 2i involving 23 soccer players diagnosed with patellar tendinopathy and referred to the Physiotherapy Service of Sports Recovery Center, CEREDE, in Barcelona. The inclusion criteria were: football players with medical diagnosis of patellar tendinopathy with more than 12 weeks of evolution, aged between 18 and 41 years (24.09 ± SD 5.12 years), who had previously received medical treatments and physiotherapy programs without achieving adequate functional recovery. Exclusion criteria included: previous infiltration with corticosteroids, treatment with nonsteroidal anti-inflammatory drugs (NSAIDs) three months prior to treatment with EPI®, previous knee surgery or any systemic pathology with symptoms in the knee. All patients who met the inclusion criteria signed an informed consent.

All patients were football players (n = 23), 23 men (100%) and no women (0%). Regarding the professional category, 13% (n = 3) were professionals, 65.2% semi-professionals (n = 15) and 21.7% amateurs (n = 5). Out of the 23 patients, the dominant lower limb was the right one in 87.0% of cases (n = 20) and the left one in 13.0% (n = 3). The right knee was injured in 8.7% of cases (n = 2), left knee in 78.3% (n = 18) and both in 13.0% (n = 3). The patellar tendon of the lower extremity of support was the most affected. The duration of disease from the onset of first symptoms was measured in weeks with an average of 70.70 weeks (SD ± 67.65, 95% CI 41.44 to 99.95) (Table 1).

Table 1. Clinical variables of the total sample (n = 23)

Time of clinic (weeks)	70.70 ± 67.65 (IC 95% 44–99.95)
Idle time sport (weeks)	31.96 ± 37.37; (IC 95% 15.8–48.12)
Injured limb	Right: 8.7% (n = 2) Left: 78.3% (n = 18)
Blazine classification 1st observation	STAGE 1: 4.3% (n = 1) STAGE 2: 13.0% (n = 3) STAGE 3: 82. 6 (n = 19)
VISA-P classification 1st observation	GROUP 1 (VISA-P ≤ 50) (worst prognosis): n = 12 (52.2%) GROUP 2 (VISA-p > 50) (best prognosis): n = 11 (47.8%)

Regarding the time players had to stay at rest from sport, unable to play due to pain in the patellar tendon, it was also measured in weeks, where the average time was of 31.96 weeks (SD ± 37.37, 95% CI 15.79–48.12).

The diagnosis of patellar tendinopathy was made in the first clinical examination, according to Blazina's Classification, a clinical and functional classification used in this type of tendinopathy. The results were: stage 1 was 4.3% (n = 1), stage 2 was 13.0% (n = 3) and stage 3 was 82.6% (n = 19). This was confirmed by ultrasound examination, showing the presence of tendon thickening, hypoechoic image, bone irregularities and/or calcifications. Evaluation was performed by color Doppler ultrasound to analyze the presence of neovascularization. In the total study population (n = 23) the clinical location corresponded to the osteotendinous junction of the lower pole of the patella, leading to the diagnosis of enthesopathy of the proximal patella.

In the first observation and according to their score on the questionnaire of the Victorian Institute of Sport Assessment-Patellar Tendon (VISA-P). The patients were classified into two groups according to the index of severity of symptoms in patients with jumper's knee: Group-1 (worse prognosis, n = 12) who scored a VISA-P ≤ 50 (52.2%) and Group-2 (better prognosis, n = 11) with VISA-P score > 50 (47.8%). The demographic descriptions and clinical variables of the groups are registered in Tables 2 and 3.

Table 2. Demographic descriptions of GROUP 1 and GROUP 2 (n = 23)

Age (years)		Total: 24.09 ± SD 5.12 años GROUP 1 (VISA-P ≤ 50): 24.75 ± 6.60 (IC 95% 20.55–28.95 GROUP 2 (VISA-P > 50): 23.36 ± 2.94 (IC 95% 21.39–25.34)
Gender	MEN	GROUP 1 (VISA-P ≤ 50): n = 12 (100%) GROUP 2 (VISA-P > 50): n = 11 (100%)
Dominant limb	RIGHT	GROUP 1 (VISA-P ≤ 50): n = 10 (83.3%) GROUP 2 (VISA-P > 50): n = 10 (90.9%)
	LEFT	GROUP 1 (VISA-P ≤ 50): n = 2 (16.7%) GROUP 2 (VISA-P > 50): n = 1 (9.1%)
Sport	Soccer	GROUP 1 (VISA-P ≤ 50): n = 12 (100%) GROUP 2 (VISA-P > 50): n = 11 (100%)
Professional category	Professional	GROUP 1 (VISA-P ≤ 50): n = 3 (25%) GROUP 2 (VISA-P > 50): n = 0 (0%)
	Semiprofessional	GROUP 1 (VISA-P ≤ 50): n = 6 (50%) GROUP 2 (VISA-P > 50): n = 9 (81.8%)
	Amateur	GROUP 1 (VISA-P ≤ 50): n = 3 (25%) GROUP 2 (VISA-P > 50): n = 2 (18.2%)

Table 3. Clinical variables of GROUP 1 and GROUP 2 (n = 23)

Time of clinic (weeks)	GROUP 1 (VISA-P ≤ 50) = 74.50 ± 88.54 (IC 95% 18.24–130.76) GROUP 2 (VISA-P > 50) = 66.5 ± 37.53 (IC 95% 41.33–91.76
Idle time sport (weeks)	GROUP 1 (VISA-P ≤ 50) = 43.50 ± 44.46 (IC 95% 15.25–71.75) GROUP 2 (VISA-P > 50) = 19.36 ± 23.76 (IC 95% 3.40–35.33)
Blazine classification 1st observation GROUP 1 (VISA-P ≤ 50)	STAGE 1: 0.0% (n = 0) STAGE 2: 0.0% (n = 0) STAGE 3: 100% (n = 12)
Blazine classification 1st observation GROUP 2 (VISA-P > 50)	STAGE 1: 9.1% (n = 1) STAGE 2: 27.3% (n = 3) STAGE 3: 63.6% (n = 7)
VISA-P classification 1st observation	GROUP 1 (VISA-P ≤ 50) (worse prognosis) = 33.67 ± 12.27 (IC 95% 25.87–41.46) GROUP 2 (VISA-P > 50) (better prognosis) = 68.09 ± 10.50 (IC 95% 61.04–75.15)

Patients were treated in supine position with the knee in slight flexion (± 20°) by placing a pad in the popliteal region. The area was disinfected with isopropyl alcohol and the standard sterilization measures were taken for these types of procedures. Treatment with EPI was carried out using a specific approved EPI device (Epiadvanced, Barcelona, Spain), certified according to Directive 93/42/CEE (Fig. 1). All patients underwent treatment with the EPI® technique that was applied with an intensity of 3 mA during a period of 10 s. In all cases, the EPI procedure was performed with the guide of a high-resolution Doppler ultrasound with a 3–14 MHz linear probe (Samsung HS50) by means of three punctures with a 0.3 mm needle in the injured area of the tendon. All patients received a weekly session of EPI combined with two weekly sessions of eccentric exercise consisting of three sets of ten repetitions at low velocity and without pain.

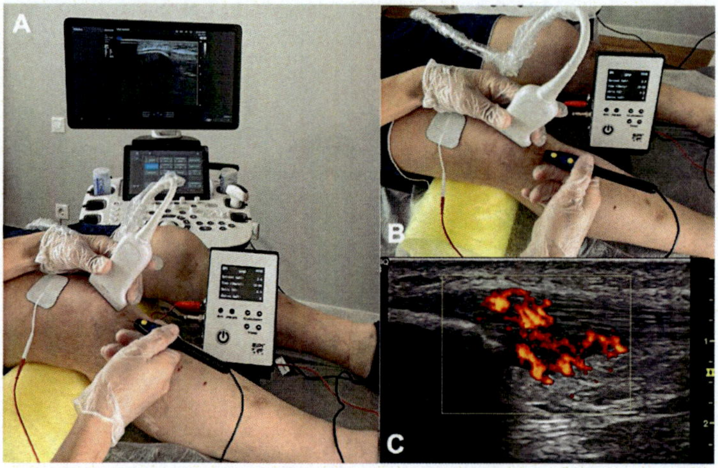

Fig. 1. A, B. Percutaneous electrolysis (EPI) procedure with specific device. **C.** Patellar tendinopathy studied with high-definition color Doppler ultrasound. Longitudinal view reveals intensive hypervascularization combined with hypoechoic zones. Important tendon thickening was detected.

3 Results

In Group 1 (n = 12), 83.3% of the patients were treated in the left knee (n = 10) and 16.7% had both knees treated (n = 2). In Group 2 (n = 11), 18.2% were treated in the right knee (n = 2), 72.75% in the left knee (n = 8) and 9.15% of them had both knees treated (n = 1). Regarding the VISA-P score, Group 1 (n = 12) in the first examination had a mean VISA-P score of 33.67 (SD ± 12.27, 95% CI 25.87–41.6) and in the last observation an average of 83.17 (SD ± 15.36, 95% CI 73.41–92.93) showing a statistically significant improvement in the results of the VISA-P questionnaire ($p \leq 0.005$) (Table 4).

Group 2 (n = 11) initially had a mean VISA-P of 68.09 (SD ± 10.50, 95% CI 61.04 to 75.15) and in the last one an average of 85.27 (SD ± 6.15, 95% CI 81.14–89.40) with statistically significant differences between the scores of the first and last observation ($p \leq 0.005$). Out of the 23 patients, only one patient (8.3%) did not show any alteration

Table 4. Results of the VISA-P score on the day of discharge

	VISA-P start	VISA-P end	Improvement VISA-P
GROUP 1 (VISA-P $\leq$ 50) (worse prognosis)) n = 12 (52.2%)	33.67 ± 12.27	83.17 ± 15.36	48.67 ± 18.45$^{(*)}$ (p $\leq$ 0.001)
GROUP 2 (VISAP > 50) (better prognosis) n = 11 (47.8%)	68.09 ± 10.50	85.27 ± 6.15	17.18 ± 13.54$^{(*)(p \leq 0.002)}$

since the first observation; the remaining patients (91.7%) improved according to the scores used in the VISA-P questionnaire (n = 22) (Table 4).

The average number of sessions performed with the EPI® technique in patients from Group 1 (n = 12) was 5 sessions of EPI® (SD ± 2.48, 95% CI 3.42–6.58) and in patients from Group 2 (n = 11) was of 2.45 sessions of EPI® (SD ± 1.29, 95% CI 1.59–3.32), with statistically significant differences in both groups ($p \leq 0.005$). Concerning the total sample, 82.6% (n = 19) were in stage III of the Blazina's Classification, which in many times is considered as criteria for surgical intervention. 13.0% of the patients were at stage II and 4.3% at stage I of the Blazina's Classification. If analyzed by groups of worst prognoses (Group 1 = VISA-P $\leq$ 50) and better prognosis (Group 2 = VISA-P > 50), in Group 1 (n = 12) 100% of the patients were in stage III of the Blazina's Classification while in Group 2 (n = 11) one patient (9.1%) was in stage I, 3 (27.3%) patients in stage II and 7 (63.6%) in stage III.

The authors performed an analysis of the survival function by Kaplan-Meier to assess the probability of a patient from the sample not be cured when receiving the treatment with EPI®, in a specific interval of time. To assess whether this was statistically significant ($p \leq 0.005$) the p by log-rank was calculated. In the worse prognosis Group (n = 12) the average healing time was 4.66 weeks (95% CI 3.25–6.08) and in the better prognosis Group (Group 2, n = 11) the time was of 2.18 weeks (95% CI 2.57–4.39), both with statistically significant differences ($p \leq 0.001$). Regarding the estimation of the number of interventions with EPI® need to achieve healing, in the Group with worse prognosis (Group 1, n = 12) the average number was 5.0 of EPI® sessions (95% 3.59–6.40) and in the Group with better prognosis (Group 2 n = 11) the mean number of sessions to achieve healing was of 2.45 EPI® (95% CI 1.69–4.74), both with statistically significant differences ($p \leq 0.005$).

4 Discussion and Conclusions

The present study shows the promising results obtained with EPI combined with eccentric exercise in the treatment of proximal patellar enthesopathy in a population of soccer players.

This study, through the analysis of Kaplan-Meier survival, allowed to estimate the time and number of sessions with EPI® technology needed to achieve cure for both the

Group with worse prognosis (Group 1) and for Group with better prognosis (Group 2), independently of the state of the patient according to the Blazina's Classification.

Functional improvement was assessed in 91.7% of the total sample (n = 22), measured by the Victorian Institute of Sport Assessment-Patellar Tendon (VISA-P) questionnaire.

In Group 1 (VISA-P $\leq$ 50) 100% of the patients were discharged from physiotherapy in 5 weeks' time, with an average of 5 sessions and a remarkable clinical improvement of VISA-P compared to the first observation.

In Group 2 (VISA-P > 50), all patients were discharged after 3 weeks, with a mean of 2.45 sessions and a greater clinical improvement of VISA-P compared to the first observation, although lower than in Group 1.

The good results are based on the local effect obtained with EPI, an electrochemical therapy that induces a short local inflammatory response with a cathodic current flow that leads to the rapid regeneration of the injured tendon [2, 3, 10, 13] combined with eccentric exercise, a type of exercise which has shown good results in the decrease in pain [3, 14].

Various treatment methods are used in the management of patellar tendinopathy including rest, activity modification, anti-inflammatory medication, injection therapies, taping, eccentric exercises, extra corporeal shock wave therapy, and surgery. Even though various treatment options are available for this type of tendinopathy, no single method has proven to result in a consistent and near complete recovery in patients [15, 16].

EPI® technology has proven to be very effective resulting in a great improvement in knee function and a rapid return to the previous level of activity after few sessions.

A major limitation of this study is the lack of a control group. The use of placebo for comparison with the control group may have reported stronger conclusions, however, most patients were referred to us after other treatments had failed. Therefore, it is unlikely that any group would accept a placebo for a sufficient period of time.

In conclusion, the ultrasound guided percutaneous electrolysis technology (EPI®) has proven to be highly effective in the treatment of proximal patellar enthesopathy in a population of soccer players, with a significant improvement in most patients from the total sample. EPI® technology has also proven to be effective in curing insertional patellar tendinopathy in soccer players regardless of the duration and evolution of the clinical symptoms.

References

1. Abat, F., et al.: Randomized controlled trial comparing the effectiveness of the ultrasound-guided galvanic electrolysis technique (USGET) versus conventional electro-physiotherapeutic treatment on patellar tendinopathy. J. Exp. Orthop. **3**, 34 (2016)
2. Abat, F., Gelber, P.E., Polidori, F., Monllau, J.C., Sanchez-Ibañez, J.M.: Clinical results after ultrasound-guided intratissue percutaneous electrolysis (EPI®) and eccentric exercise in the treatment of patellar tendinopathy. Knee Surg. Sport. Traumatol. Arthrosc. **23**, 1046–1052 (2015)
3. Castanheira, A., Peixoto, H., Machado, J.: Overcoming challenges in healthcare interoperability regulatory compliance. In: Advances in Intelligent Systems and Computing, vol. 1239 AISC, pp. 44–53 (2021)

4. Abat, F., Diesel, W.-J., Gelber, P.-E., Polidori, F., Monllau, J.-C., Sanchez-Ibanez, J.-M.: Effectiveness of the intratissue percutaneous electrolysis (EPI®) technique and isoinertial eccentric exercise in the treatment of patellar tendinopathy at two years follow-up. Muscles. Ligaments Tendons J. **4**, 188–193 (2014)
5. Gómez-Chiguano, G.F., Navarro-Santana, M.J., Cleland, J.A., Arias-Buría, J.L., Fernández-de-las-Peñas, C., Ortega-Santiago, R., Plaza-Manzano, G.: Effectiveness of ultrasound-guided percutaneous electrolysis for musculoskeletal pain: a systematic review and meta-analysis. Pain Med. **22**, 1055–1071 (2021)
6. De-Miguel-Valtierra, L., Salom-Moreno, J., Fernández-de-las-Peñas, C., Cleland, J.A., Arias-Buría, J.L.: Ultrasound-guided application of percutaneous electrolysis as an adjunct to exercise and manual therapy for subacromial pain syndrome: a randomized clinical trial. J. Pain **19**, 1201–1210 (2018)
7. Srivastava, V., Purwar, R.: An extension of local mesh peak valley edge based feature descriptor for image retrieval in bio-medical images. ADCAIJ Adv. Distrib. Comput. Artif. Intell. J. **7**(1), 77–89 (2018). https://doi.org/10.14201/ADCAIJ2018717789
8. Varela-Rodríguez, S., Cáceres-Pajuelo, J.E., Sánchez-Sánchez, J.L.: Percutaneous electrolysis in patients with musculoskeletal disorders: a systematic review. J. Mol. Genet. Med. **15**, 476 (2021)
9. García-Bermejo, P., De-La-Cruz-Torres, B., Naranjo-Orellana, J., Albornoz-Cabello, M.: Autonomic responses to ultrasound-guided percutaneous needle electrolysis: effect of needle puncture or electrical current? J. Altern. Complement. Med. **24**, 69–75 (2018)
10. Arias-Buría, J.L., Truyols-Domínguez, S., Valero-Alcaide, R., Salom-Moreno, J., Atín-Arratibel, M.A., Fernández-De-Las-Peñas, C.: Ultrasound-guided percutaneous electrolysis and eccentric exercises for subacromial pain syndrome: a randomized clinical trial. Evid. Based Complement. Altern. Med. 315219 (2015)
11. Abat, F., et al.: An experimental study of muscular injury repair in a mouse model of notexin-induced lesion with EPI® technique. BMC Sports Sci. Med. Rehabil. **7**, 7 (2015)
12. Sánchez-Sánchez, J.L., Calderón-Díez, L., Herrero-Turrión, J., Méndez-Sánchez, R., Arias-Buría, J.L., Fernández-de-las-Peñas, C.: Changes in gene expression associated with collagen regeneration and remodeling of extracellular matrix after percutaneous electrolysis on collagenase-induced Achilles tendinopathy in an experimental animal model: a pilot study. J. Clin. Med. **9**, 3316 (2020)
13. Rodríguez-Huguet, M., et al.: Effectiveness of percutaneous electrolysis in supraspinatus tendinopathy: a single-blinded randomized controlled trial. J. Clin. Med. **9**, 1837 (2020)
14. Visnes, H., Bahr, R.: The evolution of eccentric training as treatment for patellar tendinopathy (jumper's knee): a critical review of exercise programmes. Br. J. Sports Med. **41**(4), 217–23 (2007)
15. Costa, Â., Novais, P., Corchado, J., Neves, J.: Increased performance and better patient attendance in an hospital with the use of smart agendas. Logic J. IGPL **20**(4), 689–698 (2012)
16. Muaidi, Q.I.: Rehabilitation of patellar tendinopathy. J. Musculoskelet. Neuronal Interact. **20**(4), 535–540 (2020). PMID: 33265081; PMCID: PMC7716685

Check for updates

Ability-Centered Examination of People with Motor Impairments' Interaction with Television Towards More Accessible Smart Home Entertainment Environments

Ovidiu-Ciprian Ungurean and Radu-Daniel Vatavu(✉)

MintViz Lab, MANSiD Research Center, Ştefan cel Mare University of Suceava, Suceava, Romania
radu.vatavu@usm.ro

Abstract. We report results from a study with 41 people with motor impairments to understand their preferences for suitable input modalities, matched to their specific motor abilities, to interact with television. We report high preferences for voice- and smartphone-based input over hand gestures, head gestures, and eye gaze input. We discuss our findings in connection to the characteristic features of Ambient Intelligent environments, WHO levels of body functionality and disability, and a taxonomy of motor abilities, and we draw implications for more accessible home entertainment environments for users with motor impairments.

Keywords: Motor impairments · Smart home entertainment · Ambient Intelligence

1 Introduction

Home entertainment, primarily represented by smart TVs, has known many advances in display technology [9], interactive content [17], communications [26], and input modalities [16,18,22]. In this context, Ambient Intelligence (AmI) environments [4] with their many quality properties– sensitive, responsive, adaptive, ubiquitous, transparent, and intelligent—can further enhance home entertainment experiences by centering on, adapting to, and anticipating users' needs [1].

Intelligent adaptation, in particular, is essential to people with specific abilities to enable independent use of interactive systems. Prior studies [10,23] have shown that people with motor impairments experience accessibility challenges when using conventional TV remote controls. For example, in Fig. 1a, a person with Spinal Cord Injury (SCI) at vertebra C5 grasps the TV remote control with a specific hand pose, supports it against the chest, and applies pressure on the buttons with both arms; in Fig. 1b, a person with SCI at C5 uses the TV remote control from the bed, a task that needs locating the remote without visual feedback, preventing its slipping, and using both hands to reposition the remote to

V. Julián et al. (Eds.): ISAmI 2022, LNNS 603, pp. 32–43, 2023.
https://doi.org/10.1007/978-3-031-22356-3_4

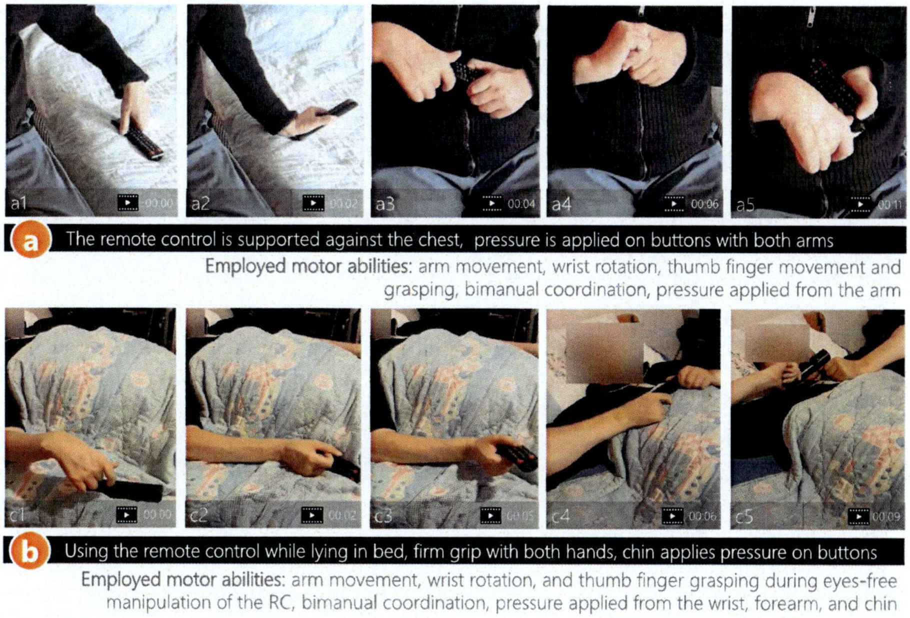

Fig. 1. Snapshots from videos collected during our study illustrating the use of TV remote controls under various motor abilities; see the text for descriptions.

press the buttons with the chin. The figure also enumerates specific motor abilities, e.g., arm movement, wrist rotation, bimanual coordination, etc., that are applied to use the conventional TV remote controls in a variety of situations.

Unfortunately, little is known about designing home entertainment systems that adapt to the specific motor abilities of the people who use them. Our practical contributions are: (1) We present results from a study with 41 people with motor impairments to understand preferences for TV input modalities matched to their motor abilities; our findings complement the results from [23] about accessibility challenges for TV input; (2) We draw implications for more accessible home entertainment systems by using AmI quality properties [4], WHO's levels of body functioning and disability [30], and a taxonomy of motor abilities [7].

2 Related Work

Prior work has introduced many innovations in smart home entertainment, from smart TV features that enable communication and social TV watching at a distance [26] to entertainment experiences that scale at room level [9] with AR [15,17]. For example, Audience Silhouettes [26] target viewers' peripheral attention by displaying cues of audience kinesics at the bottom of the TV screen, and IllumiRoom [9] extends the presentation of content from the TV screen to the wall behind it to enable new video gaming experiences. Such systems can be characterized in the context of AmI environments by being sensitive, responsive, and transparent [4], while the connection they make between AmI and AR has been highlighted [28]. We refer readers to Cook *et al.* [4] and Aarts and Encarnação [1] for overviews of AmI technology and applications.

Extensive work has been conducted to document the accessibility of interactive systems for people with motor impairments [11,12,14,23–25], and accessible interaction techniques have been proposed for mouse input [6], mobile devices [2,29], interactive surfaces [13], text entry [5], drawing [8], and wearable devices [19,21]. Şiean *et al.* [20] overviewed assistive technology at the intersection of AmI and AR/MR. However, the accessibility of interacting with television has been little examined for people with motor impairments [10,23] and, overall, users with disabilities have been little voiced in the scientific community of interactive media experiences [27]. Nevertheless, many accessibility challenges exist when interacting with television, and Ungurean and Vatavu [23] reported that people with motor impairments spend significantly more time watching television than people without impairments. In this context, it is important to understand and document user preferences and needs for suitable input modalities for interactive smart television and home entertainment systems, matched to specific motor abilities, for unrestricted access to the interactive experiences delivered by those systems. Our study, described next, addresses this need.

3 Study

We conducted interviews with people with motor impairments to elicit their preferences for TV input modalities matched to their motor abilities. Our results complement the findings from [23] about accessibility challenges for TV input.

3.1 Participants

We interviewed 41 people with various motor impairments aged between 10 and 72 years old (M=42.2, SD=13.6 years, normally distributed values according to a Shapiro-Wilk test, W=.987, p=.922).[1] The main cause of motor disability in our sample was Spinal Cord Injury (SCI), reported by 65.9% of the participants. Other conditions included Cerebral Palsy, Muscular Dystrophy, and Traumatic Brain Injury; see Table 1 and [23] for more details.

3.2 Motor Abilities

We center our scientific examination on the relationship between motor abilities and preferences for input modalities for smart TVs. To this end, we employ a taxonomy from Fleishman [7], who identified 11 psychomotor factors that account

[1] Two participants were under 18 years old. The responses of P_5 were entered by her father (online survey) and P_{22} was accompanied by a parent during the interview.

Table 1. Participants' demographics, self-reported impairments, and use of TV.

Participant	Health	Self-reported impairments‡											Cus-tom	TV functions§						Assis-tance
(age, gender)	condition†	1	2	3	4	5	6	7	8	9	10	11	RC	O	C	V	T	I	S	needed
P_1 (54 yrs., M)	SCI (C5)	·	◆	·	·	◆	·	◆	◆	◆	◆	◆	·	·	◆	◆	·	·	·	·
P_2 (37 yrs., M)	SCI (C4,C5)	·	◆	◆	·	·	·	◆	◆	·	◆	·	·	·	◆	◆	·	·	·	·
P_3 (46 yrs., M)	SCI (C5,C6)	·	·	·	·	·	·	◆	·	◆	·	·	·	·	◆	◆	·	◆	·	·
P_4 (23 yrs., F)	SCI (C6)	·	·	·	·	·	·	◆	·	·	·	·	·	·	◆	·	·	·	·	·
P_5 (10 yrs., F)	MD	◆	·	◆	·	◆	◆	◆	◆	·	◆	◆	◆	·	◆	◆	·	·	·	·
P_6 (41 yrs., M)	TBI	·	·	·	·	·	·	◆	◆	·	·	·	·	◆	◆	◆	·	·	·	·
P_7 (28 yrs., M)	SCI (C5,C6)	◆	·	◆	·	·	◆	◆	◆	·	◆	·	·	◆	◆	◆	◆	◆	◆	◆
P_8 (35 yrs., M)	SCI (C4,C5)	◆	◆	◆	·	◆	◆	◆	◆	◆	◆	◆	·	◆	◆	◆	·	·	·	·
P_9 (38 yrs., M)	SCI (C4,C5)	◆	◆	◆	·	◆	◆	◆	◆	·	◆	◆	·	◆	◆	◆	·	·	·	·
P_{10} (34 yrs., M)	SCI (C5)	◆	◆	◆	·	◆	◆	◆	◆	◆	◆	◆	·	·	◆	◆	·	·	·	◆
P_{11} (47 yrs., F)	SCI (C5,C6)	·	·	·	·	·	·	◆	◆	·	·	·	·	◆	◆	◆	·	·	·	·
P_{12} (34 yrs., F)	Ataxia	◆	·	◆	·	·	◆	·	·	·	·	·	·	·	◆	◆	·	·	·	·
P_{13} (65 yrs., M)	SCI (C6)	·	·	·	·	·	·	◆	·	·	·	·	·	·	·	·	◆	·	·	·
P_{14} (35 yrs., M)	Cerebral Palsy	·	◆	◆	·	·	·	·	·	·	·	·	·	·	·	◆	·	·	·	·
P_{15} (72 yrs., F)	Ataxia	·	·	·	·	◆	◆	·	·	·	·	·	·	·	◆	◆	·	·	·	·
P_{16} (38 yrs., M)	Ataxia	◆	·	·	·	◆	◆	·	·	·	◆	◆	·	·	◆	◆	·	◆	·	·
P_{17} (54 yrs., M)	SCI (C5)	·	◆	·	·	◆	◆	·	◆	·	·	·	·	·	◆	◆	·	·	·	·
P_{18} (34 yrs., F)	Ataxia	◆	·	·	·	◆	·	◆	·	·	◆	◆	·	·	◆	◆	·	·	·	◆
P_{19} (54 yrs., M)	TBI	◆	◆	◆	·	◆	◆	◆	◆	·	◆	◆	·	·	◆	·	·	·	·	◆
P_{20} (43 yrs., M)	SCI (C6)	·	◆	◆	·	◆	◆	◆	◆	◆	◆	·	·	·	◆	◆	·	·	·	·
P_{21} (45 yrs., M)	SCI (C6)	·	·	◆	·	◆	◆	◆	◆	◆	◆	·	·	·	◆	◆	·	·	·	·
P_{22} (15 yrs., F)	SCI (C6)	◆	◆	◆	·	◆	·	◆	◆	◆	◆	◆	·	◆	◆	◆	·	·	·	·
P_{23} (41 yrs., M)	SCI (C6,C7)	·	·	◆	·	◆	◆	◆	◆	·	·	◆	◆	·	◆	·	◆	·	·	·
P_{24} (55 yrs., M)	SCI (C5,C6)	◆	◆	◆	·	◆	◆	◆	◆	◆	◆	◆	·	·	◆	·	·	·	·	◆
P_{25} (49 yrs., M)	SCI (C3,C4)	◆	◆	◆	·	◆	◆	◆	◆	◆	◆	◆	◆	◆	◆	◆	◆	◆	◆	◆

(Continiued)

Table 1. Participants' demographics, self-reported impairments, and use of TV.

Participant	Health	Self-reported impairments‡											Cus-tom	TV functions§						Assis-tance
(age, gender)	condition†	1	2	3	4	5	6	7	8	9	10	11	RC	O	C	V	T	I	S	needed
P_{26} (20 yrs., M)	SCI (C7)	·	·	·	·	·	·	·	◆	·	◆	·	◆	·	◆	◆	·	·	·	·
P_{27} (24 yrs., F)	SCI (C5,C6)	·	·	◆	·	◆	◆	◆	◆	·	◆	·	·	·	◆	◆	·	·	·	·
P_{28} (63 yrs., M)	SCI (C2,C3)	·	·	◆	·	◆	◆	◆	◆	◆	◆	◆	·	·	◆	◆	·	·	·	◆
P_{29} (24 yrs., M)	SCI (C5)	◆	·	◆	·	◆	·	◆	◆	·	◆	◆	·	·	◆	◆	·	·	·	·
P_{30} (51 yrs., M)	SCI (C5,C6)	◆	·	◆	·	·	·	◆	·	◆	◆	◆	·	·	◆	◆	◆	·	·	·
P_{31} (46 yrs., M)	Cerebral Palsy	◆	◆	◆	◆	◆	◆	◆	◆	◆	◆	◆	·	·	·	·	·	·	·	◆
P_{32} (43 yrs., M)	SCI (C6)	·	·	·	·	◆	·	◆	◆	◆	◆	◆	·	◆	◆	◆	·	·	·	·
P_{33} (42 yrs., M)	SCI (C6,C7)	·	·	·	·	◆	·	◆	◆	·	◆	◆	·	·	◆	◆	◆	·	·	·
P_{34} (55 yrs., M)	SCI (C6,C7)	◆	·	◆	·	◆	·	◆	·	·	◆	◆	·	·	◆	◆	·	·	·	·
P_{35} (45 yrs., M)	SCI (C3,C4)	◆	◆	·	·	◆	·	◆	◆	◆	◆	◆	◆	·	◆	◆	·	·	·	·
P_{36} (57 yrs., M)	Cerebral Palsy	◆	◆	·	·	·	·	·	·	·	·	◆	·	◆	◆	◆	·	·	·	·
P_{37} (36 yrs., F)	SMA 3	◆	·	·	·	·	◆	◆	·	·	·	◆	·	·	◆	◆	·	·	◆	·
P_{38} (62 yrs., F)	Brain aneurysm	◆	◆	◆	◆	◆	◆	·	◆	·	◆	·	·	·	◆	◆	·	·	·	◆
P_{39} (43 yrs., M)	SCI (C6)	·	◆	·	·	◆	·	◆	◆	◆	◆	◆	·	·	◆	◆	·	·	◆	·
P_{40} (50 yrs., F)	Cerebral anoxia	◆	·	·	·	◆	·	·	◆	·	·	·	·	·	◆	◆	·	·	·	·
P_{41} (42 yrs., F)	Cerebral Palsy	◆	◆	·	·	◆	·	·	·	·	·	·	·	·	◆	◆	·	·	·	·
	Summary	**22**	**18**	**22**	**2**	**28**	**20**	**31**	**28**	**15**	**27**	**23**	**5**	**9**	**39**	**35**	**5**	**4**	**4**	**9**

†SCI=Spinal Cord Injury; TBI=Traumatic Brain Injury; MD=Muscular Dystrophy; The code in the parentheses denotes the affected vertebra(e), e.g., C4=the fourth cervical vertebra.

‡1=Slow movements; 2=Spasm; 3=Low strength; 4=Tremor; 5=Poor coordination; 6=Rapid fatigue; 7=Difficulty gripping; 8=Difficulty holding; 9=Lack of sensation; 10=Difficulty controlling direction; 11=Difficulty controlling distance; see [6].

§O=on/off; C=channels; V=volume; T=teletext; I=channel info; S=smart TV options

Table 2. Motor abilities required to operate the TV remote control; see also Fig. 1.

Task	Description	Required motor abilities	Correspondence with [7]
Reach	Reach for the RC	Move the arm	(1) control precision
Grasp	Pick up and seize the RC	Control the wrist, move the fingers, apply force	(7) manual dexterity
Hold	Hold the RC steadily	Apply continuous force	manual dexterity
Point	Point the RC at the TV	Move the arm, control the wrist	(1) control precision, (9) arm-hand steadiness
Press	Press buttons on the RC	Apply force, move fingers	(8) finger dexterity, (11) aiming

for the variance of human performance on a wide range of tasks: (1) control precision, (2) multilimb coordination, (3) response orientation, (4) reaction time, (5) speed of arm movement, (6) rate control, (7) manual dexterity, (8) finger dexterity, (9) arm-hand steadiness, (10) wrist/finger speed, (11) aiming. Table 2 connects abilities with the TV remote, the control condition in our study, against which we assess the perceived suitability of other input modalities; see next.

3.3 Measures

We collected the following information during our interviews:

- Assessment of the participants' health and disability conditions using Findlater *et al.* [6] eleven categories of self-reported impairments; see Table 1.
- Information about frequently used TV functions.
- Information about assistance needed for TV watching and control.
- Elicitation of preferences for input modalities for the TV, including *smartphone*, *voice*, *hand gestures*, *head gestures*, *eye gaze*, and *TV remote control;* the latter represents our control condition. These categories were informed by prior work that examined head gestures [2] and eye gaze [5] for users with motor impairments, smart TV models featuring voice [16] and hand gesture [18,22] input, and prior work [11,29] that reported effective smartphone use by people with motor impairments. Participants provided ratings for each condition using 5-point Likert scales: 1 ("not suitable at all"), 2 ("little suited"), 3 ("moderately suited"), 4 ("suited"), and 5 ("very suited").

4 Results

We found a large number (N=236) and a diversity (SD=8.03) of self-reported motor impairments, among which the most frequent were *difficulty gripping* (75.6%), *poor coordination* (68.3%), *difficulty holding* (68.3%), and *difficulty controlling direction* (65.9%); see Table 1. These impairments have a direct impact on the use of conventional TV remote controls that require gripping, holding, and pointing; see Table 2. We also found that the most frequently employed TV functions were changing channels (95.1%), adjusting volume (85.4%), and turning

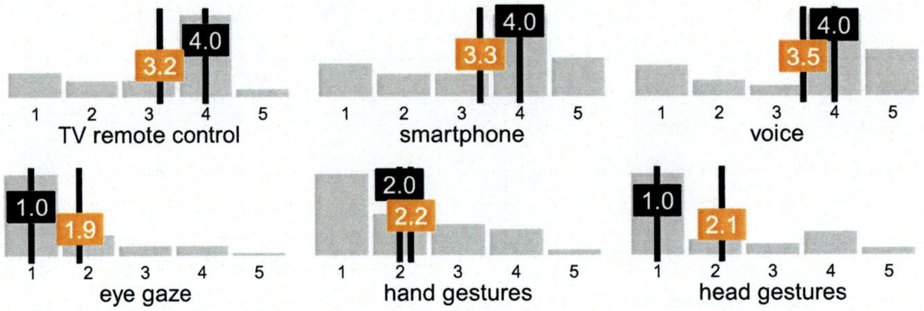

Fig. 2. Perceived suitability of input modalities for TV control. *Notes:* means are shown in orange, medians in black; larger values indicate better perceived suitability.

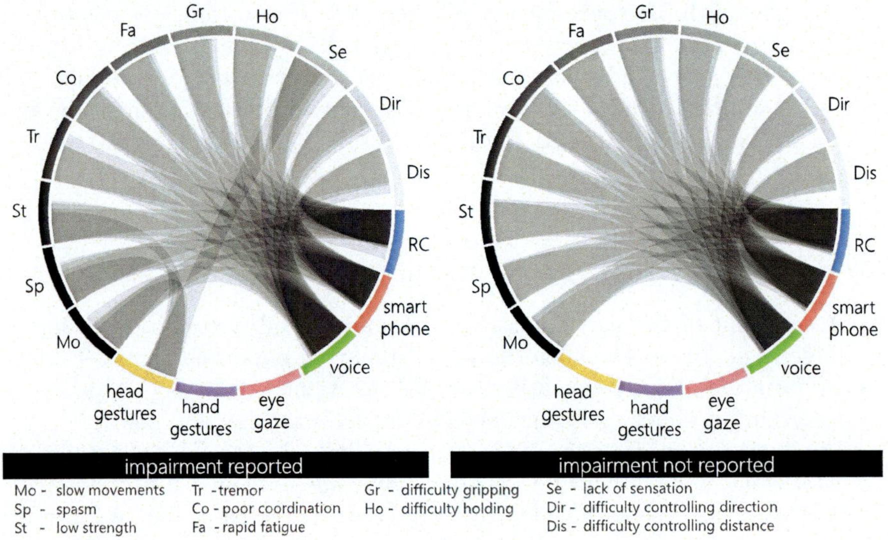

Fig. 3. The influence of specific motor impairments on the perceived suitability of TV input modalities. On the left, average suitability ratings computed for the participants that reported a specific impairment. On the right, average ratings when a specific impairment was not reported. *Note:* only arcs with average ratings above 2.5 are shown.

on/off the TV screen (22.0%). Corroborated, these findings suggest that simple designs of TV remote controls with only a few buttons might be more accessible. However, even if such solutions are commercially available, prior work [23] has reported little adoption. For example, five of our participants (12.2%) reported having used custom TV remote controls, but also mentioned the shortcomings of such solutions, including high prices. Also, a percent of 22.0% of our participants mentioned the need of assistance to control the TV.

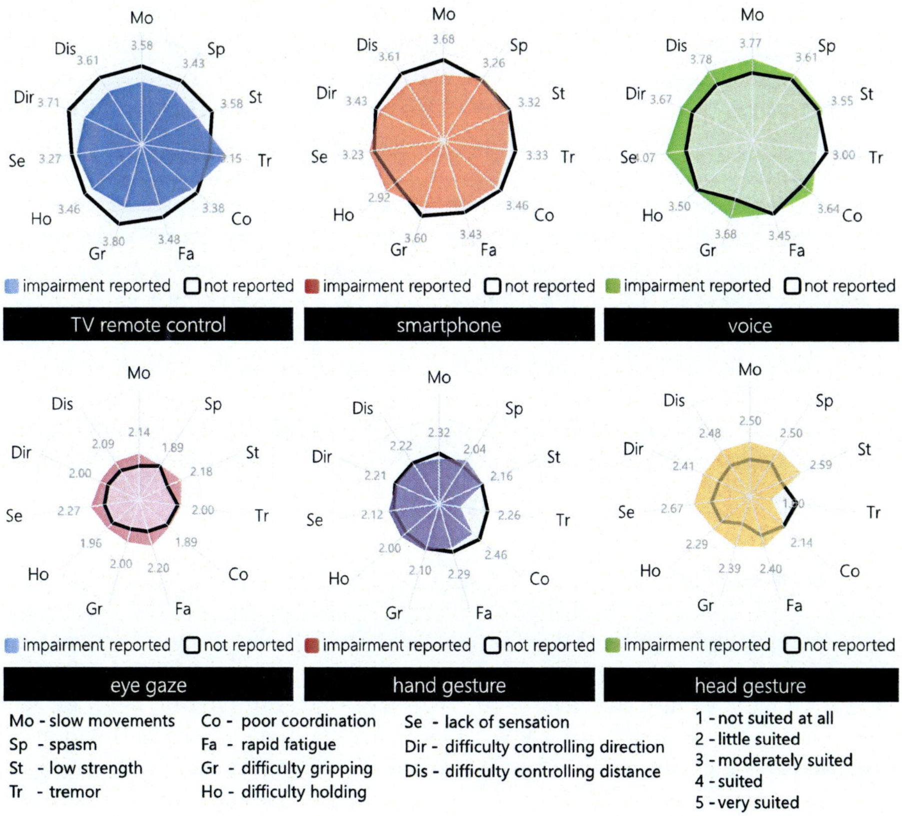

Fig. 4. Perceived suitability of the input modalities for TV control according to specific motor impairments. *Note:* larger values indicate better suitability.

Figure 2 shows participants' mean ratings of the six input modalities for TV control examined in our study. Overall, voice, smartphone, and TV remote control scored higher (Mdn=4, M$\geq$3.2) than hand, head gestures, and eye gaze (Mdn$\leq$2, M$\leq$2.2). Figure 3 shows the relationships between self-reported motor impairments and the perceived suitability of input modalities: on the left, the average suitability ratings are illustrated for each type of impairment; on the right, suitability ratings are shown for the cases where participants did not report a specific motor impairment. Several differences are visible, such as a higher preference for voice input and a preference for head gestures (Fig. 3, left) compared to the case where a specific impairment was not reported (Fig. 3, right). To understand these differences better, Fig. 4 shows radar charts with eleven dimensions according to each motor impairment category; see Table 1 for correspondences. We found higher suitability of voice (Mdn=4), smartphone (Mdn=4), and TV remote control (Mdn=4) over hand gestures (Mdn=2), head gestures (Mdn=1), and eye gaze (Mdn=1). We also found statistically significant positive correlations between TV remote control and smartphone (Spearman's $\rho_{(N=41)}$=.332, p=.05), between head gestures and eye gaze ($\rho_{(N=41)}$=.687, p=.01), and significant negative correlations between TV remote control and head gestures ($\rho_{(N=41)}=-.495$, p=.01), and between eye gaze and TV remote control ($\rho_{(N=41)}=-.438$, p=.01) and smartphone ($\rho_{(N=41)}=-.419$, p=.01).

Overall, we found that the absence of the *slow movements* motor impairment led to higher preferences for TV remote control (3.58 vs. 2.86), smartphone (3.68 vs. 3.00), and hand gestures (2.32 vs. 2.09), but not for voice, eye gaze, and head gestures; see Fig. 4, top left. Also, the absence of *difficulty gripping* led to higher ratings of th TV remote control (3.80 vs. 3.00) and smartphone (3.60 vs. 3.23), but not of the other input modalities. We also found statistically significant correlations between the preference for hand gestures and the presence of *slow movements* (Kendall's $\tau_{(N=41)}$=.307, p=.05), *low strength* ($\tau_{(N=41)}$=.313, p=.05), and *difficulty gripping* ($\tau_{(N=41)}$=.308, p=.05), and between voice and *lack of sensation* ($\tau_{(N=41)}$=.322, p=.05). We capitalize on these empirical findings to discuss implications for more accessible smart TV systems; see next.

5 Discussion

We present implications of our findings for accessible home entertainment environments that center on the smart TV. To this end, we adopt three perspectives: (1) smart home entertainment as an instance of an AmI environment, for which we connect to the quality characteristics of AmI systems [1,4]; (2) the multilevel relationship between body functioning and disability, for which we employ the biopsychosocial model [30] from the International Classification of Functioning, Disability, and Health; and (3) the relationship between users' motor abilities and their preferences for input modalities, for which we use Fleishman's [7] taxonomy (Sect. 3.2 and Table 2). Figure 5 illustrates these connections for the two highest-rated input modalities in our study, voice and smartphone input.

The experience of disability occurs at three levels of body functioning [30]: (a) at the level of a *body part*, where body structure and functions can be impaired; (b) at the level of the *whole person*, where activity limitations can occur; and (c) at the level of *the person in a social context*, where participation in life situations can be restricted; see Fig. 5, top. Each of these levels correspond to different perspectives—biological, individual, and social,—for which specific tasks in a home entertainment environment centered on the smart TV can be identified: (a) input devices are operated at the *body part* level [10], (b) independent TV watching is an activity performed at the *whole person* level [23], and (c) social television watching [26] fosters active participation and involvement in a *social context*. The two highest rated input modalities in our study facilitate the connection between these tasks and AmI quality features; see Fig. 5, middle and bottom left. For example, smartphones have been extensively examined for smart environments [3], where they act both as sensors and as information displays for the user. Also, voice input is increasingly present in the form of voice assistants in smart environments. Together, voice- and smartphone-based input

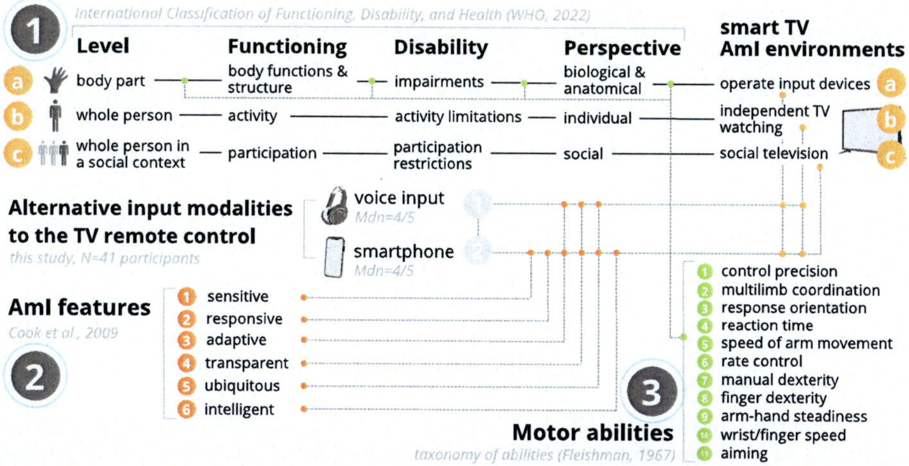

Fig. 5. Connections between body functioning and disability [30], AmI quality properties [4], and motor abilities [7] in the context of smart TV entertainment environments. *Note:* the small colored circles indicate connections between the various concepts.

enable interactions that fulfill the set of quality properties of AmI outlined by Cook *et al.* [4]. Figure 5, bottom right lists Fleishman's [7] motor abilities that connect to body functions and structure at the *body part* level, e.g., finger dexterity is reduced for people with SCI and affects use of specific input devices.

By capitalizing on the three perspectives—biological, individual, and social,—we propose several lines of future work for more accessible home entertainment environments for users with motor impairments: (i) interaction design that favors conjoint and interchangeable use of voice, smartphone, and TV remote control according to users' specific motor abilities; (ii) design of smart TV interactions by adopting the quality properties of AmI as design requirements; (iii) flexible, adaptive input via user profiles consisting of self-reported motor impairments, specific motor abilities, and input preferences; and (iv) design for social TV watching to foster social participation.

6 Conclusion

We examined the relationship between specific motor abilities and users' preferences for TV input modalities and found high perceived suitability of voice- and smartphone-based input, two input modalities compatible with voice assistants and smart mobile devices used as mediators of AmI interactions. Our findings complement prior work [23] about the accessibility challenges of TV remote controls with new results and corresponding implications for future work towards more accessible home entertainment environments centered on the smart TV.

Acknowledgements. This work was supported by a grant of the Ministry of Research, Innovation and Digitization, CNCS/CCCDI-UEFISCDI, project number PN-III-P4-ID-PCE-2020-0434 (PCE29/2021), within PNCDI III.

References

1. Aarts, E., Encarnação, J.: True Visions: The Emergence of Ambient Intelligence. Springer, Berlin (2006)
2. Cicek, M., Dave, A., Feng, W., Huang, M.X., Haines, J.K., Nichols, J.: Designing and evaluating head-based pointing on smartphones for people with motor impairments. In: Proceedings of the ASSETS '20. ACM, New York, USA (2020)
3. Clinch, S.: Smartphones and pervasive public displays. IEEE Pervasive Comput. **12**(1), 92–95 (2013)
4. Cook, D.J., Augusto, J.C., Jakkula, V.R.: Review: ambient intelligence: technologies, applications, and opportunities. Pervasive Mob. Comp. **5**(4), 277–298 (2009)
5. Feng, W., Zou, J., Kurauchi, A., Morimoto, C.H., Betke, M.: HGaze typing: head-gesture assisted gaze typing. In: Proceediongs of the ETRA '21. ACM, NY, USA (2021)
6. Findlater, L., Jansen, A., Shinohara, K., Dixon, M., Kamb, P., Rakita, J., Wobbrock, J.O.: Enhanced area cursors: reducing fine pointing demands for people with motor impairments. In: Proceedings of the UIST '10, pp. 153–162. ACM, NY, USA (2010)
7. Fleishman, E.A.: Performance assessment based on an empirically derived task taxonomy. Human Factors **9**(4), 349–366 (1967)
8. Hornof, A.J., Cavender, A.: Eyedraw: Enabling children with severe motor impairments to draw with their eyes. In: Proceedings of the CHI '05, pp. 161–170. ACM, NY, USA (2005)
9. Jones, B.R., Benko, H., Ofek, E., Wilson, A.D.: Illumiroom: immersive experiences beyond the TV screen. Commun. ACM **58**(6), 93–100 (2015)
10. Mehrotra, S.: Potmote: a TV remote control for older adults. In: Proceedings of ASSETS '18, pp. 486–488. ACM, New York, USA (2018)
11. Montague, K., Nicolau, H., Hanson, V.L.: Motor-impaired touchscreen interactions in the wild. In: Proceedings of the ASSETS '14, pp. 123–130. ACM, New York, USA (2014)
12. Mott, M., Tang, J., Kane, S., Cutrell, E., Ringel Morris, M.: "I just went into it assuming that I wouldn't be able to have the full experience": understanding the accessibility of virtual reality for people with limited mobility. In: Proceedings of ASSETS '20. ACM, New York, USA (2020)
13. Mott, M.E., Vatavu, R.D., Kane, S.K., Wobbrock, J.O.: Smart touch: Improving touch accuracy for people with motor impairments with template matching. In: Proceedings of CHI '16, pp. 1934–1946. ACM, New York, USA (2016)
14. Naftali, M., Findlater, L.: Accessibility in context: understanding the truly mobile experience of smartphone users with motor impairments. In: ASSETS '14, pp. 209–216. ACM, New York, USA (2014)
15. Popovici, I., Vatavu, R.D.: Understanding users' preferences for augmented reality television. In: Proceedings of ISMAR '19, pp. 269–278 (2019)
16. Roettgers, J.: One in Five Consumers Uses Voice to Interact with their TVs (2019). https://variety.com/2019/digital/news/tv-voice-control-usage-1203178496

17. Saeghe, P., Abercrombie, G., Weir, B., Clinch, S., Pettifer, S., Stevens, R.: Augmented reality and television: dimensions and themes. In: IMX '20, pp. 13–23 (2020)
18. Samsung: How do I use hand gestures to control my Samsung smart TV? (2020). https://www.samsung.com/ph/support/tv-audio-video/how-do-i-use-hand-gestures-to-control-my-samsung-smart-tv
19. Schipor, O.A., Bilius, L.B., Vatavu, R.D.: WearSkill: personalized and interchangeable input with wearables for users with motor impairments. In: W4A '22 (2022)
20. Siean, A.I., Bilius, L.B., Vatavu, R.D.: Assistive technology in the synchrony between ambient intelligence and mixed reality for people with motor disabilities. In: Proceedings of ISAmI '21 (2021)
21. Siean, A.I., Vatavu, R.D.: Wearable interactions for users with motor impairments: systematic review, inventory, and research implications. In: ASSETS '21 (2021)
22. Sony: Sony BRAVIA CAM (2022). https://electronics.sony.com/tv-video/televisions/television-accessories/p/cmubc1
23. Ungurean, O.C., Vatavu, R.D.: Coping, hacking, and DIY: reframing the accessibility of interactions with television for people with motor impairments. In: Proceedings of IMX '21, pp. 37–49. ACM, NY, USA (2021)
24. Ungurean, O.C., Vatavu, R.D.: Users with motor impairments' preferences for smart wearables to access and interact with ambient intelligence applications and services. In: Proceedings of ISAmI '21 (2021)
25. Ungurean, O.C., Vatavu, R.D.: "I gave up wearing rings:" insights on the perceptions and preferences of wheelchair users for interactions with wearables. In: IEEE Pervasive Computing (2022)
26. Vatavu, R.D.: Audience silhouettes: peripheral awareness of synchronous audience kinesics for social television. In: Proceedings of TVX '15, pp. 13–22. ACM, NY, USA (2015)
27. Vatavu, R.D.: Accessibility of interactive television and media experiences: users with disabilities have been little voiced at IMX and TVX. In: IMX '21 (2021)
28. Vatavu, R.D.: Are ambient intelligence and augmented reality two sides of the same coin? Implications for human-computer interaction. In: CHI EA '22 (2022)
29. Vatavu, R.D., Ungurean, O.C.: Stroke-gesture input for people with motor impairments: empirical results & research roadmap. In: Proceedings of CHI '19, pp. 1–14 (2019)
30. World Health Organization: ICF (2022). https://icd.who.int/dev11/l-icf/en

Discovery of Profitable Stock Price Movement Patterns from Various High Utility Pattern Mining

Mohammad Iqbal(✉), Edwina Simanungkalit, Safira Nur Latifa, Nurul Hidayat, and Imam Mukhlash

Department of Mathematics, Faculty of Science and Data Analytics, Institut Teknologi Sepuluh Nopember, Sukolilo, Surabaya 60111, Indonesia
iqbal@matematika.its.ac.id , nurul_hdy@matematika.its.ac.id,
imamm@matematika.its.ac.id

Abstract. Understanding stock price movements in a few moments consider a pivotal step for investors to decide their next actions. To read the movements easily, we can offer stock price movement patterns to the investor. Existing studies on mining stock price movement patterns mostly focused on the frequent ones. In this work, we aim to assist the investors to see stock price movement patterns that bring profit. Additionally, we consider profitable stock price movements pattern that also frequently occurs in the dataset. To be specific, this work applies two different data mining methods; fuzzy high utility itemset mining and skyline frequent utility pattern mining. Further, this work formulates profit stock price returns to help us recognise the benefit ones. Also, this work addresses the uncertainty movements from the fuzzy set. In this work, we analyzed fifteen companies in Indonesia. We listed a few suggestions like portfolios including the company, the company sector, and the whole listed company that is most profitable.

Keywords: Stock price · Profit · High utility mining · Data mining

1 Introduction

Investors need to read the stock price movements sharply to decide their next action whether to sell their stocks. To ease the movements readings', we can provide the stock price movements patterns from the historical stock price dataset by using data mining techniques. A few studies have been conducted to analyze the stock price movement patterns using association rules mining [1,13] and sequential pattern mining [10,12]. The previous studies typically focused on finding the stock price movement patterns that frequently occur on the given historical stock price dataset without showing the profitable ones. Intuitively, the investors want to ask the most profitable stock price. Hence, this work aims to deliver the profitable stock price movement patterns by adopting the idea

V. Julián et al. (Eds.): ISAmI 2022, LNNS 603, pp. 44–53, 2023.
https://doi.org/10.1007/978-3-031-22356-3_5

of utility measurement in the mining process from [4]. In this work, we define a profit utility stock price measurement by calculating the return and the volume of stock price. According to the profit utility measurement, we search the profitable stock price movement patterns by implementing high utility itemset mining (HUIM) method.

As numerous HUIM methods have been developed [5,7,11], this work attempts to utilize two different HUIM methods in order to offer various profitable stock price movement patterns. More specifically, we can obtain profitable stock price movement patterns by applying Fuzzy High Utility Itemset Mining (FHUIM) in [8]. To have more concise patterns and efficient mining processes, this work employs Skyline Frequent Utility Patterns Mining (SFUPM) [9] to get frequent and profitable stock price movement patterns. In addition, this work analyzed both patterns in three different profiles, which may be seen as portfolio-like. First, we analyze the pattern for each company. Then, we explore the company sectors depending on the listed companies in this work. Lastly, we want to find the patterns of the whole listed companies.

In a summary, this work contributions can be described as follows:

- A profit utility that fits with the stock price application is formulated;
- Two types of stock price movement patterns that mostly focused on the profit ones are discovered;
- A few portfolios-like are designed to offer the company profiles that brings profitable stock price movement patterns.

The next following organization of this paper is described as follows: Sect. 2 reviews a few related studies. Section 3 explains the high utility itemsets mining approaches that we focus on. Section 4 discusses the analysis of both patterns and their pros and cons. Section 5 summarizes this work results.

2 Past Works

We review some past works on stock price movements analysis using data mining techniques. In this work, we categorize the analysis into two parts: stock price movements patterns and predictions. For the predictions, the studies generally works on different domain applications, such as: FOREX market [6], China's commercial bank [2]. However, this work is more interested on stock price movement patterns analysis with the aim of investors' understandings on the overall stock price movement patterns. Several studies provide the patterns based on association rules mining (ARM) or sequence pattern mining (SPM). Based on ARM, Arafah *et al.* [1] analyzed sixty companies on Indonesia, Syu [14] studied to a self-risk management portfolio on Taiwan stock market, Cheng [3] recommended stock investment decision model which focused on sport and entertainment fields. But, the previous studies ignored the interval time and the order of the stock price movements. Then, Mukhlash et al. [12] continued to extract the stock price movement patterns using sequential pattern mining, thus the generated patterns followed the movement order. Also, they considered the interval time of the movement based on fuzzy set. Despite of that, our previous studies did not offer the stock price movement patterns indicating profit. Accordingly, this work employs high utility pattern mining concept to bring profitable stock price movement patterns; more specifically, this work adopts FHUIM and SFUPM methods.

3 Methodology

In this work, we want to offer stock price movements which benefit at the moments. In general, this work has three main steps to discover the patterns. First, we transform each return of stock price as the movement pattern from fuzzy set. Second, we implement two data mining approaches: FHUIM [8], and SFUPM [9], to have various profitable stock price movement patterns. Third, we analyze the patterns in terms of each company, inter-companies and sectors. To explain the details of the steps, we shall describe some necessary definitions in the next section.

3.1 Definitions

Let $\mathbf{s} = (s_1, s_2, \ldots, s_T)$ be a historical stock price sequence of one company with time-stamps $1, 2, \ldots, T = t$, such that $s_t \in \mathbb{R}$. When we speak on a few companies, we collect a set of historical stock price sequences, $\mathcal{S} = \{\mathbf{s}_1, \mathbf{s}_2, \ldots, \mathbf{s}_p\}$ with p companies and $\mathbf{s}_p = (s_{p_1}, \ldots, s_{p_T})$. Next, we ought to calculate the returns of each company, as follows:

$$r_{p_i} = \frac{s_{p_{i+1}} - s_{p_i}}{s_{p_i}}, \quad i = 1, 2, \ldots, T-1; \tag{1}$$

Subsequently, we have a set of return sequences $R = \{\mathbf{r}_1, \ldots, \mathbf{r}_p\}$ with $\mathbf{r}_p = (r_{p_1}, r_{p_2}, \ldots, r_{p_{T-1}})$. In general, we categorize each return following $r > 0$ with *Profit* (P), $r < 0$ with *Loss* (L), and $r = 0$ with *Stable* (S). In this work, we want to have more diverse categorical representation to mention the quantity from fuzzy set. Moreover, we utilize trapezoidal membership function by allowing the linguistic terms: *low* (ℓ), *medium* (m) and *high* (h). Let xy or x and y be two terms from (P, L, S) and (ℓ, m, h), respectively. The membership function can be written below.

$$\mu_{xy}(r_{p_i}) = \begin{cases} 0, & r_{p_i} \le a, \\ \frac{r_{p_i} - a}{b - a}, & a < r_{p_i} < b, \\ \frac{c - r_{p_i}}{c - b}, & b \le r_{p_i} < c, \\ 1, & r_{p_i} \ge c, \end{cases} \tag{2}$$

for any $x \in$ (P, L), and $y \in$ (ℓ, m, h), also $a, b, c, d \in R$. Further we set three time intervals of returns, such as: *long* (l), *normal* (n), and *short* (s) from triangular membership function, as follows:

$$\mu_z(i) = \begin{cases} 0, & i \leq e, \\ \frac{i-e}{f-e}, & e < i < f, \\ 1, & i \geq f, \end{cases} \tag{3}$$

with $e, f, g \in i$ and $z \in$ (l, n, s). As the results, we obtain 21 distinct items in a set $I = \{$Pℓ_l, Pℓ_n, Pℓ_s, Pm_l, Pm_n, Pm_s, ..., S_l, S_n, S_s$\}$. For an example, Pm_s means that stock price in the company showed *Profit* with a *medium* value in a *short* time.[1]

This work also considers the volume of stock price. Let say $V = \{\mathbf{v}_1, \ldots, \mathbf{v}_p\}$ with $\mathbf{v}_p = (v_{p_1}, \ldots, v_{p_T})$. In this work, we formulate external utility u; that often known profit in the customer transaction case, with multiplication between the volume of item e $\overline{v}$ and its returns $\overline{r}$, as follows:

$$u(e) = \overline{r}(e) \times \overline{v}(e) = \sum_j (r_j(e) \times v_j(e)); \tag{4}$$

where e is an item in I, and j is any indexes where item e can be found in R or V. $r_j(e)$ is a return value of item e at index j. $v_j(e)$ is a volume value of item e at index j. Following the nature of high utility mining approaches, we construct a quantity set of R based on u, i.e., $Q = \{(e_k, u_n(e_k)), k = 1, \ldots, |I|, n = 1, \ldots, p, e_k \in I\}$. Then, we extract two stock price movements patterns from Q: profitable stock price movements patterns $\mathbf{s}_p$, and further with frequent ones $\mathbf{s}_{\mathrm{fp}}$. Next, we define the two patterns from two high utility mining methods that we focus on.

3.2 The Focused High Utility Mining Methods

Two stock price movement patterns are analyzed in this work. First, a profitable stock price movement pattern $\mathbf{s}_p$ from fuzzy high utility itemset mining. Second, a frequent and profitable stock price movement pattern $\mathbf{s}_{\mathrm{fp}}$ from skyline frequent utility pattern mining. The brief details from $\mathbf{s}_p$ and $\mathbf{s}_{\mathrm{fp}}$ mining concepts in the next few paragraphs.

Fuzzy High Utility Itemset Mining or FHUIM [8] is applied to find profitable stock price movement pattern $\mathbf{s}_p$ from Q. The FHUIM constructs any high fuzzy utility upper bound itemsets and then sort out the high fuzzy utility itemsets ones or $\mathbf{s}_p$. A pattern $\mathbf{s}'_p$ is said a high fuzzy utility upper bound itemsets (hfuubi) when its utility upper bound value (fuub) is greater than or equal to min_util given by:

$$\begin{aligned} \mathrm{fuub}_{\mathbf{s}'_p} &= \sum_{\mathbf{s}'_p \subseteq Q} \mathrm{mtfu}_{\mathbf{s}'_p} \geq \mathrm{min\_util}; \\ &= \sum_{\mathbf{s}'_p \subseteq Q} \mathrm{mtfu}_{\mathbf{s}'_p} \geq \delta \times \sum_{k=1}^{|Q|} \mathrm{mtfu}_k; \end{aligned} \tag{5}$$

[1] Note that we follow the range value of each item based on the given data.

with $\delta \in [0,1]$ is user-predefined parameter. To avoid the ambiguity, the $\text{mtfu}_{\mathbf{s}'_p}$ is the maximal fuzzy utility of an itemsets $\mathbf{s}'_p$ in Q and mtfu_k is the maximal fuzzy utility of itemset e_k in each row of Q. Then, the maximal fuzzy itemset of $\mathbf{s}'_p$ is given by:

$$\text{mtfu}_{\mathbf{s}'_p} = \sum_{\mathbf{s}'_p \subseteq Q} \sum_{e_k \subseteq \mathbf{s}'_p} \text{mfu} = \sum_{\mathbf{s}'_p \subseteq Q} \sum_{e_k \subseteq \mathbf{s}'_p} \max \left\{ \text{fu}_{e_1}, \text{fu}_{e_2}, \ldots \text{fu}_{e_k} \right\}; \quad (6)$$

with $\text{fu}_{e_k} = \mu_v \times u_n$, where μ_v is the triangular membership function to describe the volume of the returns into three linguistic terms lots of volume (ℓv), normal of volume (nv) and small of volume (sv). Then, we search any $\mathbf{s}'_p$ that having the actual utility value afu is greater or equal to min_util which later called as high fuzzy utility itemset (hfui) $\mathbf{s}_p$, as follows:

$$\text{afu}_{\mathbf{s}_p} = \sum_{k=1}^{|Q|} \text{fu}_k = \sum_{k=1}^{|Q|} \min \left\{ \text{fu}_{k_1}, \text{fu}_{k_2}, \ldots \right\} \geq \text{min\_util}; \quad (7)$$

with $\text{fu}_k = \mu_k \times \sum_{e_k \in \mathbf{s}_p} u_n(e_k)$. Next, we explain how to obtain a frequent and profitable stock price movement from skyline frequent utility patterns mining.

Skyline Frequent Utility Patterns Mining or SFUPM [9] is used here to search a frequent and profitable stock price movement pattern $\mathbf{s}_{\text{fp}}$ in Q. Let $\mathbf{s}_{\text{fp}}$ and $\mathbf{s}'_{\text{fp}}$ be two itemsets in I. A pattern $\mathbf{s}_{\text{fp}}$ is said an SFUP by satisfying $(f(\mathbf{s}_{\text{fp}}) \geq u(\mathbf{s}_{\text{fp}})) \succeq (f(\mathbf{s}'_{\text{fp}}), u(\mathbf{s}'_{\text{fp}}))$, where f is the occurrence number of itemset $\mathbf{s}_{\text{fp}} \in Q$. In general, the SFUPM construct an utility list to help us finding the potential skyline frequent utility patterns (PSFUP). A pattern $\mathbf{s}'_{\text{fp}}$ is called PSFUP by satisfying the following criteria:

$$u(\mathbf{s}'_{\text{fp}}) \geq \text{utilmax} = \max \left\{ u(\mathbf{s}'_{\text{fp}}) \mid f(\mathbf{s}'_{\text{fp}} \geq j) \right\}; \quad (8)$$

with $j \in \{1, 2, \ldots, |Q|\}$. Further, the PFSUP can be called as SFUP by fulfilling the following criteria:

$$u(\mathbf{s}_{\text{fp}}) + \sum_{\mathbf{s}'_{\text{fp}} \subseteq Q, \mathbf{s}'_{\text{fp}} \neq \mathbf{s}_{\text{fp}}} u(\mathbf{s}_{\text{fp}}) \geq \text{utilmax}; \quad (9)$$

A pattern $\mathbf{s}_{\text{fp}}$ will be extended under depth-first search technique whenever $\mathbf{s}_{\text{fp}}$ meets the above-criteria. Conversely, the extension will be stopped when $\mathbf{s}_{\text{fp}}$ cannot satisfy the above-criteria. In the next section, we will discuss the profitable stock price movements pattern $\mathbf{s}_p$ and the frequent and profitable stock price movements pattern $\mathbf{s}_{\text{fp}}$.

4 Experiments

4.1 Data and Pre-processing

We collected the historical stock price sequences of fifteen companies in Indonesia, for one year from https://finanace.yahoo.com/. This work focused on fifteen public limited companies (in abbreviation): ACES, ADHI, ANTM, ASII, BMRI, BBNI, BBRI, INDF, ISAT, MNCN, PGAS, BBCA, TLKM, UNVR, WSKT. In the experiments, we analyzed two patterns $\mathbf{s}_p$ from FHUIM and $\mathbf{s}_{\text{fp}}$ from SFUPM. Further, we set $\delta = \{0, 0.05, 0.2, \ldots, 1\}$ to help us searching $\mathbf{s}_p$. In the discussion, this work manages three portfolios-like: (i) patterns for each company profile, (ii) patterns for different company sector profiles, and (iii) patterns for the whole companies profile. In this work, the fifteen listed companies can be categorized into five sectors: basic materials and utilities, consumer cyclical, consumer defensive, industrial, financial services, and communication services.

4.2 Discussions

Each company profile. We start to discuss both $\mathbf{s}_p$ and $\mathbf{s}_{\text{fp}}$ for each listed company in this work. From Fig. 1a, the number of patterns $\mathbf{s}_p$ will decrease when the parameter δ is increased. According to the various volume values, UNVR has the highest number of profitable stock price movement patterns yet BBNI showed the lowest ones. For illustration, we set $\delta = 0.3$.[2] We discovered that ANTM has the highest profit with afu value $= 6.44 \times 10^8$ for $\mathbf{s}_p$: (Pm_s_nv, Pℓ_s_ℓv, Pℓ_n_ℓv, Pℓ_s_sv). In other words, the stock price movement of ANTM showed *medium profits* in a *short time* (for 1–2 days) with *normal volume*, then *low profits* in a *short time* (for 1–2 days) with *lots of volume*, then *low profits* in a *normal time* (for 3–6 days) with *lots of volume*, then *low profits* in a *short time* (for 1–2 days) with *small of volume* by having total profits 64.4 million. In fact, UNVR has the smallest afu value 25 million with $\mathbf{s}_p$: (Pm_s_nv, Pℓ_s_ℓv, Ph_s_sv, S_n_sv) or *medium profits* in a *short time* with *normal volume*, then *low profits* in a *short time* (for 1–2 days) with *lots of volume*, then *high profits* in a *short time* (for 1–2 days) with *small volume*, then *stable profits* in a *normal time* (for 3–6 days) with *small of volume*.

Table 1. The discovered patterns $\mathbf{s}_{\text{fp}}$ for each listed company in this work.

The pattern $\mathbf{s}_{\text{fp}}$	Company	$u(\mathbf{s}_{\text{fp}})$
Pℓ_n Pm_s	ANTM	867163900
Pℓ_n	PGAS	393665920
Pℓ_l Pm_s	BBRI	289972224
...	...	...
Pm_s	BBCA	22134600
Pm_s	UNVR	18194368

Unlike FHUIM, the SFUPM has no parameter to find the patterns $\mathbf{s}_{\text{fp}}$. That makes, SFUPM is more efficient compared to FHUIM. However, in this application, SFUPM seems difficult to provide rich stock price movement patterns since

[2] Due to the limitation space, we neglect the details information of the results.

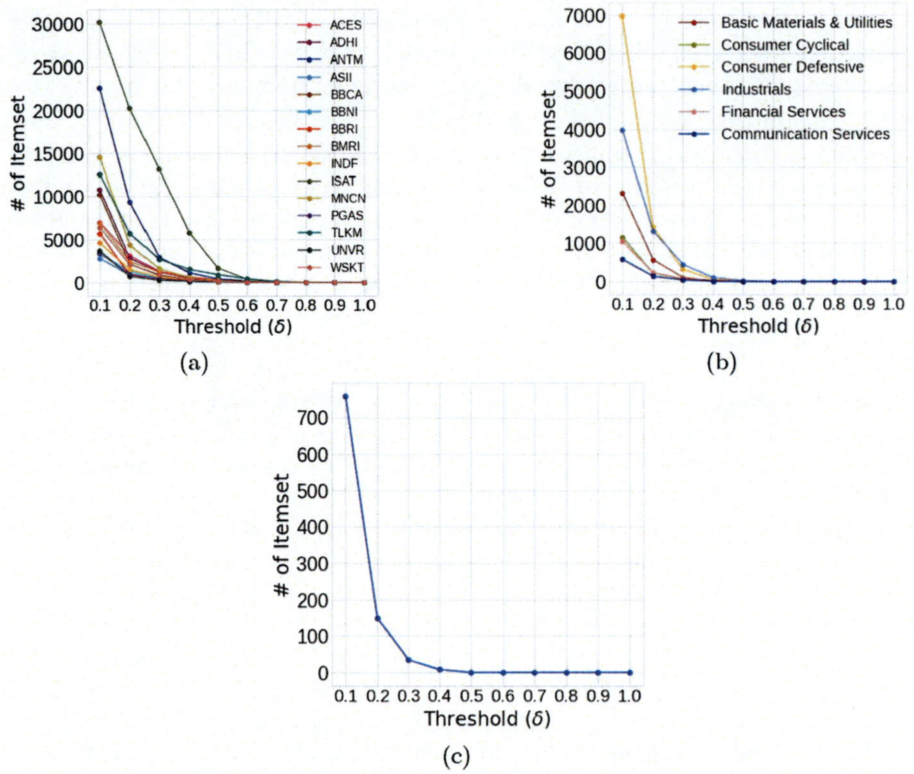

Fig. 1. The number of profitable stock price movement patterns (a) for each company, (b) for each company sector, and (c) for the whole listed company against parameter δ.

we need to verify that the stock price movement patterns are not only profitable (with the high utility) but also frequently occur in the dataset. We found that each company showed a single pattern $\mathbf{s}_{\text{fp}}$ on Table 1. The discovered patterns $\mathbf{s}_{\text{fp}}$ are more concise than $\mathbf{s}_p$. In general, we agreed to have the ANTM as the most profitable company here and the UNVR as the lowest profitable company. The stock price movement pattern of ANTM $\mathbf{s}_{\text{fp}}$: (Pℓ_n, Pm_s) or *low profits* in a *normal time* (for 3–6 days), then *medium profits* in a *short time* (for 1–2 days) with the total expected profits around eight hundred million (without fuzzification process for the volume feature). The $\mathbf{s}_{\text{fp}}$ for UNVR is *medium profit* in a *short time* (for 1–2 days) with the total expected profits around one hundred million.

The company sector profile. Then, we discuss both $\mathbf{s}_p$ and $\mathbf{s}_{\text{fp}}$ on each company sector. Similar to the previous results, if the parameter δ is getting bigger, then the number of the generated $\mathbf{s}_p$ will be smaller, see on Fig. 1b. We also can see that the consumer defensive has the largest number of $\mathbf{s}_p$ as its volume companies has the highest variance. Meanwhile, the generated patterns $\mathbf{s}_p$ on the

Table 2. The discovered patterns $\mathbf{s}_{\text{fp}}$ for each company sector.

Company Sector	Pattern $\mathbf{s}_{\text{fp}}$	Companies	$u(\mathbf{s}_{\text{fp}})$
Basic Materials and Utilities Sector	Pℓ_n	ANTM, PGAS	1.14×10^9
Communication Services Sector	Pℓ_s	ISAT, MNCN, TLKM	4.78×10^8
Financial Services Sector	Pm_s	BMRI, BBRI,	3.66×10^8
Industrial Sector	Pℓ_n Pm_s	ADHI , WSKT	2.47×10^8
Consumer Cylical Sector	Pm_s	ACES , ASII	1.04×10^8
Consumer Defensive Sector	Pm_s	INDF , UNVR	3.6×10^7

communication services sector is the smallest ones. Let us setting the parameter $\delta = 0.3$ for each company sector. Again, we discuss patterns $\mathbf{s}_p$ with the highest afu on each company sector.

The basic materials and utilities sector showed the highest total profits around one billion. It came from two companies ANTM and PGAS with the stock price movement patterns *low profits* in a *short time* with *small volume*, then *low profits* in a *normal time* with *small volume*, then *high profits* in a *short time* with *lots of volume*. The financial services sector had the second place with the total profits around five hundred million. It occurred from BBRI company. In the third place, the communication services sector showed the total profits around four hundred million from TLKM company. The fourth place is the industrial sector with the total profits around two hundred million from WSKT company. The fifth place is the consumer cyclical sector with the total profits around one hundred million from ASII company. The last one is the consumer defensive sector with the total profits around thirty million from INDF and UNVR companies.

By using SFUPM, we fail to see the stock price movement pattern in almost all company sectors (see on Table 2) as we expected in the early discussion. We can only see the stock price movement pattern in the industrial sector from ADHI and WSKT companies, such as: *low profits* in a *normal time*, then *medium profits* in a *short time* with total profits around two hundred million. Despite that, we generally obtained the most profitable coming from the basic material and utilities sector and the lowest one from the consumer defensive sector, which can be seen on Table 2.

All listed company profiles. At last, we discuss both $\mathbf{s}_p$ and $\mathbf{s}_{\text{fp}}$ for the whole listed companies in this work. Within $\delta \geq 0.4$, we cannot find any patterns of $\mathbf{s}_p$ by focusing on all companies as depicted in Fig. 1 c. By setting parameter $\delta = 0.2$, we found that several companies will leads to *low profits* or even the stock price movement patterns is getting drop in this case. We can found the stock price movement pattern increasing and remain stable at the end if the companies from the basic material and utilities sector; more specifically, ANTM and PGAS companies as what we discovered in all part of our discussions. Speaking on SFUPM, all companies showed a same pattern $\mathbf{s}_{\text{fp}}$: (Pℓ_n) with the utility value 2.7×10^9 or *low profit* in a *short time* (for 1–2 days) with the total expected profits more than 2 billion. From here, we cannot have any insight useful stock price movement patterns. Subsequently, we may say that the SFUPM cannot handle to local optima issue when mining the patterns of the whole companies due to the utilmax usage for all candidate patterns (see on Table 3).

Table 3. The discovered patterns $\mathbf{s}_p$ for all listed companies in this work when $\delta = 0.2$

Pattern $\mathbf{s}_p$	Company	afu
Pm_s_ℓv, Pℓ_s_nv, Pℓ_n_nv	ACES, ADHI, ANTM, ASII, BBNI, BBRI, BMRI, INDF, ISAT, MNCN, PGAS, WSKT	1.8×10^9
Pℓ_s_nv, Pℓ_n_nv, S_s_ℓv	ACES, ADHI, ANTM, ASII, BBCA, BBNI, BBRI, BMRI, INDF, ISAT, MNCN, PGAS, TLKM, UNVR, WSKT	1.7×10^9
Pm_s_ℓv, Pℓ_s_nv, Pℓ_n_nv, Ph_s_ℓv, S_n_ℓv	ANTM, PGAS	1.3×10^9
...	...	...

5 Conclusions

This work attempts to deliver profitable stock price movement patterns and then frequent ones given the historical stock price sequences and their volume by applying fuzzy high utility itemset mining and skyline frequent utility patterns mining. In this work, we formulated the profit stock price utility and the interval time of the stock price movement. Moreover, we inherit three portfolios-like by analyzing each company profile, each company sector profile and the whole listed companies profile. In our findings, frequent and profitable stock price patterns from SFUPM mostly fail to deliver the movement information which is caused by the local optima issue from the utilmax. Hence, we suggest to implement FHUIM instead of SFUPM for this work application. Following the listed companies in this work, we agreed to have the basic material and utilities sector as the most profitable stock price movements, especially from ANTM and PGAS companies. For future work, we will consider another feature on the historical stock price dataset, such as: high or low stock price data.

Acknowledgements. This research is partially supported by the Institute for Research and Community Development of Institut Teknologi Sepuluh Nopember, Surabaya through Lab Based Education scheme with Funding Agreement Letter of 1102/PKS/ITS/2019.

References

1. Arafah, A.A., Mukhlash, I.: The application of fuzzy association rule on co-movement analyze of Indonesian stock price. Proc. Comput. Sci. **59**, 235–243

(2015). https://doi.org/10.1016/j.procs.2015.07.541. International Conference on Computer Science and Computational Intelligence (ICCSCI 2015)
2. Chen, Y., Wu, J., Wu, Z.: China's commercial bank stock price prediction using a novel k-means-lstm hybrid approach. Expert Syst. Appl. **202**, 117370 (2022). https://doi.org/10.1016/j.eswa.2022.117370
3. Cheng, K.C., Huang, M.J., Fu, C.K., Wang, K.H., Wang, H.M., Lin, L.H.: Establishing a multiple-criteria decision-making model for stock investment decisions using data mining techniques. Sustainability **13**(6) (2021). https://doi.org/10.3390/su13063100
4. Fournier-Viger, P., Chun-Wei Lin, J., Truong-Chi, T., Nkambou, R.: A Survey of High Utility Itemset Mining, pp. 1–45. Springer International Publishing, Cham (2019). https://doi.org/10.1007/978-3-030-04921-8_1
5. Fournier-Viger, P., Wu, C.W., Zida, S., Tseng, V.S.: Fhm: Faster high-utility itemset mining using estimated utility co-occurrence pruning. In: Andreasen, T., Christiansen, H., Cubero, J.C., Raś, Z.W. (eds.) Foundations of Intelligent Systems, pp. 83–92. Springer International Publishing, Cham (2014). https://doi.org/10.1007/978-3-319-08326-1_9
6. Kocian, A., Chessa, S.: Auto regressive integrated moving average modeling and support vector machine classification of financial time series. In: Bucciarelli, E., Chen, S.H., Corchado, J.M. (eds.) Decision Economics. Designs, Models, and Techniques for Boundedly Rational Decisions, pp. 1–8. Springer International Publishing, Cham (2019)
7. Krishnamoorthy, S.: Hminer: efficiently mining high utility itemsets. Expert Syst. Appl. **90**, 168–183 (2017). https://doi.org/10.1016/j.eswa.2017.08.028
8. Lan, G.C., Hong, T.P., Lin, Y.H., Wang, S.L.: Fuzzy utility mining with upper-bound measure. Appl. Soft Comput. **30**, 767–777 (2015). https://doi.org/10.1016/j.asoc.2015.01.055
9. Lin, J.C.W., Yang, L., Fournier-Viger, P., Hong, T.P.: Mining of skyline patterns by considering both frequent and utility constraints. Eng. Appl. Artif. Intell. **77**, 229–238 (2019). https://doi.org/10.1016/j.engappai.2018.10.010
10. Liu, H., Du, H.: Stock sequence pattern mining method based on swi-gsp algorithm. In: Proceedings of the 2017 International Conference on Data Mining, Communications and Information Technology. DMCIT '17. Association for Computing Machinery, New York, USA (2017). https://doi.org/10.1145/3089871.3089875
11. Liu, M., Qu, J.: Mining High Utility Itemsets Without Candidate Generation, pp. 55–64 (2012). https://doi.org/10.1145/2396761.2396773
12. Mukhlash, I., Iqbal, M., Wulandari, M.N., Arafah, A.A.: Mining fuzzy time intervals of fuzzy stock price co-movements patterns. ICIC Express Lett. B: Appl. **10**, 859–866 (2019). https://doi.org/10.24507/icicelb.10.10.859
13. Raval, H.R., Kaushik, D.: Efficient Mining of Fast Frequent Itemset Discovery from Stock Market Data (2015)
14. Syu, J.H., Yeh, Y.R., Wu, M.E., Ho, J.M.: Self-management portfolio system with adaptive association mining: a practical application on Taiwan stock market. Mathematics **9**(10) (2021). https://doi.org/10.3390/math9101093

Automatic Detection of Oil Spills from SAR Images Using Deep Learning

Krishna Patel[1](✉), Chintan Bhatt[2], and Juan M. Corchado[3,4,5]

[1] Department of Computer Science and Engineering, Devang Patel Institute of Advance Technology and Research (DEPSTAR), Charotar University of Science And Technology (CHARUSAT), Charusat Campus, Changa 388421, India
krishnapatel.ce@charusat.ac.in

[2] Department of Computer Science and Engineering, School of Technology, Pandit Deendayal Energy University, Gandhinagar, India
chintan.bhatt@sot.pdpu.ac.in

[3] BISITE Research Group, University of Salamanca, 37007 Salamanca, Spain
corchado@usal.es

[4] Air Institute, IoT Digital Innovation Hub, 37188 Salamanca, Spain

[5] Department of Electronics, Information and Communication, Faculty of Engineering, Osaka Institute of Technology, Osaka 535-8585, Japan

Abstract. Large tankers, ships, and pipeline cracks pour oil on sea surfaces, causing significant damage and devastation to the maritime ecosystem. From the synthetic aperture radar (SAR) images, the target scenarios are ships, sea, land surfaces, look-alikes, and oil spills. Oil spill detection and segmentation using SAR pictures is critical for leak cleaning and environmental protection. The paper presents a deep learning-based framework for the oil spill identification on the dataset generated by the European Maritime Safety Agency (EMSA). In the first step, a 23-layer Convolutional Neural Network detects oil spills by classifying the patches into less than 0.5% and more than 0.5% oil spill pixels. The second step uses a U-Net architecture to apply the semantic segmentation on the generated patches. The results evidenced that the deep learning framework and segmentation models outperform the identification of an oil spill from SAR images. The present research makes a substantial contribution to future research on oil spill detection and the processing of SAR images.

Keywords: Deep learning framework · Marine surveillance · Oil spill detection · Convolutional neural network · SAR images

1 Introduction

An oil spill event is a hazard for the natural environment and has negative consequences for air, water, and land bodies. It also has the potential to pollute sea surfaces, fish, and other aquatic organisms, and harm bird species. Accidents involving oil tankers are the primary cause of oil spills. Fuel, crude oil, and gasoline are transported via ships and pipelines. Oil slicks must be removed from the body of water to keep the environment

V. Julián et al. (Eds.): ISAmI 2022, LNNS 603, pp. 54–64, 2023.
https://doi.org/10.1007/978-3-031-22356-3_6

clean and safe, and to maintain the safety of aquatic life. SAR (Synthetic Aperture Radar) is a type of radar that is commonly deployed on planes or satellites to collect data such as surface images of the sea and land. SAR's sensors emit radio waves, through which surfaces are reflected and a visual image of the targeted surface is created, including ships, sea, land surfaces, look-alikes, and oil spills, and may all be captured in SAR images. There are some images of the environmental phenomena that look-alikes could portray such as grease ice, sea wave, and shadows. In SAR photos, both look-alikes and oil spills resemble dark dots when reflected in the radio waves, which makes it difficult to differentiate between them. The literature overview of the techniques or algorithms for identifying oil spills from SAR images is presented in Table 1.

Table 1. Review of literature on the detection of oil spills from SAR images.

Paper	Dataset	Framework/algorit hm	Evaluation metrics
[12]	Database generated by European space agency (ESA)	Deep convolutional neural networks (DCNN)	Accuracy—93%
[13]	Database generated by European space agency (ESA) and EMSA	Deep Convolutional neural networks (DCNN)	Precision—86.21%
[15]	SAR images with VV polarization in NOWPAP database	Adversarial f—divergence learning	Accuracy— 96%
[20]	Oil spill images of Dongying city, China (March 2016 to December 2016)	Deep learning algorithm	mAP—99.30%, F1-Score—98.5%
[21]	Database generated by European space agency (ESA)	Deep learning based hyperspectral remote sensing technology	Accuracy 89.7%

Deep learning [1–4] and neural network classifiers have been used to categorize and segment oil spills in recent studies [5–32]. All these approaches mostly rely on the manual extraction of features i.e. semi-automatic classification or a Jaccard score to classify oil spills with moderate accuracy. The paper presents a 2-stage CNN which categorizes and segments the oil spill from the SAR images. The 23-layer Convolutional Neural Network is used in the first step to identify oil spills with over 97% specificity, accuracy and sensitivity in classifying the patches for less than 0.5% and more than 0.5% oil spill pixels. The reason behind categorizing into less than or greater than 0.5% is that some studies have already identified oil spill for a greater percentage. However, in this paper, we tried to analyze the oil spill result with a lesser patch size i.e. greater or less than 0.5%. A substantial oil spill has been detected in the second stage. The framework presented in the paper achieves 96% of accuracy, 87% of precision, and 80% of Dice loss. The proposed study improves Dice score and precision compared to cutting-edge techniques. The remainder of the paper is organised as follows. The dataset that has been used is explained in Sect. 2 along with the proposed methodology, the results of the proposed algorithm are presented in Sect. 3, and the experimental research and its findings are discussed in Sect. 4. In Sect. 5, the conclusions are drawn from the conducted research.

2 Dataset and Methods

2.1 Oil Spill Dataset Description

A fundamental limitation that the research community must overcome is that very few common datasets are available for oil spill detection. In previous research [13, 33–36] authors limited their activities to datasets that were altered depending on the approach. However, because each approach uses a distinct dataset, the reported results were not comparable. Thus, there is no common point of comparison. This subsection contains a detailed description and analysis of the dataset.

Using the Copernicus Open Access Hub (https://scihub.copernicus.eu/) database [37], the European Space Agency (ESA) acquires the satellite SAR images [40] of oil-polluted maritime areas.

The European Maritime Safety Agency (EMSA) released information on the pollution event's geographic coordinates and timestamps via the CleanSeaNet service. As a result, the EMSA data indicate that the black patches represented in the SAR photos are oil spills, resulting in a subset of solid ground truth data. The SAR images of oil pollution data are captured using the European Satellite programme Sentinel-1.

Ships, sea, land surfaces, look-alikes, and oil spills are all classes of interest represented in the images, with the latter always being viewed as a background class. The mask corresponding to each type and the number of pixels is shown in Table 2.

Table 2. Class label and number of pixels for each class.

Class	3-D mask	Number of pixels
Land	Green	9,942,113
Look-alike	Red	13,585,532
Oil spill	Cyan	3,188,485
Sea	Black	225,071,583
Ship	Brown	87,287

The oil spill dataset is openly accessible on the lab's website (https://mklab.iti.gr/results/oil-spill-detection-dataset/) [37–39] enabling the scientific community to utilize a common evaluation basis.

2.2 Semantic Segmentation

As shown in Fig. 1, the input images were first given to a frost filter which is used to remove the speckle noise present in the background from SAR images.

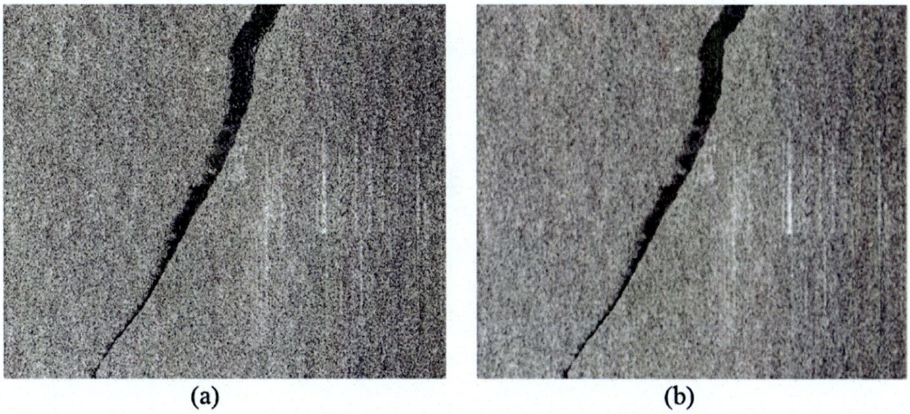

Fig. 1. (a) Original image prior to filter application (b) Resultant image following the application of the filter.

Applying the filter only removes the speckle noise and does not modify the dark object pixels (i.e. oil spill). However, while reducing the speckle noise from SAR images it might also reduce the other dark pixels which leads to confusion with the detection of oil spill instances. The application of the filter leads to better and more accurate oil spill detection.

After that, the filtered images were divided into $64 \times 64 \times 3$ patches. Furthermore, patches are categorized into two groups: one with less than 0.5% oil spill pixels and the other with more than 0.5% oil pixels. Patches with a 0.5% oil leak had about 40 pixels with a "0.5" label out of a total of 4096 pixels that can be called background patches. Furthermore, when combined with other patches, the dataset's imbalance is increased due to the patches with less than 0.5% oil spill, diminishing the model's accuracy. Figure 2 shows the deep learning architecture in which patches are given as input. The 23-layer CNN, Rectified Linear Unit (ReLU), pooling layer, SoftMax layer, and fully connected layers are used in the first stage.

Table 3. Layers of the proposed algorithm.

Layer name	Kernel size	Total no of layers
Input layer	$64 \times 64 \times 3$	1
ReLU/convolutional layer	9×9	4
Pooling layer	2×2	1
ReLU/convolutional layer	7×7	4
Pooling layer	2×2	1
Fully connected layer	2	1

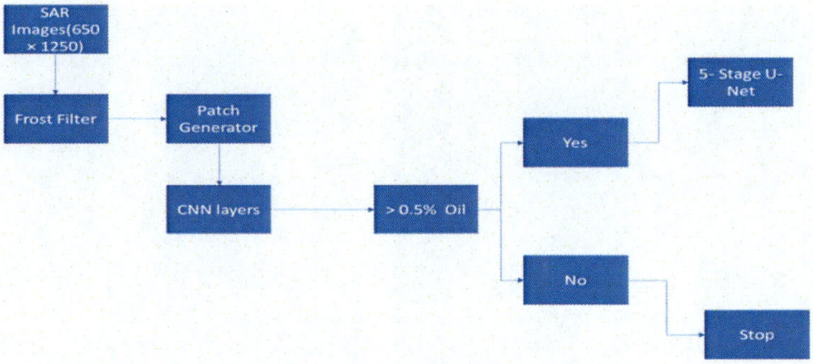

Fig. 2. Proposed architecture.

In recognised patches with more than 0.5% oil leak incidents, oil spill pixels are detected and segmented using the five-stage U-Net [41].

The bottleneck bridge, the five-stage contracting path, and the five-stage expansive path make up the U-Net, as shown in Table 3. In each contracting step, a single 2 × 2 max-pooling layer with a stride of 2 and two 3 × 3 unpadded convolutional/ ReLU layers are used. The five-channel 64, 128, 256, 512 and 1024 are included in the five steps of the contracting path. In the expansion path, every level is made up of two 3 × 3 convolutional/ReLU layers and 2 × 2 upsampling. Matching masks of 64 by 64 by 1 pixel are produced using a final 1 × 1 convolutional layer.

2.3 Evaluation Metrics

The accuracy, sensitivity, and specificity of the initial classification stage were examined in this work.

$$\text{Accuracy} = (\text{TP} + \text{TN})/(\text{TP} + \text{TN} + \text{FP} + \text{FN}) \tag{1}$$

$$\text{Recall(Sensitivity)} = \text{TP}/(\text{TP} + \text{FN}) \tag{2}$$

$$\text{Specificity} = \text{TN}/(\text{TN} + \text{FP}) \tag{3}$$

The false negative, true positive, true negative, and false-positive measures are FN, TP, TN, and FP, respectively, and are used for classification of patches of more than or less than 0.5% oil spill pixels.

The proposed model evaluates the pixel-wise precision, recall, accuracy, and Dice score (i.e., f1 score) for the detection of oil spill cases within each patch in the second stage. The following are the equations for precision and Dice scores:

$$\text{Precision} = \text{TP}/(\text{TP} + \text{FP}) \tag{4}$$

$$\text{Dice} = 2\text{TP}/(2\text{TP} + \text{FP} + \text{FN}) \tag{5}$$

where, the false negative, true positive, true negative, and false positive are the pixel-wise classification FN, TP, TN, and FP, respectively.

3 Results

The 210 images of resolution 1250 × 650 × 3 were divided into the number of patches of size 64 × 64 × 3. The generated patches are categorized into two groups, first patches with more than 0.5% oil spill pixels, and the second with less than 0.5% oil spill pixel patches. As a result, it helps to identify the patches with a significant amount of oil spill distribution. A total of 299,890 patches were chosen, and it was discovered that less than 0.5% of them had oil spills and the remaining patches had an oil spill of more than 0.5%.

Table 4 illustrates the accuracy, sensitivity, and specificity of a cross-validation model with five-fold (mean) and five-fold. The above-mentioned performance measures' standard deviation was calculated.

Table 4. Oil spill detection performance on 5-stage convolutional neural network.

Measures	Training accuracy	Testing accuracy	Sensitivity	Specificity
5-Fold (Mean)	1	0.97	0.98	0.96
5-Fold (Deviation)	7.1×10^{-5}	1.4×10^{-4}	2.4×10^{-4}	2.1×10^{-4}

Table 4 shows that the proposed CNN model outperforms the other approaches in terms of validation accuracy, sensitivity, and specificity by 97%. As a result, we predict that all patches with large oil dispersion are recognized and segmented further into the U-Net stage.

To create a segmentation mask, the five-stage U-Net will have input patches with an oil leak of more than 0.5%. To maintain a balanced pixel-wise classification and to avoid biased pixel-wise classification, the patches used to train the U-Net were of 64× 64 × 3 with 40–60% oil spill pixels.

To train the U-Net model, 212,294 patches were generated from 168 SAR images (or almost 80% of the SAR dataset). Testing is done on the remaining images of around 42 images of 684 patches (approximately 20% of the SAR Dataset). The experimental results of semantic segmentation for both the training and testing patches are presented in Table 4.

Table 5. Training and testing oil spill detection performance on second stage U-net.

Measures	Accuracy	Precision	Recall	Dice
Training Values	0.96	0.89	0.84	0.87
Testing Values	0.97	0.87	0.79	0.8

As shown in Table 5 Testing values, the proposed framework generated segmenting patches of 64 × 64 × 3 with a 97% accuracy, 87% precision, and an 80% Dice score. The performance of the testing phase is almost similar to the performance of training.

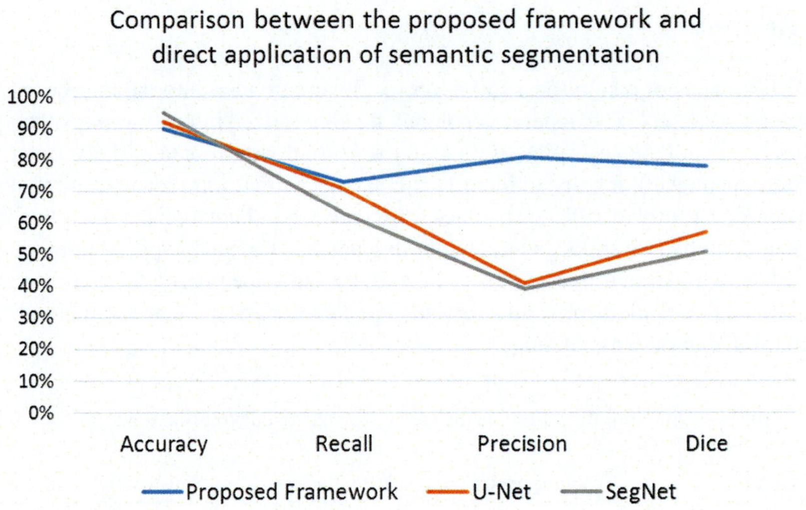

Fig. 3. Comparative graph analysis of the direct application of semantic segmentation and of the proposed framework

By using the U-Net and SegNet, the accuracy and recall were higher than those of the proposed framework. GDL loss reduction was used to update the model, which led to a significant false-positive rate, resulting in low precision and Dice loss, as shown in Fig. 3.

Figure 4 shows the detected labels and ground truth masks generated by the deep learning framework. The model proposed in the paper was successfully able to detect oil spills from SAR images.

At the same time, it was unable to precisely detect the data patches of higher background distribution as presented in Fig. 4c, and a large amount of oil spill distribution, as presented in Fig. 4d.

4 Discussion

We used deep-learning architectures to overcome the problem of recognizing and segmenting oil spills from SAR images. The work presented by Krestenitis et al. [18] used semantic segmentation, U-Net, fully convolutional CNN which failed to detect the oil spill, and simply applying the U-Net on the SAR dataset gained a maximum Jaccard score of 0.5. Recent deep learning-based algorithms [9–29] used patches that minimize the background speckle noise in the tested images, where both feature extraction and classification of SAR images can be done using the CNN structure. Several pre-trained models were introduced such as VGG-16, ResNet 101, and GAN for the categorization of patches, however, their performance was poor. The authors of [23] presented the improvement in their experiments by using two techniques, namely, VGG-16 and a generation of dark batch on SAR images. The authors discovered that removing areas with a minute presence of oil spill, using the CNN structure proposed in the paper, improved the framework's performance in semantic segmentation. The precision was increased

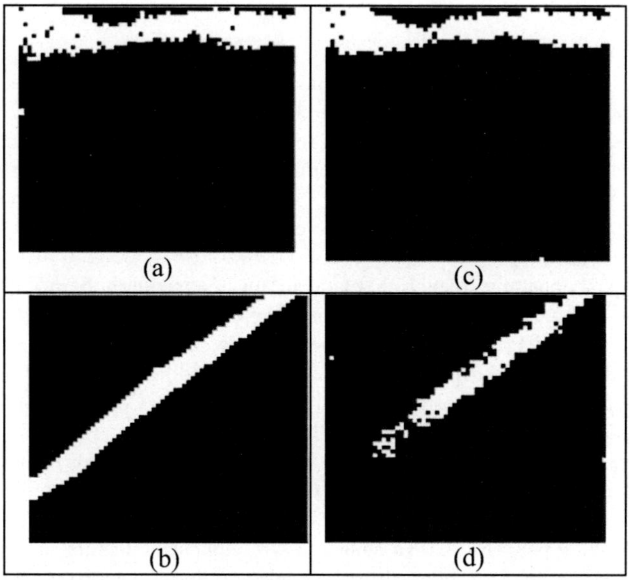

Fig. 4. Ground truth labels (a, b); mask predicted with more than 93% of accuracy (c); mask predicted with more than 78% of accuracy (d).

to 89% and the f1 score was increased to 80% when we trained the five-stage U-Net on data that had, approximately, 40–60% oil patches and testing was performed on the patches which had been generated by the 23-layer CNN.

5 Conclusion

In order to detect oil spills from SAR images, a deep learning-based framework has been presented in this paper. The framework begins with the creation of a patch wherein each image of 1250 × 650 × 3 is divided into 64 × 64 × 3 patches, the speckle noise in the image can be removed by using the frost filter, which leads to overcoming the false detection of an oil spill. To detect the oil spill from the patches, frost filter applied patches are then supplied to 23-layer CNN which is already 70 epochs of training on 80% of the training patches and received five-fold and ten-fold cross-validation (i.e., oil spill detected if more than 0.5% in the patches). The proposed framework outperforms the other model with an accuracy, sensitivity, and specificity of approximately 97%. Furthermore, the proposed framework successfully detects patches of oil spill of more than 0.5% using a five-stage U-net with a semantic segmentation stage and a Dice loss. The proposed framework had 87% precision and outperforms the state-of-the-art CNN algorithm. The following are some of the limitations of the proposed framework: (1) the deep learning framework improves the classification of oil spills and background, but it is ineffective for the detection of other classes such as sea, ship, and look-alikes. Nevertheless, the relevant research work, including ours, was focused on identifying oil spills. This will help oil leaks to be repaired and marine ecosystems to be protected.

References

1. Menegotto, A., Lopes Becker, C., Cazella, S.: Computer-aided hepatocarcinoma diagnosis using multimodal deep learning. Adv. Intell. Syst. Comput. **1006**, 3–10 (2020)
2. Alonso, R. Deep symbolic learning and semantics for an explainable and ethical artificial intelligence. Advances in Intelligent Systems and Computing, 1239 AISC, pp. 272–278 (2021)
3. Hernández, G., Rodríguez, S., González, A., Corchado, J., & Prieto, J. Video analysis system using deep learning algorithms. Advances in Intelligent Systems and Computing, 1239 AISC, pp. 186–199 (2021)
4. Shoeibi, N. cooperative deeptech platform for innovation-hub members of disruptive. Advances in Intelligent Systems and Computing, 1239 AISC, pp. 298–304 (2021)
5. Calabresi, G.; Del Frate, F.; Lichtenegger, J., Petrocchi, A.; Trivero, P.: Neural networks for oil spill detection using ERS-SAR data. In: Proceedings of the IEEE 1999 International Geoscience and Remote Sensing Symposium. IGARSS'99 (Cat. No.99CH36293), Hamburg, Germany, 28 June–2 July 1999; Volume 38, pp. 2282–2287
6. De Souza, D.L.; Neto, A.D.D.; da Mata, W.: Intelligent system for feature extraction of oil slick in SAR images: Speckle filter analysis. In: Proceedings of the International Conference on Neural Information Processing, Hong Kong, China, 3–6 October (2006)
7. Stathakis, D.; Topouzelis, K.; Karathanassi, V.: Large-scale feature selection using evolved neural networks. In: Proceedings of the Image and Signal Processing for Remote Sensing XII, Stockholm, Sweden, 11–14 September (2006)
8. Topouzelis, K., Karathanassi, V., Pavlakis, P., Rokos, D.: Detection and discrimination between oil spills and look-alike phenomena through neural networks. ISPRS J. Photogramm. Remote Sens. **62**, 264–270 (2007)
9. Singha, S.; Bellerby, T.J.; Trieschmann, O.: Satellite Oil Spill Detection Using Artificial Neural Networks. IEEE J. Sel. Top. Appl. Earth Obs. Remote Sens., 6, 2355–2363 (2013)
10. Song, D., Ding, Y., Li, X., Zhang, B., Xu, M.: Ocean Oil Spill Classification with RADARSAT-2 SAR Based on an Optimized Wavelet Neural Network. Remote Sens. **9**, 799 (2017)
11. Chen, G., Li, Y., Sun, G., Zhang, Y.: Application of Deep Networks to Oil Spill Detection Using Polarimetric Synthetic Aperture Radar Images. Appl. Sci. **7**, 968 (2017)
12. Gallego, A.-J., Gil, P., Pertusa, A., Fisher, R.B.: Semantic Segmentation of SLAR Imagery with Convolutional LSTM Selectional AutoEncoders. Remote Sens. **11**, 1402 (2019)
13. Orfanidis, G.; Ioannidis, K.; Avgerinakis, K.; Vrochidis, S.; Kompatsiaris, I.: A deep neural network for oil spill semantic segmentation in Sar images. In: Proceedings of the 25th IEEE International Conference on Image Processing (ICIP), Athens, Greece, 7–10 October (2018)
14. Nieto-Hidalgo, M., Gallego, A.-J., Gil, P., Pertusa, A.: Two-Stage Convolutional Neural Network for Ship and Spill Detection Using SLAR Images. IEEE Trans. Geosci. Remote Sens. **56**, 5217–5230 (2018)
15. Yu, X., Zhang, H., Luo, C., Qi, H., Ren, P.: Oil Spill Segmentation via Adversarial f - Divergence Learning. IEEE Trans. Geosci. Remote Sens. **56**, 4973–4988 (2018)
16. Yin, L., Zhang, M., Zhang, Y., Qiao, F.: The long-term prediction of the oil-contaminated water from the Sanchi collision in the East China Sea. Acta Oceanol. Sin. **37**, 69–72 (2018)
17. Gallego, A.-J., Gil, P., Pertusa, A., Fisher, R.B.: Segmentation of Oil Spills on Side- Looking Airborne Radar Imagery with Autoencoders. Sensors **18**, 797 (2018)
18. Guo, H., Wei, G., An, J.: Dark Spot Detection in SAR Images of Oil Spill Using Segnet. Appl. Sci. **8**, 2670 (2018)
19. Li, Y., Zhang, Y., Yuan, Z., Guo, H., Pan, H., Guo, J.: Marine Oil Spill Detection Based on the Comprehensive Use of Polarimetric SAR Data. Sustainability **10**, 4408 (2018)
20. Jiao, Z., Jia, G., Cai, Y.: A new approach to oil spill detection that combines deep learning with unmanned aerial vehicles. Comput. Ind. Eng. **135**, 1300–1311 (2019)

21. Zhu, X.; Li, Y.; Zhang, Q.; Liu, B.: Oil Film Classification Using Deep Learning-Based Hyperspectral SAR Technology. ISPRS Int. J. Geo-Inf. 2019, 8, 181. Sensors, 21, 2351 15 of 15 (2021)
22. Krestenitis, M., Orfanidis, G., Ioannidis, K., Avgerinakis, K., Vrochidis, S., Kompatsiaris, I.: Oil Spill Identification from Satellite Images Using Deep Neural Networks. Remote Sens. **11**, 1762 (2019)
23. Qiao, F., et al.: Modelling oil trajectories and potentially contaminated areas from the Sanchi oil spill. Sci. Total. Environ. **685**, 856–866 (2019)
24. Yang, J.-F., Wan, J.-H., Ma, Y., Zhang, J., Hu, Y.-B., Jiang, Z.-C.: Oil Spill Hyperspectral Remote Sensing Detection Based on DCNN with Multi-Scale Features. J. Coast. Res. **90**, 332–339 (2019)
25. Park, S.-H., Jung, H.-S., Lee, M.-J., Lee, W.-J., Choi, M.-J.: Oil Spill Detection from PlanetScope Satellite Image: Application to Oil Spill Accident near Ras Al Zour Area, Kuwait in August 2017. J. Coast. Res. **90**, 251–260 (2019)
26. Liu, B., Li, Y., Li, G., Liu, A.: A Spectral Feature Based Convolutional Neural Network for Classification of Sea Surface Oil Spill. ISPRS Int. J. Geo-Inf. **8**, 160 (2019)
27. Zeng, K., Wang, Y.: A Deep Convolutional Neural Network for Oil Spill Detection from Spaceborne SAR Images. Remote. Sens. **12**, 1015 (2020)
28. Yekeen, S.T., Balogun, A., Yusof, K.B.W.: A novel deep learning instance segmentation model for automated marine oil spill detection. ISPRS J. Photogramm. Remote Sens. **167**, 190–200 (2020)
29. Yekeen, S.T.; Balogun, A.L.: Automated Marine Oil Spill Detection Using Deep Learning Instance Segmentation Model. Int. Arch. Photogramm. Remote Sens. Spatial Inf. Sci, XLIII-B3, 1271–1276 (2020)
30. Bianchi, F., Espeseth, M., Borch, N.: Large-Scale Detection and Categorization of Oil Spills from SAR Images with Deep Learning. Remote Sens. **12**, 2260 (2020)
31. Zhang, J., Feng, H., Luo, Q., Li, Y., Wei, J., Li, J.: Oil Spill Detection in Quad- Polarimetric SAR Images Using an Advanced Convolutional Neural Network Based on SuperPixel Model. Remote Sens. **12**, 944 (2020)
32. Baek, W., Jung, H., Kim, D.: Oil spill detection of Kerch strait in November 2007 from dual-polarized TerraSAR-X image using artificial and convolutional neural network regression models. J. Coast. Res. **102**, 137–144 (2020)
33. Bohara, M., Patel, K., Patel, B., & Desai, J.: An AI-Based Web Portal for Cotton Price Analysis and Prediction. In: 3rd International Conference on Integrated Intelligent Computing Communication & Security (ICIIC 2021), Atlantis Press, pp. 33–39 (2021)
34. Kothadiya, D., Chaudhari, A., Macwan, R., Patel, K., & Bhatt, C.: The Convergence of Deep Learning and Computer Vision: Smart City Applications and Research Challenges. In: 3rd International Conference on Integrated Intelligent Computing Communication & Security (ICIIC 2021), Atlantis Press, pp. 14–22 (2021)
35. Konik, M., Bradtke, K.: Object-oriented approach to oil spill detection using ENVISAT ASAR images. ISPRS J. Photogramm. Remote Sens. **118**, 37–52 (2016)
36. Topouzelis, K., Psyllos, A.: Oil spill feature selection and classification using decision tree forest on SAR image data. ISPRS J. Photogramm. Remote Sens. **68**, 135–143 (2012)
37. Oil Spill Detection Dataset, https://mklab.iti.gr/results/oil-spill-detection-dataset/
38. Krestenitis, M., Orfanidis, G., Ioannidis, K., Avgerinakis, K., Vrochidis, S., Kompatsiaris, I.: Oil spill identification from satellite images using deep neural networks. Remote Sensing **11**(15), 1762 (2019)
39. Krestenitis, M., Orfanidis, G., Ioannidis, K., Avgerinakis, K., Vrochidis, S., Kompatsiaris, I.: Early Identification of Oil Spills in Satellite Images Using Deep CNNs. In: Kompatsiaris, I., Huet, B., Mezaris, V., Gurrin, C., Cheng, W.-H., Vrochidis, S. (eds.) MMM 2019. LNCS,

vol. 11295, pp. 424–435. Springer, Cham (2019). https://doi.org/10.1007/978-3-030-05710-7_35

40. Skobelev, P., Simonova, E., Zhilyaev, A., & Travin, V. Swarm of satellites: Multi-agent mission scheduler for constellation of earth remote sensing satellites. Lecture Notes in Computer Science (including subseries Lecture Notes in Artificial Intelligence and Lecture Notes in Bioinformatics), 10978 LNAI, pp. 348–352 (2018)
41. Ronneberger, O., Fischer, P., Brox, T.: U-Net: Convolutional Networks for Biomedical Image Segmentation. In: Navab, N., Hornegger, J., Wells, W.M., Frangi, A.F. (eds.) MICCAI 2015. LNCS, vol. 9351, pp. 234–241. Springer, Cham (2015). https://doi.org/10.1007/978-3-319-24574-4_28

Intelligent Appliances for AAL Environments*

Ricardo Costa[1(✉)], Pedro Borges[2], João Cerdeira[2], and Luís Lima[1]

[1] School of Management and Technology, Polytechnic of Porto, Felgueiras, Portugal
rcosta@estg.ipp.pt, llima@estg.ipp.pt
[2] LOQR S.A., Felgueiras, Portugal
p@loqr.io, j@loqr.io
https://www.estg.ipp.pt/ , https://loqr.io/

Abstract. Healthcare projects intend to decrease the economic and social costs of the ageing population phenomenon through the relocation of healthcare services delivery and management to the home. The VirtualECare project was one of those Ambient Assisted Living environments in which a step forward was taken with the introduction of proactive techniques for better adapting to its user's needs, namely elderly or chronic patients, by learning through interaction based on contexts. The VirtualECare2 project aimed to extend the VirtualECare platform by extending its usage from the user's home to institutions like elderly care facilities and hospitals and providing real-time information to caregivers and relatives. All this learning and information sharing, however, requires data that must comply with current and future GDPR and be locally and remotely accessible by the different modules and deployments of the project. This document exposes the initial approach of the VirtualECare2 project to the creation of an Appliance, both physical and virtual, that will allow more granular control of all this data.

Keywords: Ambient Assisted Living · e-Health · Active ageing · Privacy · Security

1 Introduction

With the evolution of technology in all sectors of society, it becomes more usual to find applications of that technology in healthcare projects intended to decrease the economic and social costs of the ageing population phenomenon [10] through the de-localisation of healthcare services delivery and management to the home. The VirtualECare project [7], initially presented in 2008, was one of those Ambient Assisted Living environments. However, again because of the fast technological evolution, the initial project was able to evolve by introducing proactive techniques for better adapting to its user's needs [4], namely elderly or chronic patients, by learning through interaction based on contexts.

* Supported by Organization LOQR S.A

V. Julián et al. (Eds.): ISAmI 2022, LNNS 603, pp. 65–73, 2023.
https://doi.org/10.1007/978-3-031-22356-3_7

1.1 VirtualECare and VirtualECare2

The VirtualECare project (see Fig. 1) envisions a new and effective way of providing healthcare. The treatment will no more be institution centred but, instead, will shift to be user-centred, resulting in a better and cheaper service for both the user and the provider. Although this may seem a contradiction at first glance, the lower costs are possible by using new, low-cost devices and technologies that allow the monitoring and assessment of patient vital data and location in real-time. This same data, after adequate processing, can conduct alarms, recommendations or even actions to assist the needed patient while he maintains his everyday, day-to-day life. These new healthcare provision approximations bring significant advantages that may also be extended to relatives and friends of the patient, since they can be informed, in real-time, of his actual condition, allowing them to also intervene in case of necessity [7,8].

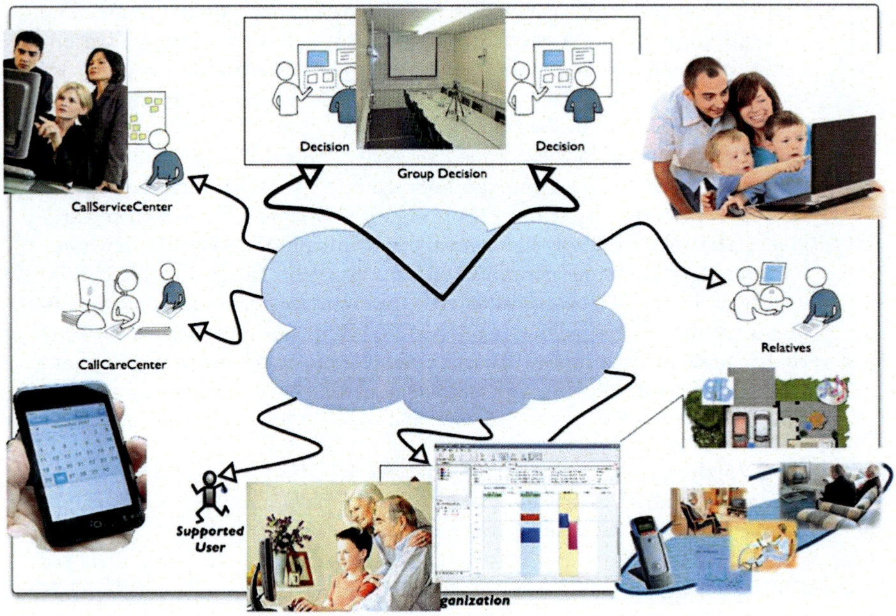

Fig. 1. The VirtualECare project.

The VirtualECare system architecture (see Fig. 2) was idealised to be a modular one to allow the addition of new modules in the future for system evolution. This project can receive new modules based on the OSGi technology and decentralised systems, either locally or in the central system (cloud). Implemented under the Multi-Agent paradigm, all the modules and agents in the system communicate with each other, creating an intricate communication and information sharing system. Each module can have several agents and control them independently, but it keeps connection with the remaining modules in a meta-communication form. Several modules are already working in both system types: sensor monitoring system, user intelligent calendar organiser, and user recognition, among others [3,5,6].

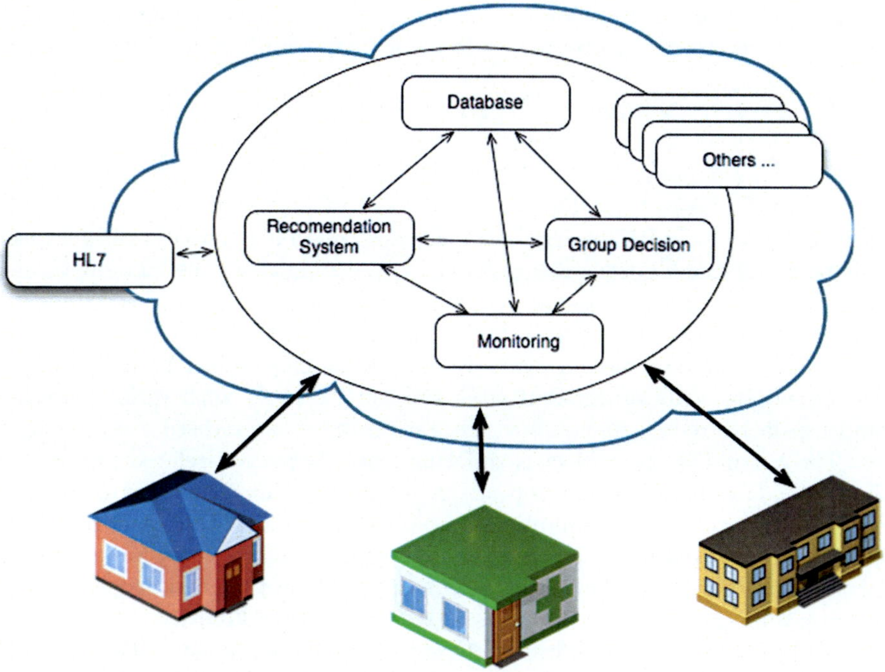

Fig. 2. VirtualECare modular architecture.

The evolution of the VirtualECare project, VirtualECare2, represents the first step in introducing new user interfaces and hardware based on TV and mobile platforms (e.g., smart TVs, tablets and smartphones). The objective is to use hardware already present in the user's traditional habitat (e.g., home). We believe that this approach will help surpass the usual user initial renitence [6].

However, all this learning and information sharing requires data and must comply with current, and future EU General Data Protection Regulation (GDPR) [1,9,11]. All the information and data sharing must be locally and remotely accessible by the different modules and deployments of the project. Additionally, these deployments are usually made by not so technical people opening the need for the process to be as smooth as possible. These were the main reason behind our Intelligent Appliance approach.

2 Intelligent Appliances

Cybersecurity is one of the main areas of activity where there must be a substantial investment. It serves as a pillar for the sustained and secure growth of the remaining areas. Data integrity and information security (among others) are essential for developing digital services, considering the current requirements resulting from data protection legislation, which obliges organisations to adequately protect the information belonging to the individuals with whom they interact [2,12].

The drive for data protection and information management is not new in the cybersecurity world and for cybersecurity professionals. However, the detail and requirements imposed by the GDPR take it to an all-new level. The GDPR is a regulatory framework for data protection and privacy that was enacted on 25th May 2018. It includes 99 articles or clauses covering virtually every aspect of business and information management—everything from the consent to collect and process information to the "right to be deleted".

There is a need for solutions that ensures the development of applications and services for different sectors, with other technological solutions that will adapt to the requirements of different sizes and must comply with various regulatory and legislative requirements in a safe, easy and dematerialised way.

The VirtualECare project is suffering from these data-related problems. We believe it is essential to develop an "Appliance" (something easy to deploy on the client site) that will enable a simplified plug-and-play installation process across the different habitats of multiple users without the need for specific skills or training. This will allow data to remain stored on the client-side and only be accessed temporarily, making it easier to comply with GDPR.

Additionally, and to address different segments, we believe the "Appliance" must have two different, but complementary, versions:

- virtual, available in the marketplaces of the leading clouds (Azure, AWS), aimed at home and small elderly care institutions, enabling the VirtualECare platform to be activated in a matter of minutes;
- physical, aimed at more prominent elderly care institutions and hospitals, which only need an electrical and internet connection to activate the VirtualECare platform after receipt by each location.

After activation, the "Appliance" will be easily configured centrally to provide access to the various VirtualECare modules in any of its formats. Additionally, these "Appliances" must also support remote diagnostics and update processes, significantly reducing the time needed to identify and resolve situations that require support services, thus reducing downtime.

2.1 Virtual Appliance

Requirements We believe the technical requirements necessary for the implementation of the Virtual Appliance regarding its compatibility with AWS and Azure clouds, would have to fulfil the following criteria:

- Packaging the different microservices developed by the VirtualECare platform in Docker containers;
- Use of a Kubernetes cluster to deploy the different VirtualECare platform microservices and some support services (e.g., Redis, Conductor and MongoDB, among others);
- Compatibility with ElasticSearch for indexing information and storing metrics;
- Persistence of structured information in SQL relational database and binary objectives in blob storage;
- Use of a load balancer to ensure fault tolerance and high availability;
- Use of an application firewall to allow content filtering following OWASP Top 10 best practices;
- Connection via Virtual Private Network to the VirtualECare central platform to access the existing remaining modules.

Proof of Concept The development of a proof of concept (PoC) for the Virtual Appliance was needed to assess the feasibility of implementing them within the scope of AWS and Azure clouds. For this purpose, specific accounts were created in those clouds to carry out experimentation activities with the available resources to identify and validate which of them fit the defined requirements. Considering the growing diversity and scope of cloud services, this work was of considerable complexity, requiring the deployment of different services and analysis and customisation of their configurations. At the end of this process, it was possible to define and validate a technical architecture for the implementation of the Virtual Appliance in each of these clouds (see Fig. 1), including:

- IAM, SSO and MFA mechanisms (e.g., AWS Control Tower) to ensure centralised management of access and the application of good security and authentication practices for users with access to the Virtual Appliance's cloud infrastructure;
- managed Kubernetes application clusters (e.g. AWS EKS) to ensure the existence of a secure and resilient control plane and the possibility of dynamic self-scaling of the infrastructure to cope with any volume of use, automatically increasing the number of nodes in the cluster when demand increases and decreasing them when they are no longer needed;
- data indexing system (e.g. AWS OpenSearch), to allow a quick search for information, essential to allow timely location (e.g. while the user is waiting) of a given process amid large volumes of accumulated data;
- persistence of information in a structured way, with support for data normalisation, searches, automated backups and geo-replication (e.g. AWS RDS), to store data collected from users efficiently and securely;
- support for storing binary blobs with support for large volumes of data reliably, with low latency and availability (e.g. AWS S3) to allow the archiving and sharing of data;
- support for high availability and fault tolerance using load balancers (e.g. AWS ALB), with monitoring of health checks to detect unavailability of nodes in an application cluster;

- possibility of filtering OSI layer 7 content (e.g. AWS WAF) with traffic inspection and HTTPS/TLS support;
- secure interconnection with VirtualECare central platform via private and authenticated IPSEC VPN (e.g. AWS PrivateLink) to allow connectivity with core modules;
- integration with a DNS management system (e.g. AWS Route 53) to allow the automated registration of specific Internet subdomains for each deployment through the Marketplace;
- possibility of managing digital certificates (e.g. AWS Certificate Manager) to allow the automated issuance of an SSL/TLS certificate for the Internet domains created for each deployment that joins the platform through the Marketplace;
- support for connecting the nodes of the application cluster via NAT protocol (e.g. AWS NAT Gateway) and from a single point (e.g. AWS Internet Gateway) to allow connection to the Internet from private addresses and facilitate the application of IP whitelisting on target services;
- storage in a shared file system (e.g. AWS EFS) in a reliable and scalable way to allow the sharing of configuration and cache files between several nodes of the application cluster so that if one node becomes unavailable, one of the others can continue processing without loss of context of requests that were being handled by the failing node;
- integration with logging systems to allow the recording of logs related to the activity of cloud services and resources (e.g. AWS CloudWatch) and of all actions/events (e.g. AWS CloudTrail) that take place on the platform associated with each customer.

2.2 Physical Appliance Requirements

Requirements Regarding the Physical Appliance, we have identified, in addition to the need to observe the generality of the points previously listed for the Virtual Appliance (to support the application layer used adequately), the existence of minimum hardware characteristics to ensure an effective and efficient operation. Thus, in the specific case of the Physical Appliance, the following additional requirements were also identified:

- rack-mountable enclosure with support for hot-swapping of disks to allow the resolution of faults in the same without stopping the operation of the services;
- Redundant power supplies;
- 512 GB of RAM (minimum);
- Xeon E5-4610 v2 8-Core 2.3 GHz CPU (or higher);
- 300 GB of disk space (minimum);
- Support for RAID1;
- Support for iDRAC technology to allow complete remote management of the machine (including formatting, rebooting, etc.);

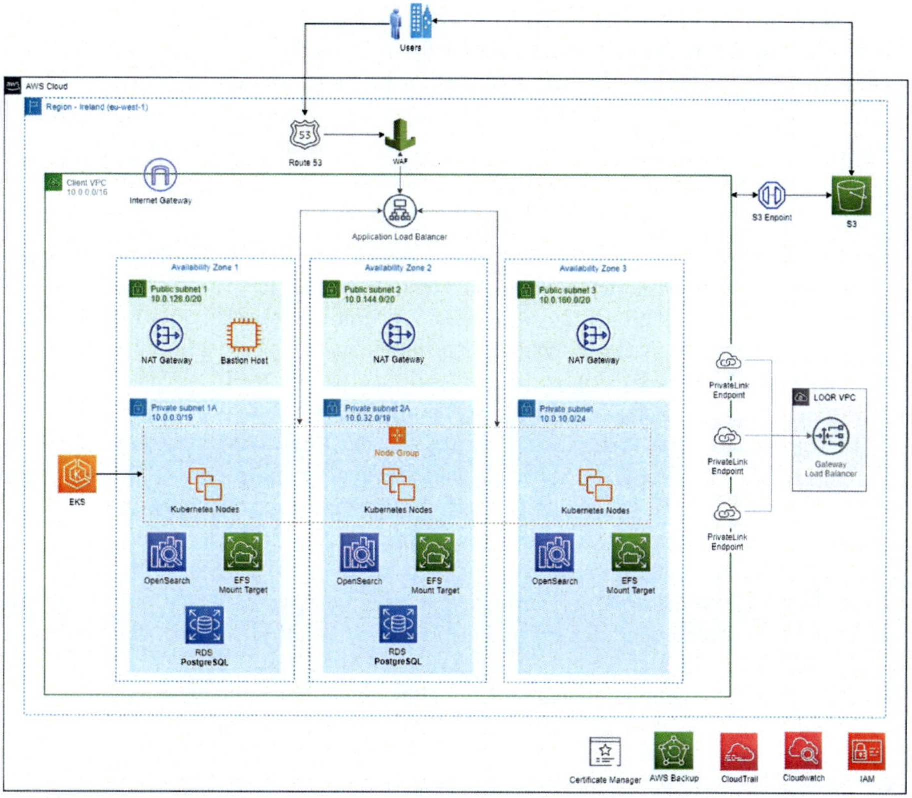

Fig. 3. Virtual appliance architecture on AWS cloud.

- to respond to growth scenarios in which large volumes of digital identities are stored (including the respective documentation, video interviews, etc.), the Physical Appliance should be able to be connected to an external SAN storage module with:
  - the rack-mountable enclosure with support for hot-swapping of disks, allowing up to 3 disks to fail simultaneously;
  - connection via fibre channel with HBA transceivers to support high levels of throughput (minimum of 100 Mb/s);
  - capacity of (at least) 32 TB;
  - support for 2.5" SAS, SATA or SSD disks;
  - support for fast volume rebuilding;
  - support for adaptive RAID mode.

Proof of Concept After completing the definition of the requirements for the Physical Appliance, a search was done for components, parts and accessories that could be used in the development of a proof of concept of the same.

Once this process was completed, some units were purchased that would allow for a configuration compatible with the defined requirements, followed by the respective assembly according to the respective manufacturer's instructions. However, to find an optimised, reliable and easily replicable configuration for future production, it was necessary to carry out iterative experimentation of different combinations and assembly options of the various components, proceeding (among others) with the following validations:

- Physical Appliance started and detected all components (processor, memories, disks, etc.) that were physically connected to it;
- It was possible to monitor and detect any failure remotely, as well as remote access and manage the Physical Appliance, to diagnose and solve any problems that might occur by updating its operating system/components, rebooting or, in extreme cases, complete reinstallation;
- In the initial tests carried out, it was found that, after exchanging several disks, some of the stored data on the Physical Appliance were lost, contrary to expectations. However, after further investigation, this was due to a RAID mode configuration error. After correcting it, the data persisted and a toleration to degradation/breakage/removal of up to 3 disks simultaneously was achieved, allowing enough time to proceed with the scheduling of an intervention to replace the defective component without shutdown the system and without causing any service degradation or data loss;
- Data transfer throughput to storage (measured for large files) reached 246.78 Mb/s, mainly exceeding the 100 Mb/s considered as a minimum requirement.

3 Conclusions and Future Work

This paper presented a much-needed evolution for the VirtualECare project to deal with the new privacy regulations and added ease of deployment. We believe introducing this Intelligent Appliance approach will open new horizons and allow this and other similar environments to quickly address these issues, allowing them to continue to grow in the target communities. In a relatively short amount of time, it was possible to start and complete this initial investigation activity, comprising four tasks, with the following results:

- Software requirements for the Virtual Appliance to be developed;
- Hardware and software requirements for the Physical Appliance to be developed;
- Proof of concept of the Virtual Appliance instantiation in a cloud environment and observing the previously identified requirements;
- Proof of concept of assembly of the Physical Appliance in a laboratory environment and fulfilling the previously identified requirements.

The successful conclusion of this activity and the achievement of these results allowed us to validate the feasibility of the proposed objectives and the approach followed. We will now start the development activities of both the Physical and Virtual Appliances to systematise the obtained information and produce optimised and economically viable versions.

References

1. Agyei, E.E.Y.F., Oinas-Kukkonen, H.: Gdpr and systems for health behavior change: a systematic review. In: International Conference on Persuasive Technology, pp. 234–246. Springer, Heidelberg (2020)
2. Al-Issa, Y., Ottom, M.A., Tamrawi, A.: ehealth cloud security challenges: a survey. J. Healthcare Eng. (2019)
3. Almeida, A., Costa, R., Lima, L., Novais, P.: Non-obstructive authentication in aal environments. In: Workshop Proceedings of the 7th International Conference on Intelligent Environments, pp. 63–73. IOS Press (2011)
4. Carneiro, D., Novais, P., Costa, R., Neves, J.: Case-based reasoning decision making in ambient assisted living. In: International Work-Conference on Artificial Neural Networks, pp. 788–795. Springer, Heidelberg (2009)
5. Costa, Â., Laredo, J.L., Novais, P., Corchado, J.M., Neves, J.: igenda: an event scheduler for common users and centralised systems. In: 5th International Workshop on Soft Computing Models in Industrial and Environmental Applications (SOCO 2010), pp. 55–62. Springer, Heidelberg (2010)
6. Costa, R., Calçada, L., Jesus, D., Lima, L., Lima, L.C.: Ami: monitoring physical activity. In: Ambient Intelligence-Software and Applications, pp. 233–239, Springer, Heidelberg (2014)
7. Costa, R., Novais, P., Lima, L., Carneiro, D., Samico, D., Oliveira, J., Machado, J., Neves, J.: Virtualecare: intelligent assisted living. In: International Conference on Electronic Healthcare, pp. 138–144. Springer, Heidelberg (2008)
8. Costa, R., Novais, P., Lima, L., Cruz, J.B., Neves, J.: Virtualecare: group support in collaborative networks organizations for digital homecare. In: Handbook of Digital Homecare, pp. 151–178. Springer, Heidelberg (2009)
9. Larrucea, X., Moffie, M., Asaf, S., Santamaria, I.: Towards a gdpr compliant way to secure European cross border healthcare industry 4.0. Comput. Standards Interfaces **69**, 103408 (2020)
10. McNicoll, G.: World population ageing 1950–2050. Population Dev. Rev. **28**(4), 814–816 (2002)
11. Orel, A., Bernik, I.: Gdpr and health personal data; tricks and traps of compliance. In: EFMI-STC, pp. 155–159 (2018)
12. Sahama, T., Simpson, L., Lane, B.: Security and privacy in ehealth: is it possible? In: 2013 IEEE 15th International Conference on e-Health Networking, Applications and Services (Healthcom 2013), pp. 249–253. IEEE (2013)

Time Series Forecasting for Improving Quality of Life and Ecosystem Services in Smart Cities

Raúl López-Blanco[1](✉), Juan Herranz Martín[1], Ricardo S. Alonso[1,2], and Javier Prieto[1]

[1] BISITE Research Group, University of Salamanca Edificio Multiusos I+D+i, Calle Espejo 2, 37007 Salamanca, Spain
raullb@usal.es, pkjuan@usal.es, ralorin@usal.es, ralonso@air-institute.com, javierp@usal.es

[2] AIR Institute—Deep Tech Lab IoT Digital Innovation Hub, Salamanca, Spain
https://bisite.usal.es , https://air-institute.com

Abstract. Quality of life is one of the factors that most influence the mood of citizens. As many studies have shown, one of the ways to increase the perception of quality of life are the actions on the Green Infrastructure of cities. Some studies have resorted to LSTM and ARIMA networks to make environmental predictions, however, as will be shown in this article, the seasonality of these models is a brake on the predictions. In order to perform efficient actions, an application case is presented, which has made use of cutting-edge methodologies thanks to IoT technology, Big Data and Artificial Intelligence to collect environmental data in order to perform time series prediction processes with them using GAM models, which have proven to be the most efficient during the tests carried out. Thanks to this work, it has been possible to obtain information on future environmental scenarios in order to make the best decisions on the influence that urban actions implemented by local authorities will have on citizens.

Keywords: Time series forecasting · Smart Cities · Pollutants analysis · Ecosystem services · Life quality

1 Introduction

Urban living, especially since the 1960s, has become the dominant trend to the detriment of rural life [30]. This trend is expected to continue in the coming years, with 68% of the population expected to live there by 2050 [35].

This progressive overcrowding of the cities [19] makes guaranteeing the quality of life of all citizens an arduous and complex job facing environmental and social problems, having to protect tangible cultural resources, as well as other more intangible ones such as innovation and business [36].

V. Julián et al. (Eds.): ISAmI 2022, LNNS 603, pp. 74–85, 2023.
https://doi.org/10.1007/978-3-031-22356-3_8

Taking into account all social situations makes many different perspectives blend together, where technological innovations, circular economy and environmental performance take special relevance [11] which focuses us more and more on the incipient Smart Cities [43].

All these considerations seek to improve the quality of life of citizens. One of the most used ways to improve the perception of quality of life is the implementation of green areas near the areas frequented by the city's inhabitants [23,40]. These types of actions are closely related to the United Nations Sustainable Development Goals (SDGs) [34]. These actions seek to improve the quality of life in the city through the development of municipal management models as a strategy for adaptation to climate change and in the search for the reduction of the urban heat island effect [17].

To achieve these objectives, innovative technology is used with IoT sensor-based stations increasingly used in all types of environments from agriculture [24] to energy saving [25], Cloud Computing and Artificial Intelligence to provide data, analysis and forecasts for current and future scenarios [9] with low-cost solutions. Aligned with Climate Change [44] mitigation and adaptation strategies [12] and essential for the improvement of citizens' quality of life, cutting-edge technology has been used to provide the best of solutions to the problem posed. IoT devices are used to capture heterogeneous [7,21] data sources, guaranteeing the security of these data through secure protocols, although other types of techniques based on Blockchain can be used [6,32]. Another of the technologies used for data management is Big Data [37], the use of which is becoming a trend for sustainability, comfort and climate change problems [28] due to the dimension of these. With the processed in numerous studies are applied on them Machine Learning techniques such as classification in climatic zones [4]. In order to communicate the results, these are transmitted to the general public in graphs and panels [15].

Combining the data collected by these technologies and applying machine learning models to make predictions on the evolution of pollutants, this work aims to achieve more accurate results using GAM models [38], which, as will be seen throughout the article, yield better results than other models traditionally used in this type of predictions, such as LSMT [10] and ARIMA [27].

The rest of the article is divided into several sections; first, there is Sect. 2 where related papers appear. Section 3 explains the data set and compares different prediction algorithms. Section 4 displays a GAM model with the collected data. Finally, in Sect. 5, the conclusions of the application of these prediction techniques and possible lines of future work are explained.

2 Related Works

In this section, the relationships that the work has with the field of sensorization and Smart Cities are discussed, and other works with which it is related due to the similarity in the treatment of data are explained. Other works with similar techniques in terms of Climate Change mitigation and Green Infrastructure are also studied. Finally, we discuss works that use prediction techniques based on Artificial Intelligence (AI).

Prediction of future situations is one of the objectives sought by Smart Cities [22], that is why data collected from IoT devices in Smart Cities [13] undergo processing to become part of datasets managed by Big Data tools [37]. These processing can be applied both at the edge [41] and in the cloud [2].

Another important aspect is data handling, as the amount of data received from IoT devices [5] and other open data sources can be a problem in the environmental sector due to the Earth being a very changeable system [20]. For data storage, a platform has been used that is able to organize data from different sources in a modular way [14] depending on their origin and respecting this modularity, which Smart Cities should count on [29].

Determinants for the solution are also the studies that have managed to demonstrate that the quality of life of citizens is closely related to the perception they have of their environment and more specifically with the state of Green Infrastructure in their area of influence as can be seen in [23,40]. This way of acting, optimizing the Green Infrastructure, allows to palliate in a simple way [1] some effects of Climate Change [3] while improving the perception of the citizens standard of living [16].

After processing, the data are used for the application of technologies such as Artificial Intelligence and Machine Learning [33] widely used for Climate Change predictions. Time series data, as is the current case, are widely used to make predictions, whether in electricity consumption [26] or to monitor the traffic of a website [8].

From the studies reviewed, it can be deduced that the LSTM and ARIMA [18] networks produce improved results, as reflected by Chang in [10]. This study aims to confirm better results for the GAM models [38] in the prediction of the evolution of pollutants than those provided by LSTM and ARIMA.

3 Forecasting Methods for Climate Data

This section is divided in two; Sect. 3.1 where a description of the dataset used in the study is made, to continue in Sect. 3.2 with a comparison of models to make time series predictions, showing their performance when acting.

3.1 Description of the Dataset

The pollutants used in the study are: CO, NO_2, O_3 and $PM_{2.5}$. The temporal evolution of these pollutants in the air of the city of Salamanca (Spain) can be seen in Fig. 1. All of them are measured in $\mu g/m^3$, except CO which is measured in mg/m^3.

A series of long periods of empty values is observed in the data set for CO and $PM_{2.5}$, namely 2128 and 2519 days, respectively. The recorded data correspond to the period between the years 1997 and 2020 (both included) for NO_2, O_3 y $PM_{2.5}$; and between 1997 and 2019 for CO.

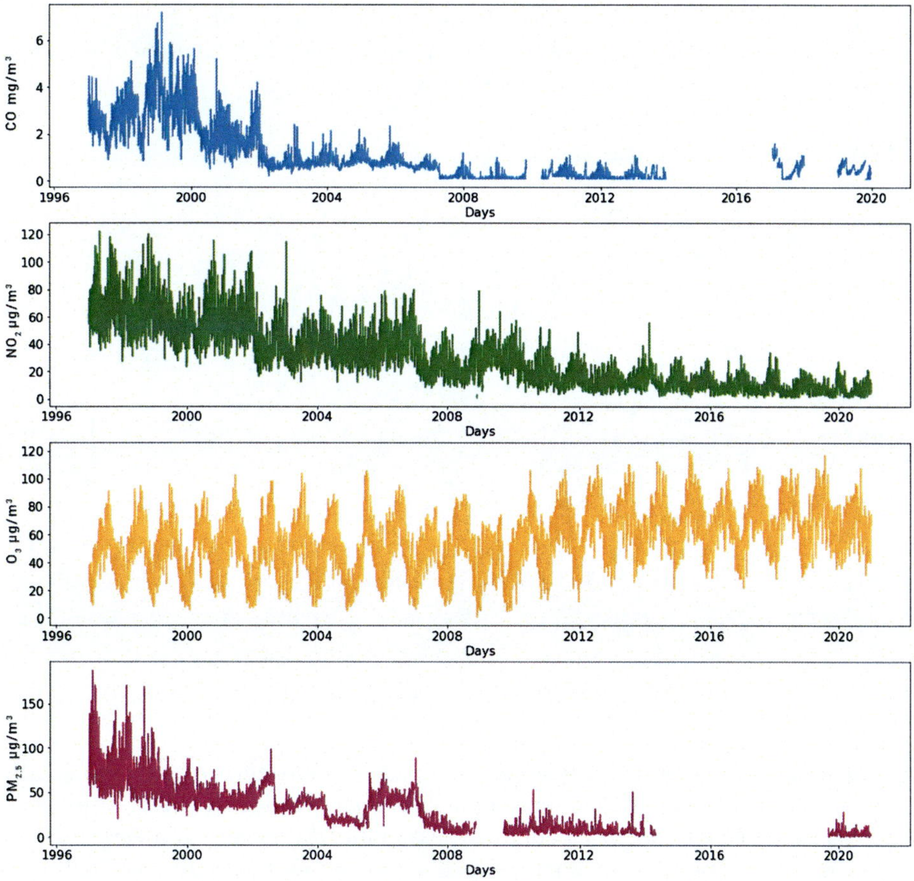

Fig. 1. Historical evolution of pollutants in the city of Salamanca (Spain).

3.2 Algorithms for Time Series Prediction

The prediction of time series requires data sets that vary over time, for this there are different types of algorithms capable of undertaking the task, among them we will study some of the most common ones such as Long Short Term Memory (LSTM) networks, Autoregressive Integrated Moving Average (ARIMA) models and Generative Additive Models (GAM).

Firstly, LSTM networks are a type of recurrent networks whose main characteristic is that information can persist in the layers of the network, generating loops that allow remembering previous states, thus generating a long-term memory, which makes them ideal for learning from situations and making predictions. However, it requires data with a very marked seasonality [42], which does not occur in the current data set. The results obtained from their evaluation do not provide predictive capacity, since the networks either return a trivial result in the form of a constant or impute the value of the previous day depending on the time window considered, as can be seen in Fig. 2.

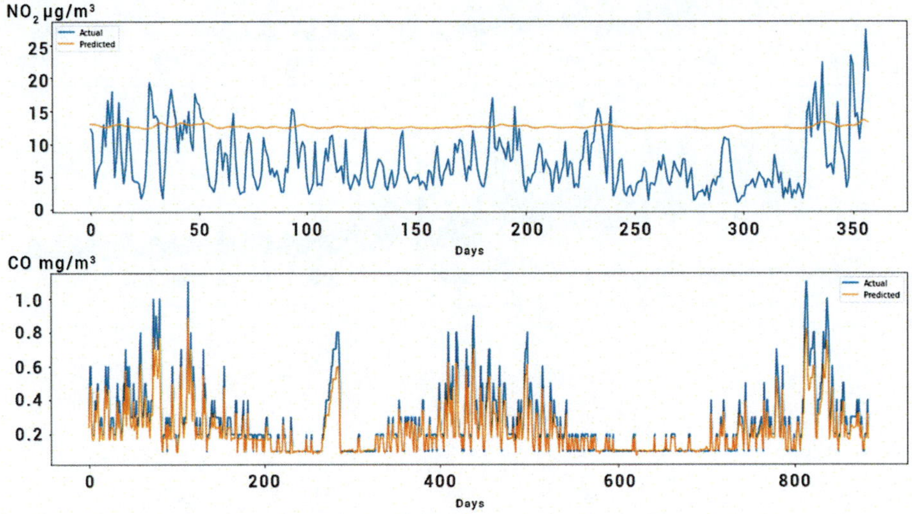

Fig. 2. Evaluation of LSTM networks with the pollutants NO_2 and CO.

The second algorithm studied is the Autoregressive Integrated Moving Average (ARIMA) model. This model is used in statistics for time series prediction and is very efficient when calculating seasonality by means of a series of components (p autoregressive, d integrated, q moving average). In addition, it has already been used in some climate studies as a [18]. Despite these characteristics and the good results of this algorithm when modeling seasonality and trends, ARIMA can fail in the face of extreme [27] values, which negatively affect the stationarity of the series. This type of extreme values can occur occasionally when we talk about environmental and air quality values because they vary a lot depending on certain human behaviors [31].

The last of the algorithms analyzed for this case study was the Generative Additive Model (GAM). This model is defined by means of the by the Eq. [39], where $g(t)$ is the trend models non-periodic changes, such as growths over time $s(t)$, seasonality presents the periodic changes (weekly, monthly, yearly), $h(t)$ adds the effects of vacations (covering potentially irregular schedules $geq1$ day(s)) and $e(t)$ which deals with idiosyncratic changes not accommodated by the model.

$$y(t) = g(t) + s(t) + h(t) + e(t) \tag{1}$$

The sum of these equations broken down below composes the mathematical equality that results in the model that is responsible for making the predictions that will be seen in Sect. 4.

$$g(t) = \frac{C}{1 + exp(-k(t-m))} \tag{2}$$

The equation Eq. 2 is the trend and is composed of the parameters C which is the carrying capacity, k which is the growth rate and m is a compensation parameter.

$$s(t) = \sum_{n=1}^{N} \left(a_n \cos\left(\frac{2\pi nt}{P}\right) + b_n \sin\left(\frac{2\pi nt}{P}\right)\right) \tag{3}$$

The equation Eq. 3 represents the stationarity which is based on Fourier series which makes the model much more adaptive by including periodicity.

The two remaining equations $h(t)$ and $e(t)$ refer respectively to the marked and predictable holiday periods and to the idiosyncratic changes that the model does not manage to add in its predictions. Due to its versatility, this algorithm has been chosen to make the predictions on the data.

4 Experiment and Results

Having defined the algorithm to be used, it is necessary to comment that the Generative Additive Model (GAM) implementation of the Prophet package has been chosen, which is very useful for data sets that contain an extended time period of detailed historical observations, with very marked seasonalities that incur important but irregular previously known events and with data points that have large outliers whose nonlinear growth trends approach a limit.

A series of predictions have been made with this package and are seen in Figs. 1 and 2. This series of predictions has used the NO_2 value as an additional regressor to predict the empty data intervals of $PM_{2.5}$ and CO, since they have a strong linear correlation (Pearson's correlation coefficient of 0.77 and 0.76, respectively). With this consideration, data are obtained to make the one-year prediction. In addition, the data set with the lowest noise and the longest possible length is selected, cutting at the beginning of the data, which is where there is the most noise and where the data differ the most from the most current levels. The first 14 years have also been removed for NO_2 and O_3, and the first 5 years for particulate matter less than 2.5 mm and CO.

To build the models, different seasonalities are used, the impact of predefined Spanish vacations, although it does not seem to have an impact on the prediction, and wind speed as an additional regressor with higher impact, testing and selecting the model with the lowest root mean square error (RMSE). The annual seasonality has been tested with different Fourier orders, and the one that best reflects the historical seasonality (lower RMSE) at annual level has been selected.

The results obtained show, as mentioned in Sect. 3.2, that LSTM and ARIMA networks are not capable of identifying temporal patterns due to the nature of the data and the fact that they deal with non-stationary and non-stationary

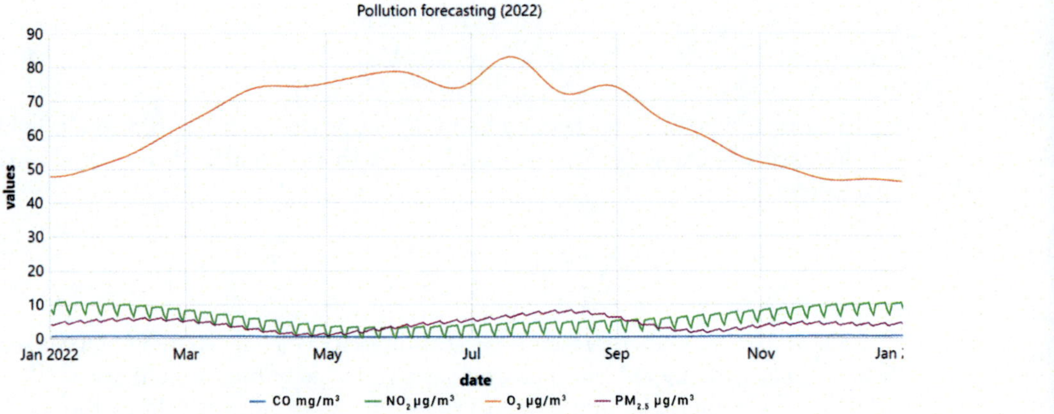

Fig. 3. Prediction of pollutant levels for the year 2022 with GAM models.

Table 1. GAM model results.

Pollutant	Mean	Standard deviation	Max/Min
NO_2	5.917	2.850	11.072/0.000
O_3	64.968	11.894	83.216/46.242
$PM_{2.5}$	4.419	1.786	8.463/0.768
CO	0.635	0.118	0.841/0.414

series. All this, together with their computational cost, means that they cannot be used as control models. Instead, a trivial reference model, corresponding to the mean value of the previous year, is used as a comparator to evaluate the RMSE obtained with the GAM model. The RMSE values obtained are those shown in Table 2.

Table 2. Comparison of the RSME between the GAM model and a trivial model.

Pollutant	RMSE	
	GAM model	Trivial model
NO_2	3.461	4.620
O_3	13.264	14.742
$PM_{2.5}$	3.536	3.550
CO	0.240	0.217

Finally, weekly seasonality has been added, also taking into account whether pollution can be directly due to the impact of traffic, since traffic seems to have weekly seasonality, being higher on daily days and decreasing on weekends, which is reflected, for example, in the levels of weekly NO_2.

From the realization of the model the error margins are extracted as shown in Fig. 4 and broken down in Table 3. In it, the error margin graphs on the prediction can be seen, where the black dots are the real history, the dark blue line is the temporal pattern it learns and uses to predict, and in lighter blue the error margin where the black dots do not appear are the two years of prediction.

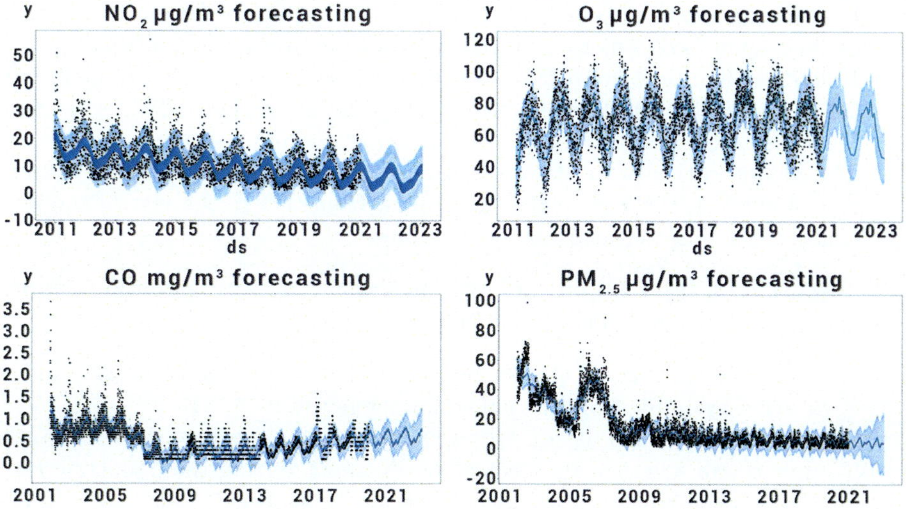

Fig. 4. Margins of error for individual pollutants.

From the RMSE values obtained, from Fig. 4 and from the results of Fig. 3 it is extracted that the predictions of NO_2 and O_3 have a clearer seasonality and less extreme events than particles and CO, therefore, they obtain better results and the error does not go up so much. For the pollutant evolution series, we can see that they are not stationary, and CO is the only one that tends to increase.

Table 3. Error margins and trends.

Pollutant	Mean	Standard deviation	Max/Min	Trend
NO_2	10.231	4.379	22.743/0.000	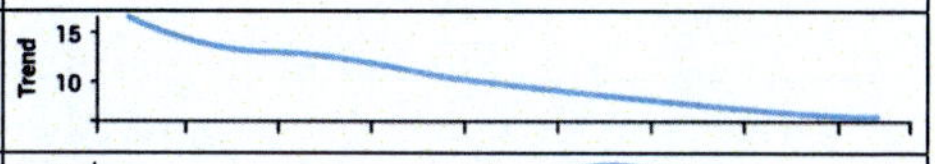
O_3	67.624	12.119	90.186/38.520	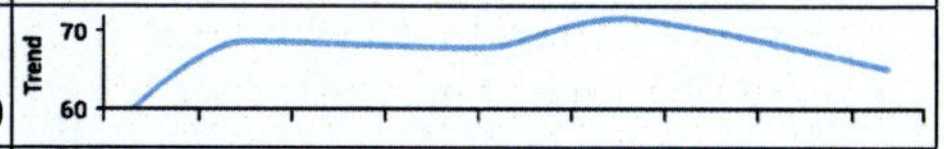
$PM_{2.5}$	15.100	13.736	55.789/0.768	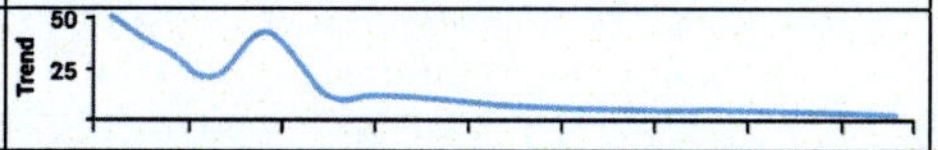
CO	0.492	0.264	1.221/0.000	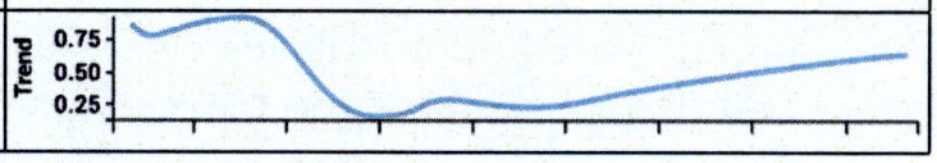

5 Conclusions and Future Work Lines

The conclusions drawn from this work highlight the usefulness of the predictions for some environmental values, as they improve the idea of the possible values of the parameters being monitored and open up lines of future work.

- Predictions and decision support systems are important for decision making, as they allow decisions to be made with all the information available.
- Understanding how prediction algorithms work is relevant to understanding the outputs they provide and why they reach conclusions. This will help to incorporate cognitive processes into decision making not previously applied.
- The conclusion is drawn that ozone levels follow an inverse seasonality (annual and weekly) to the rest of the pollutants. Contrary to the rest of the pollutants, it rises on weekends and in summer.

Regarding future lines of work, it is necessary to investigate the revision of the predictions of the model, in particular, it is needed:

- To check the accuracy of the predictions made by the algorithm with the real data when available from the collecting agency.
- To feed the algorithm with a larger training set and reduce its uncertainty.
- To investigate whether an effect of confinement during the pandemic on air quality is occurring. The prediction of the NO_2 usually remains below the highest peaks or conforming to the actual values; however, between March and May 2020 the actual values remain significantly below the prediction.

Acknowledgements. This work has been partially supported by the Institute for Business Competitiveness of Castilla y León, and the European Regional Development Fund under grant CCTT3/20/SA/0002 (AIR-SCity project).

References

1. Alkama, R., Forzieri, G., Duveiller, G., Grassi, G., Liang, S., Cescatti, A.: Vegetation-based climate mitigation in a warmer and greener world. Nat. Commun. **13**(1), 1–10 (2022)
2. Alonso, R.S., Sittón-Candanedo, I., Casado-Vara, R., Prieto, J., Corchado, J.M.: Deep reinforcement learning for the management of software-defined networks in smart farming. In: 2020 International Conference on Omni-layer Intelligent Systems (COINS), pp. 1–6. IEEE (2020)
3. Bühne, H.S., Tobias, J.A., Durant, S.M., Pettorelli, N.: Improving predictions of climate change-land use change interactions. Trends Ecol. Evol. **36**(1), 29–38 (2021)
4. Carvalho, M., Melo-Gonçalves, P., Teixeira, J., Rocha, A.: Regionalization of europe based on a k-means cluster analysis of the climate change of temperatures and precipitation. Phys. Chem. Earth, Parts A/B/C **94**, 22–28 (2016)
5. Casado-Vara, R., Chamoso, P., De la Prieta, F., Prieto, J., Corchado, J.M.: Non-linear adaptive closed-loop control system for improved efficiency in iot-blockchain management. Inf. Fusion **49**, 227–239 (2019)

6. Casado-Vara, R., de la Prieta, F., Prieto, J., Corchado, J.M.: Blockchain framework for iot data quality via edge computing. In: Proceedings of the 1st Workshop on Blockchain-enabled Networked Sensor Systems, pp. 19–24 (2018)
7. Casado-Vara, R., Martin-del Rey, A., Affes, S., Prieto, J., Corchado, J.M.: Iot network slicing on virtual layers of homogeneous data for improved algorithm operation in smart buildings. Future Gener. Comput. Syst. **102**, 965–977 (2020)
8. Casado-Vara, R., Martin del Rey, A., Pérez-Palau, D., de-la Fuente-Valentín, L., Corchado, J.M.: Web traffic time series forecasting using lstm neural networks with distributed asynchronous training. Mathematics **9**(4), 421 (2021)
9. Chamoso, P., González-Briones, A., Rodríguez, S., Corchado, J.M.: Tendencies of technologies and platforms in smart cities: a state-of-the-art review. In: Wireless Communications and Mobile Computing 2018 (2018)
10. Chang, Y.S., Chiao, H.T., Abimannan, S., Huang, Y.P., Tsai, Y.T., Lin, K.M.: An lstm-based aggregated model for air pollution forecasting. Atmos. Pollution Res. **11**(8), 1451–1463 (2020)
11. Circular Cities: Cities of Tomorrow (2018)
12. European Commission: Attitudes of Europeans Towards the Environment (Mar 2020). https://europa.eu/eurobarometer/surveys/detail/2257
13. Corchado, J.M.: Blockchain and its applications on edge computing, industry 4.0, iot and smart cities. Dieleman, S (2014)
14. Corchado, J.M., Chamoso, P., Hernández, G., Gutierrez, A.S.R., Camacho, A.R., González-Briones, A., Pinto-Santos, F., Goyenechea, E., García-Retuerta, D., Alonso-Miguel, M., et al.: Deepint. net: a rapid deployment platform for smart territories. Sensors **21**(1), 236 (2021)
15. Corchado, J.M., Pinto-Santos, F., Aghmou, O., Trabelsi, S.: Intelligent development of smart cities: Deepint. net case studies. In: Sustainable Smart Cities and Territories International Conference, pp. 211–225. Springer, Heidelberg (2021)
16. Corchado, J.M.: Technologies for Sustainable Consumption—researchgate.net (Apr 2021)
17. Deilami, K., Kamruzzaman, M., Liu, Y.: Urban heat island effect: a systematic review of spatio-temporal factors, data, methods, and mitigation measures. Int. J. Appl. Earth Observation Geoinf. **67**, 30–42 (2018)
18. Dimri, T., Ahmad, S., Sharif, M.: Time series analysis of climate variables using seasonal arima approach. J. Earth Syst. Sci. **129**(1), 1–16 (2020)
19. Duranton, G., Puga, D.: The growth of cities. Handb. Econ. Growth **2**, 781–853 (2014)
20. Faghmous, J.H., Kumar, V.: A big data guide to understanding climate change: the case for theory-guided data science. Big data **2**(3), 155–163 (2014)
21. Fan, T., Chen, Y.: A scheme of data management in the internet of things. In: 2010 2nd IEEE International Conference on Network Infrastructure and Digital Content, pp. 110–114. IEEE (2010)
22. Gharaibeh, A., Salahuddin, M.A., Hussini, S.J., Khreishah, A., Khalil, I., Guizani, M., Al-Fuqaha, A.: Smart cities: a survey on data management, security, and enabling technologies. IEEE Commun. Surv. Tutorials **19**(4), 2456–2501 (2017)
23. Giannico, V., Spano, G., Elia, M., D'Este, M., Sanesi, G., Lafortezza, R.: Green spaces, quality of life, and citizen perception in European cities. Environ. Res. **196**, 110922 (2021)
24. González-Briones, A., Castellanos-Garzón, J.A., Mezquita Martín, Y., Prieto, J., Corchado, J.M.: A framework for knowledge discovery from wireless sensor networks in rural environments: a crop irrigation systems case study. In: Wireless Communications and Mobile Computing 2018 (2018)

25. González-Briones, A., Chamoso, P., De La Prieta, F., Demazeau, Y., Corchado, J.M.: Agreement technologies for energy optimization at home. Sensors **18**(5), 1633 (2018)
26. González-Briones, A., Hernández, G., Corchado, J.M., Omatu, S., Mohamad, M.S.: Machine learning models for electricity consumption forecasting: a review. In: 2019 2nd International Conference on Computer Applications Information Security (ICCAIS), pp. 1–6 (2019). https://doi.org/10.1109/CAIS.2019.8769508
27. Grogan, M.: Limitations of Arima: Dealing with Outliers (Sep 2020). https://towardsdatascience.com/limitations-of-arima-dealing-with-outliers-30cc0c6ddf33
28. Hassani, H., Huang, X., Silva, E.: Big data and climate change. Big Data Cogn. Comput. **3**(1), 12 (2019)
29. Kitchin, R.: The promise and peril of smart cities. Comput. Law: J. Soc. Comput. Law **26**(2) (2015)
30. Llorent-Bedmar, V., Palma, V.C.C.D., Navarro-Granados, M.: The rural exodus of young people from empty Spain. Socio-educational aspects. J. Rural Stud. **82**, 303–314 (2021)
31. Ma, J., Ding, Y., Cheng, J.C., Jiang, F., Tan, Y., Gan, V.J., Wan, Z.: Identification of high impact factors of air quality on a national scale using big data and machine learning techniques. J. Cleaner Prod. **244**, 118955 (2020)
32. Mezquita, Y., Casado, R., Gonzalez-Briones, A., Prieto, J., Corchado, J.M., AETiC, A.: Blockchain technology in IoT systems: review of the challenges. In: Annals of Emerging Technologies in Computing (AETiC), Print ISSN pp. 2516–0281 (2019)
33. Milojevic-Dupont, N., Creutzig, F.: Machine learning for geographically differentiated climate change mitigation in urban areas. Sustain. Cities Soc. **64**, 102526 (2021)
34. United Nations: United Nations Sustainable Development (2015). https://www.un.org/sustainabledevelopment/
35. United Nations: Around 2.5 Billion More People will be Living in Cities by 2050 (May 2018). https://www.un.org/development/desa/en/news/population/2018-world-urbanization-prospects.html
36. Oppio, A., Bottero, M., Stanghellini, S.: Integrated Evaluation for the Management of Contemporary Cities (2016)
37. Parra Domínguez, J., Rodríguez González, S., Prieto Tejedor, J., Manuel Corchado, J., Marreiros, G., Ramos, C., et al.: Actas del iii taller de tecnologías de la información y la comunicación disruptivas para la innovación y la transformación digital: 18 de diciembre de 2020, online (2022)
38. Querejeta, M.U., Alonso, R.S.: Modeling air quality and cancer incidences in proximity to hazardous waste and incineration treatment areas. In: Second International Workshop on Data Engineering and Analytics (WDEA 2019), pp. 108–122 (2019)
39. Robson, W.: The Math of Prophet (Nov 2020). https://medium.com/future-vision/the-math-of-prophet-46864fa9c55a#:~:text=Prophet%20is%20a%20procedure%20for,several%20seasons%20of%20historical%20data
40. Santos, T., Silva, C., Tenedório, J.A.: Promoting citizens' quality of life through green urban planning. In: International Conference on Geographical Information Systems Theory, Applications and Management, pp. 153–175. Springer, Heidelberg (2017)
41. Sittón-Candanedo, I., Alonso, R.S., Corchado, J.M., Rodríguez-González, S., Casado-Vara, R.: A review of edge computing reference architectures and a new global edge proposal. Future Gener. Comput. Syst. **99**, 278–294 (2019)

42. Wang, Y., Zhu, S., Li, C.: Research on multistep time series prediction based on lstm. In: 2019 3rd International Conference on Electronic Information Technology and Computer Engineering (EITCE), pp. 1155–1159. IEEE (2019)
43. Yigitcanlar, T., Butler, L., Windle, E., Desouza, K.C., Mehmood, R., Corchado, J.M.: Can building "artificially intelligent cities" safeguard humanity from natural disasters, pandemics, and other catastrophes? an urban scholar's perspective. Sensors **20**(10), 2988 (2020)
44. Zhongming, Z., Wei, L., et al.: Urban Adaptation to Climate Change in Europe 2016-Transforming Cities in a Changing Climate (2016)

Check for updates

Security of IoT Application Layer: Requirements, Threats, and Solutions

Mahmoud Abbasi, Marta Plaza-Hernández(✉), and Yeray Mezquita

BISITE Research Group, University of Salamanca, 37007 Salamanca, Spain
mahmoudabbasi@usal.es, martaplaza@usal.es, yeraymm@usal.es

Abstract. Nowadays, communication systems and networks, become an integral part of not only our everyday life but also even in industry, critical infrastructure, business, etc. Current trends and paradigms, such as Internet of Things (IoT), promise improved quality of life, better business opportunities, cost-effective production, and careful operation management via ubiquitous connectivity and effective deployment of smart physical objects. IoT networks capture, gather, process, and transmit a huge amount of security- and privacy-critical data, then becomes them an available option for intruders. Given the fact that IoT networks deal with the fundamental aspects of human lives and critical infrastructures, cybersecurity in such networks is crucially important. The unique characteristics, large scale, and complexity of these networks bring about further vulnerabilities. The goal of this paper is to shed light on the security requirements of the IoT application layer, its threats, and potential countermeasures.

Keywords: Internet of Things · Security · Privacy · Security requirements · Taxonomy

1 Introduction

Internet of Things (IoT) paradigm refers to the ever-growing network of smart-physical devices that can sense and act on their environment, process information, communicate, and share data to fulfill their ultimate objectives [1,2]. More specifically, IoT systems play an active part in various sections, including industry, transportation, healthcare, smart grids [3], business [4], agriculture, etc. The successful implementation of IoT-powered systems in different sections lead to a substantial increase in the number of connected objects that is predicted to reach several billion in the upcoming years [5].

Accordingly, many industries and organizations are developing IoT-enabled products, services, and solutions to break into and dominate the market [6]. On the other side, one of the key objectives of IoT is to change the way we live today through creating smart devices and services that can carry out our daily tasks. Smart cities [7], smart farming [8], smart grids [9], smart healthcare [10],

V. Julián et al. (Eds.): ISAmI 2022, LNNS 603, pp. 86–100, 2023.
https://doi.org/10.1007/978-3-031-22356-3_9

smart environment [11] etc. are the concepts that are introduced in connection with IoT technology.

Despite these advantages, there are still several challenges to ensuring the full and effective involvement of IoT in the real world. One of the main challenges that IoT faces and must be overcome is security [12]. Many studies have reported evidence of security and privacy issues in IoT-enabled systems, such as authentication, access control [13], Denial-of-Service (DoS) attacks, and information leakage [14]. Indeed, not only the number of IoT security threats is ever-increasing, but also the complexity and the number of potential intruders is growing daily [15].

In the context of IoT security, several survey papers have been published [14,16]. However, few studies have been conducted to investigate the security aspects of IoT layers individually. In an attempt to fill this gap and in response to concerns about the security of the IoT application layer, our aim in this paper is to investigate a structural survey of the application layer security by introducing the key security requirements, threats, and the existing solutions. Moreover, open issues and future directions in the IoT application layer security are presented.

The paper is organized as follows: Sect. 2 provides the necessary background and motivation for our survey. Section 3 pursues a careful investigation into the necessary security requirements in the IoT application layer. Regarding the provided classification, security threats and proposed solutions in the IoT application layer is discussed in Sect. 4. Section 5 presents the challenges and future research directions in the IoT application layer. Finally, our paper is concluded in Sect. 6.

2 Background and Motivation

The description of the most adopted IoT architectures and the definition of the IoT layers and their functions are key steps towards understanding the IoT systems. Various architectures have been proposed for IoT systems by the scientific community and industry. In this paper, a three-layer architecture is selected as a reference for the definition of the IoT layers and their functions since it is the most common IoT architecture (see Fig. 1) [17]. In addition, to narrow down our search and investigate the topic as carefully as possible, our primary focus is on the IoT application layer in this architecture.

Application layer is designed as the top layer in the three-layer IoT architecture. The application layer accepts the network-level data from the second layer and utilizes the data to deliver desired services or operations. As an example, the application layer is capable of providing the data analysis service to discover useful details to forecast the condition of physical devices in the future [18]. Considering *network layer*, it is implemented as the middle layer in the three-layer IoT architecture, and is also named the transmission layer [19]. One of the main functions of the network layer is to route the processed data provided by the perception layer. In other words, this layer sends the data to the IoT devices,

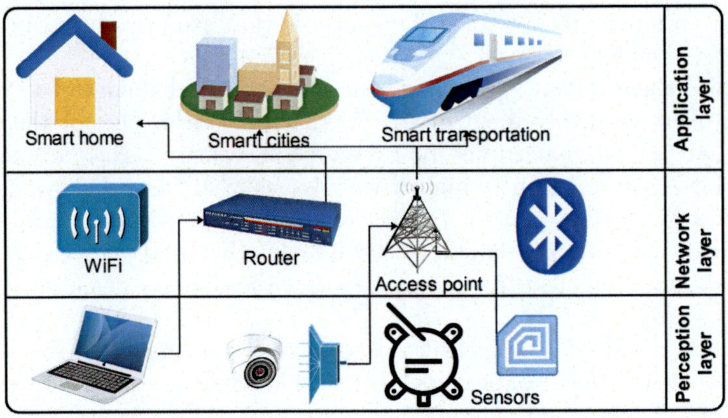

Fig. 1. Three-layer IoT architecture.

services, etc. through the communication network. *Sensor layer* is another name for *perception layer*. The perception layer is designed as the bottom layer in the three-layer IoT architecture and can interact with physical objects and entities in an IoT system via smart devices such as RFID tags and various sensors.

2.1 Motivation

There is an increasing number of IoT devices integrated into security- and safety-critical applications, such as smart cities, industrial automation, smart vehicles [20], and smart grids. Furthermore, IoT devices are capable of collecting, pre processing, and transmitting security-critical and sensitive private data, and hence, they are interesting targets for different attackers [21]. Hence, IoT security is of paramount importance. In a nutshell, IoT systems are more prone to cyber-attacks than the traditional communication systems due for the following reasons:

- Most of the IoT systems use wireless protocols for communications where a malicious actor could acquire confidential data by eavesdropping on the wireless channel [22,23].
- Most IoT devices are resource-constrained in terms of power, computation, memory, security algorithms, etc. [24].
- The growing complexity and heterogeneity of IoT systems also complicates the security and privacy issues fronted by such systems [25].
- Most IoT systems are usually large-scale and distributed in nature [25,26].

As mentioned, in this paper we consider the three-layer architecture for IoT systems. Based on this architecture, the main focus of the paper is on the application layer and providing a taxonomy of security requirements, security threats, and the associated solutions. Towards this aim and taking right steps towards our study, the security of IoT application layer is investigated from IoT use cases point of view (e.g., smart grids and cities).

3 Security Requirements in IoT Application Layer

Before the study of security threats of the IoT application layer, it is essential to investigate the requirements this layer must meet for the proper operation of the IoT systems. Failure to meet a security requirement may bring about a possible security challenge to the system. The key security requirements in the IoT application layer are listed below. These requirements are identified by careful investigation of the papers related to the security of IoT use cases. IoT and security, IoT application layer and security, IoT application layer protocols and security, IoT and privacy, etc., are the keywords that were used to search the literature in our paper. The literature was extracted from major scientific databases/publishers, including IEEE, ScienceDirect, MDPI, Elsevier, ACM, Springer, etc. [27–35].

3.1 Confidentiality

Confidentiality refers to protecting information from unauthorized access or who are not allowed to view it. Confidentiality may also refer to preserving the IoT devices and equipment from unauthorized access.

When considering the IoT use cases, confidentiality protection is challenging due to the different involved devices and components [36,37]. For example, in an Intelligent Transportation System (ITS), there are different devices such as smartphones, vehicles, roadside stations, cameras, and sensors. In some IoT use cases (e.g., IIoT and smart grids), the lack of confidentiality countermeasures can lead to the loss of customer data and vendors' data, and intellectual property such as trade secrets.

3.2 Integrity

Data/message integrity means that a message was not changed over its life cycle (i.e., between sending and receiving). In other words, it refers to the consistency, accuracy, and validity of data over workflow [38]. In IoT systems, integrity can safeguard the system against the unapproved spread, destruction, or changing of messages. In IoT use cases (e.g., healthcare, smart transportation and agriculture), it is crucially important to ensure the data integrity regarding communication and computation between different system entities, such as various sensors, actuators, controllers, human agents, etc. This is mainly due to the fact that these entities can collect massive amounts of important data. For example, in smart agriculture scenario, there are many IoT sensors and smart meters that gather different types of data, e.g., humidity, temperature, and water data. The altering of this data can lead to serious damages to different involved operations, e.g., pH of agricultural water and the applied nutrient solution for plants.

3.3 Availability

Availability is vital in IoT systems and guarantees that service and network will stay operating even in the condition of faults or malicious activities. Availability needs not only security but also the fault management process (i.e., fault detection, isolating, and then correcting the abnormal condition of the network) [39]. For IoT systems, especially safety- and mission-critical IoT systems, such as smart grids and ITS, it is important to guarantee availability of the systems. Because these systems deal with the safety of the users and the real-time functional requirements of such systems. For example, to guarantee the safety of passengers, ITS's involved devices need to be able to operate and communicate with each other [40]. The forecasting of potential bottlenecks and providing bandwidth need to be considered.

3.4 Authentication and Authorization

It is one of the principal requirements for any communication system and ensures that the right users (e.g., patients and physicians in a smart healthcare system) or devices (e.g., nodes and aggregators) can get access to the resources or do something [41]. The resources may refer to data, hardware equipment, and the services provided by a IoT network. For example, granting access to the electronic health records and patient records. In the vast majority of IoT applications, e.g., vehicular networks and ITSs, the authentication of all users and messages is vital as it can prevent serious security threats such as Sybil attacks.

4 Security Threats and Solutions in IoT Application Layer

The security of the IoT application layer is an integral part of the system design. IoT application layer protocols are the foundation for the communications among various IoT use cases, devices, and running services. In other words, IoT application layer protocols serve as an interface between the IoT use cases and end-users [42]. Hence, considering the important role of the application layer in all of the IoT use cases, security at this layer is crucial. Moreover, the intruders in the IoT application layer are probably going to disturb privacy through different attacks, such as injection attacks, unauthorized access, cross-site scripting attacks, etc.

4.1 Focus on IoT Use Cases

After extensive review and analysis, we identify five important IoT applications, including smart grids, smart healthcare, ITS, smart agriculture, IIoT, and smart cities. In the following sections, we discuss the security aspects of these applications.

Smart grids The main security goals in smart grids are: confidentiality, integrity, and availability [43]. Concerning these security requirements, one can refer to the following security threats in smart grids.

Threats: There are several types of attacks that target *confidentiality* in smart grids, including password-pilfering attacks, traffic analysis attacks, eavesdropping attacks, unauthorized access, false data injection attacks, and password theft attacks. The main objective of these attacks is to gain the desired information. There is another group of attacks that try to destroy the *integrity* of smart grids, such as data tampering attacks, wormhole attacks, data injection attacks, spoofing attacks, data manipulation attacks, man-in-the-middle attacks, and masquerading attacks [27]. The main goal of these types of attacks is to change the original payload of data. The *availability* of smart grids also can be endangered through the availability-related attacks, such as jamming, wormhole, DoS attacks (e.g., teardrop, LDoS, puppet, and smurf), buffer overflow, masquerading, man-in-the-middle attacks, and spoofing attacks [44].

In addition, using monitoring technologies, such as Advanced Metering Infrastructure (AMI) may cause privacy violation risks for users (privacy issues) [28]. For example, extracting habitual patterns information by adversaries or dissemination of industrial information. Moreover, the massive number of deployed devices and the heterogeneity of devices can raise key scalability issues for security providing.

Solutions: To deal with the security threats that target the *confidentiality* of smart grids, several methods have been proposed [45]. For example, one can use data encryption against password theft attacks. Deploying authentication mechanisms can prevent eavesdropping attacks, unauthorized access, and false data injection attacks. Moreover, using encryption protocols can prevent traffic analysis attacks. To cope with data *integrity* attacks, some solutions have been introduced. Cryptography techniques and algorithms, and authenticity are among the most used methods to prevent data *integrity* attacks [46]. Moreover, methods such as power fingerprinting techniques, strategies based on trusted network connect, and volt-var control algorithms also have been developed. Using security gateways to encrypt the traffic can be a remedy for man-in-the-middle attacks. In addition, to reduce the consequences of the data injection attack, spoofing attack, and data manipulation attack, using end-to-end encryption and authentication mechanisms is crucially important. The following measures have been made to cope with the *availability* attacks. For mitigation of DoS attacks, traffic filtering technologies, anomaly detection methods, and air gapping are promising solutions. Given jamming attacks, anti-jamming techniques can be adopted, such as [47].

Smart healthcare Regarding the applications of IoT in healthcare, there are two main concerns/requirements, including security and privacy. When it comes to security, the key requirements are confidentiality, integrity, authentication, and authorization.

Threats: Data *confidentiality* in smart healthcare systems can be endangered through unauthorized users and eavesdropping attacks [48]. Furthermore, during data transmission, adversary users and accidental communication mistakes can destroy data *integrity* in such systems.

In the smart health systems, the *authenticity* of the users (e.g., patient and physicians) and devices (e.g., nodes and aggregators) should be ensured in order to prevent from masquerading attacks against electronic health record and patient health records [49]. Moreover, authorization ensures that the right users (e.g., patients and physicians) or devices (e.g., nodes and aggregators) can get access to electronic health records and patient health records.

Besides the challenges related to security, wearable devices in smart health systems can be used for measuring data about blood pressure, temperature, heart rate, blood sugar, etc. This data usually is stored in a cloud server in the form of Personal Health Record (PHR) for further processing and analysis by physicians. As this data is vital and personal, privacy concern is the most critical security issue in healthcare-related IoT applications.

Note that some literature also refers to *data freshness* as a security requirement in smart healthcare [50]. Repeat/replay attacks is one of the often mentioned challenges endangered data freshness.

Solutions: Using cipher algorithms for data encryption is a remedy to deal with data *confidentiality* security challenges. Considering the security challenges related to data *integrity*, ensuring data integrity by means of cryptography algorithms such as AES128/256 and SHA is a solution [29].

To deal with *authentication* security challenges, different authentication mechanisms should be utilized, such as digital signatures, key based and certificate-based authentication. Additionally, to ensure *authorization* in a smart healthcare system, the access control mechanisms should be used to define the right access for each user in the system. Moreover, to address the privacy issues in smart healthcare applications, developing secure access control approaches for wearables and PHR should be considered. Furthermore, as PHRs are stored in cloud servers, using cryptographic primitives to improve the authentication protocols of PHRs is possible. In the healthcare systems, when one accesses the information, the authentication mechanisms should be human-machine authentication, while for updating the collected data in the server machine-machine authentication will work.

One of the ways to mitigate repeat/replay attacks is to assure *data freshness* through verifying the data collected from the devices (e.g., sensors). The verification can be done by different factors, such as is how up-to-date data is, non-duplication, and order of data.

Smart transportation systems (ITSs) Regarding security in ITSs, the key requirements are confidentiality, integrity, availability, and authentication/identification [51]. Indeed, the different security threats in ITSs can be classified from these requirements point of view.

Threats: *Confidentiality* protection in ITSs is challenging because there are different types of devices in an ITS, such as smartphones, vehicles, roadside stations, and IoT devices. Hence, there is a wide range of attacks against the involved devices that can destroy confidentiality. These attacks are man-in-the-middle attacks, eavesdropping attacks, model identification attacks against used machine learning techniques, parameter inference attacks against controllers [52]. Moreover, in ITSs, it is crucially important to ensure the data *integrity* regarding communication and computation between different system devices, such as vehicles, traffic controllers, and roadside infrastructures. There are different potential security risks against data integrity in ITSs, including spoofing attacks, timing attacks [53], Sybil attacks, man-in-the-middle attacks, attacks against machine learning with adversarial examples, and data poisoning, and policy manipulation attacks.

To guarantee the safety of passengers, it is crucial for ITS's involved devices to be able to operate and communicate with each other. Different attacks can restrict the *availability* of devices in ITS, such as DoS, spoofing attack, timing attack, jamming attack, man-in-the-middle attack, policy manipulation attacks, and data poisoning [30]. Regarding *authentication/identification*, it is crucially important for an ITS to properly identify and authenticate the users who want to participate in communication and data transmission [54]. This is because many security threats are posed through different types of attacks, including spoofing, timing attack, Sybil attacks, and man-in-the-middle attack.

Solutions: To alleviate *confidentiality-related* security challenges, a couple of techniques have been proposed, including symmetric cryptography, asymmetric cryptography, and a secure steganographic algorithm. Each of them has its own pros and cons. When considering data *integrity*, Message Authentication Code (MAC) is one of the main approaches to ensure data integrity in ITSs [55]. However, using this technique can cause additional computational overhead.

To cope with the *availability-related* security challenges, signature-based authentication techniques have been proposed [31]. The most important problem with this method is that it needs additional infrastructure. In addition, challenge-response protocols and message authentication codes are provided for security challenges related to *authentication* and *identification*. These methods can pose overhead in terms of time and computation.

Smart agriculture One can classify the security risks in smart agriculture into five main categories, including threats against privacy, authentication, data confidentiality and integrity, and availability.

Threats: In smart agriculture applications, there are many IoT sensors and smart meters that collect different types of data, e.g., humidity, temperature, and water quality monitoring [32]. The collected data is sensitive as the analysis of this data can disclose valuable information (e.g., the applied nutrient solution for plants and the locations of sensors) to a third party. Hence, it is important to preserve this *private information* from unauthorized access and security threats such as insider data leakage and cloud data leakage. As for *authentication-related* security challenges, a malicious user (or program) tries to forge an identity in order to enter the system as an authorized node [56]. Toward this end, the malicious may carry out different attacks, such as impersonation, spoofing, replay attack, and masquerade attack.

When it comes to data *confidentiality*, the main goal of an attacker is to stand in an ideal place to eavesdrop on the communication between IoT devices or IoT devices with an access point. There are different types of eavesdropping attacks in smart agriculture, including brute-force attacks, tracing attacks, known-key distinguishing attacks, and false data injection attacks [57]. As the name implies, the main goal of the attacks against *availability* is to services become unavailable in a smart agriculture system. DoS and jamming attacks are the main types of threats in this category [58].

Last but not least, smart agriculture systems are also subjected to data *integrity* attacks [59]. This type of attack let unauthorized entities access and modify sensitive information, such as the pH of agricultural water. A man-in-the-middle attack, forgery attack, biometrics attack, and Trojan can be included in this category.

Solutions: Different solutions have been proposed to deal with *privacy-related* challenges, including privacy-preserving techniques during the data aggregation process in a smart agriculture system, location privacy solutions, content-oriented protection, data anonymization techniques, and privacy-preserving trust evaluation methods. To reduce the threats related to data *integrity*, some solutions have been proposed, such as label-based access control technique [60], content integrity verification [61], and message authentication codes.

To provide *authentication*, different solutions have been proposed. For example, RFID authentication methods alleviate the situation when one uses RFID tags in smart agriculture [62], delegated authentication, label-based access control, and blockchain-based access control. Access control algorithms based on cipher text is one of the solutions to preserve *confidentiality* in smart agriculture. Moreover, blockchain-based access control mechanisms can be adopted in smart agriculture systems.

Industrial IoT Based on [33, 63], the main security requirements in IIoT are authentication, data/traffic flow confidentiality, integrity, and availability.

Threats: In IIoT, *authentication* is an important security requirement in order to preserve the legality of data access, and consequently, to guarantee data confidentiality. In an IIoT system with an ineffective authentication mechanism, false data injection and spoofing attacks can be launched. These types of attacks can inject adversarial code and commands into the system [64] for different purposes, such as controlling industrial machinery and performing unsafe operations.

In the context of industrial IoT systems, *confidentiality* refers to ensuring data/traffic flow access only by authorized entities. The lack of confidentiality countermeasures in an industrial system can lead to the loss of customers' data and vendors' data and intellectual property such as trade secrets. Malware is one of the security attacks that can threaten the confidentiality of an IIoT system through the disclosure of information [65]. Furthermore, in industrial IoT, there is a possibility that a malicious entity (e.g., man-in-the-middle, malware, and worms) manipulates data without detection, and consequently destroys the *integrity* of data [66]. The lack of data integrity in an industrial environment can lead to damaging consequences, such as hiding and altering crucial details related to the safety parameters of industrial pieces of machinery or standards, degradation of product quality, and industrial machinery breakdown.

Security threats may also focus on the *availability* of industrial systems, to make them unable to do their typical tasks through overloading [34]. Different types of physical and cyber-attacks can threaten the availability of an IIoT system, such as DoS attacks, DDoS attacks, Mirai botnet, BrickerBot, and Reaper.

Solutions: To deal with security challenges in IIoT systems that threaten *authentication*, different authentication techniques have been adopted, including trust-based authentication, proximity-based authentication, and edge-assisted device authentication [67]. Moreover, using authentication and verification methods, such as user key sets, digital signatures, and certificates, can mitigate security risks related to unauthorized access to the system.

Applying cryptographic techniques is one of the common countermeasures to deal with *confidentiality-* and *integrity-related* attacks in IIoT systems. Moreover, the security of cloud computing and big data components, third parties and vendors should be considered [68].

When considering the *integrity* of IIoT systems, one of the proposed solutions is to use Manufacturing Security Enforcement Device (MSED) for encryption [35]. In addition, using control and report filters after sensors, defining secure data exchange channel between IoT devices, IoT devices authorization through digital certificates/Public key Infrastructure (PKI), and data monitoring to identify possible unauthorized modifications. The key measure to increase the *availability* of IIoT systems is to protect these systems against DoS attacks. Toward this end, various approaches have been proposed such as SDN-based and distributed approaches, and the real-time availability monitoring of IoT devices.

5 Open Issues and Future Directions

In this section, we provide a few potential critical open issues and future directions based on our findings.

5.1 The Lack of Comprehensive Security- and/or Privacy-Protecting Frameworks

We review and analyze several papers related to IoT security, especially application layer security [14,16,25,27,38,42,47]. However, in all of these papers, there is no thorough framework that guarantees security or/and privacy in IoT for a wide range of use cases. To fill this gap, there is a growing need to establish a comprehensive lightweight framework to ensure the security or/and privacy in IoT environments.

5.2 Insecure Interfaces

IoT devices collect and process data, communicate, and share this data to achieve their objective. Toward this end, these devices employ user interfaces (e.g., Apps and web interfaces), services, and various communication protocols. Multiple IoT security threats arise from insecure interfaces. More specifically, these security vulnerabilities include the lack of device authentication/identification and weak encryption. For example, in home automation [69], a web interface exploitation attack may be launched by an internal intruder or an external intruder.

5.3 Blockchain

IoT systems are usually large-scale and distributed in nature. These features turn security and privacy into a key challenge in such systems. In other words, IoT environments call for scalable, decentralized, and lightweight security and privacy protection. At the same time, blockchain technology has the ability to respond to the above-mentioned challenges through providing distributed, secure, and private mechanisms. In addition, Ethereum blockchain developed a new feature, named smart contracts, that can perform a crucial function in managing, controlling, and especially securing IoT devices. Generally speaking, based on our understanding of blockchain technology and IoT security, we can refer to the following items as the roles that this technology can play for IoT security: 1) Data integrity and authentication, 2) Access control and privacy, 3) Secure communications, 4) Data assurance, and 5) IoT network self-regulation.

5.4 Machine Learning

Considering the fact that the number of IoT attacks is increasing at an exponential rate, it is necessary to provide solutions that can combine state-of-the-art methods and technologies from machine learning and Big Data. These potential solutions can especially work for intrusion and anomaly detection purposes. However, for the design of machine learning- and Big Data-based solutions, one should consider the following points: 1) The scalability of the solution, 2) Selecting the right dataset for training, 3) Continuous model training and data labeling, and 4) The computational complexity of the model.

6 Conclusion

In this survey, we studied the IoT application layer security. Toward this goal, we have first provided a background to IoT and its security. Afterward, we categorized and discussed the key security requirements of the IoT application layer, threats, and potential solutions. To take the right direction and conduct a systematic review, our study is based primarily on IoT use cases.

In this study, we also discussed some open issues and future directions based on our findings. In our future work, we will investigate the security requirements and threats of IoT application layer protocols. Moreover, we will continue to investigate more IoT applications, e.g., smart cities.

Acknowledgements. This work has been performed under the IoTalentum project within the framework of Marie Skłodowska-Curie Actions ITN-ETN (grant number 953442) which is funded by the European Union Horizon 2020 research and innovation program.

References

1. Atzori, L., Iera, A., Morabito, G.: The internet of things: a survey. Comput. Netw. **54**(15), 2787–2805 (2010)

2. Gupta, S., Ranga, V., Agrawal, P.: Epilnet: a novel approach to IoT based epileptic seizure prediction and diagnosis system using artificial intelligence. arXiv preprint arXiv:2111.03265 (2021)
3. de Alba, F.L., González-Briones, A., Chamoso, P., Pinto, T., Vale, Z., Corchado, J.M.: A P2P electricity negotiation agent systems in urban smart grids. In: International Symposium on Distributed Computing and Artificial Intelligence, pp. 97–106. Springer, Berlin (2020)
4. Unzueta, M., Bartolomé, A., Hernández, G., Parra, J., Chamoso, P.: System for recommending financial products adapted to the user's profile. In: International Symposium on Ambient Intelligence, pp. 117–126. Springer, Berlin (2020)
5. Cisco, U.: Cisco Annual Internet Report (2018–2023) White Paper. Cisco, San Jose, CA, USA (2020)
6. Ali, I., Hussain, T., Khan, K., Iqbal, A., Perviz, F.: The impact of IEEE 802.11 contention window on the performance of transmission control protocol in mobile ad-hoc network. ADCAIJ: Adv. Distrib. Comput. Artif. Intell. J. 9(3), 29–48 (2020)
7. Chamoso, P., González-Briones, A., De La Prieta, F., Venyagamoorthy, G.K., Corchado, J.M.: Smart city as a distributed platform: toward a system for citizen-oriented management. Comput. Commun. **152**, 323–332 (2020)
8. González-Briones, A., Castellanos-Garzón, J.A., Mezquita Martín, Y., Prieto, J., Corchado, J.M.: A framework for knowledge discovery from wireless sensor networks in rural environments: a crop irrigation systems case study. Wirel. Commun. Mob. Comput. 2018 (2018)
9. Abrishambaf, O., Faria, P., Gomes, L., Spínola, J., Vale, Z., Corchado, J.M.: Implementation of a real-time microgrid simulation platform based on centralized and distributed management. Energies **10**(6), 806 (2017)
10. Gonçalves, C., Rebelo, D., Silva, F., Analide, C.: Sleep performance and physical activity estimation from multisensor time series sleep environment data. In: International Symposium on Ambient Intelligence, pp. 166–176. Springer, Berlin (2020)
11. Casado-Vara, R., Martin-del Rey, A., Affes, S., Prieto, J., Corchado, J.M.: IoT network slicing on virtual layers of homogeneous data for improved algorithm operation in smart buildings. Future Gener. Comput. Syst. **102**, 965–977 (2020)
12. Da Xu, L., Lu, Y., Li, L.: Embedding blockchain technology into IoT for security: a survey. IEEE Internet Things J. **8**(13), 10452–10473 (2021)
13. Raufa, A., Abdullaha, A.H., Mateenb, A., Ashrafb, M.: Secure data access control with perception reasoning. ADCAIJ: Adv. Distrib. Comput. Artif. Intell. J. 13 (2018)
14. Swessi, D., Idoudi, H.: A survey on internet-of-things security: threats and emerging countermeasures. Wirel. Pers. Commun. 1–36 (2022)
15. Shahraki, A., Abbasi, M., Haugen, Ø.: Boosting algorithms for network intrusion detection: a comparative evaluation of real adaboost, gentle adaboost and modest adaboost. Eng. Appl. Artif. Intell. **94**, 103770 (2020)
16. Zhang, Z.-K., Cho, M.C.Y., Wang, C.-W., Hsu, C.-W., Chen, C.-K., Shieh, S.: IoT security: ongoing challenges and research opportunities. In: 2014 IEEE 7th International Conference on Service-Oriented Computing and Applications, pp. 230–234. IEEE (2014)
17. Sobin, C.: A survey on architecture, protocols and challenges in IoT. Wirel. Pers. Commun. **112**(3), 1383–1429 (2020)
18. Hernández, G., Rodríguez, S., González, A., Corchado, J.M., Prieto, J.: Video analysis system using deep learning algorithms. In: International Symposium on Ambient Intelligence, pp. 186–199. Springer, Berlin (2020)

19. Nauman, A., Qadri, Y.A., Amjad, M., Zikria, Y.B., Afzal, M.K., Kim, S.W.: Multimedia internet of things: a comprehensive survey. IEEE Access **8**, 8202–8250 (2020)
20. Fatras, N., Ma, Z., Jørgensen, B.N.: System architecture modelling framework applied to the integration of electric vehicles in the grid. In: International Symposium on Distributed Computing and Artificial Intelligence, pp. 205–209. Springer, Berlin (2020)
21. Sakarkar, G., Kolekar, M.K.H., Paithankar, K., Gaurav, G.P., Dutta, P., Chaturvedi, R., Kumar, S., et al.: Advance approach for detection of DNS tunneling attack from network packets using deep learning algorithms (2021)
22. Plaza-Hernandez, M., Sittón-Candanedo, I., Alonso, R.S., Martínez-de Iturrate, L.C., Prieto, J., Kravari, K., Kosmanis, T., Katranas, G., Silva, M.P., Corchado, J.M.: Edge computing and internet of things based platform to improve the quality of life of the silver economy on leisure cruise ships. In: 2021 International Symposium on Computer Science and Intelligent Controls (ISCSIC), pp. 159–163. IEEE (2021)
23. Salazar, G., Figueiredo, L., Ferreira, N.: Towards the development of IoT protocols. In: International Symposium on Ambient Intelligence, pp. 146–155. Springer, Berlin (2020)
24. Zahoor, S., Mir, R.N.: Resource management in pervasive internet of things: a survey. J. King Saud Univ.-Comput. Inf. Sci. 33(8), 921–935 (2021)
25. Omolara, A.E., Alabdulatif, A., Abiodun, O.I., Alawida, M., Alabdulatif, A., Arshad, H., et al.: The internet of things security: a survey encompassing unexplored areas and new insights. Comput. Secur. **112**, 102494 (2022)
26. Alonso, R.S.: Low-power distributed AI and IoT for measuring lamb's milk ingestion and predicting meat yield and malnutrition diseases. In: International Symposium on Ambient Intelligence, pp. 251–257. Springer, Berlin (2020)
27. Gunduz, M.Z., Das, R.: Cyber-security on smart grid: threats and potential solutions. Comput. Netw. **169**, 107094 (2020)
28. Gough, M.B., Santos, S.F., AlSkaif, T., Javadi, M.S., Castro, R., Catalão, J.P.: Preserving privacy of smart meter data in a smart grid environment. IEEE Trans. Ind. Inform. **18**(1), 707–718 (2021)
29. Alshehri, F., Muhammad, G.: A comprehensive survey of the internet of things (IoT) and AI-based smart healthcare. IEEE Access **9**, 3660–3678 (2020)
30. Hemalatha, R., et al.: A survey: security challenges of VANET and their current solution. Turk. J. Comput. Math. Educ. (TURCOMAT) **12**(2), 1239–1244 (2021)
31. Ali, I., Chen, Y., Faisal, M., Li, M.: Certificateless signature-based authentication scheme for vehicle-to-infrastructure communications using bilinear pairing. In: Efficient and Provably Secure Schemes for Vehicular Ad-Hoc Networks, pp. 91–119. Springer, Berlin (2022)
32. Yang, X., Shu, L., Chen, J., Ferrag, M.A., Wu, J., Nurellari, E., Huang, K.: A survey on smart agriculture: development modes, technologies, and security and privacy challenges. IEEE/CAA J. Automatica Sin. **8**(2), 273–302 (2021)
33. Gebremichael, T., Ledwaba, L.P.I., Eldefrawy, M.H., Hancke, G.P., Pereira, N., Gidlund, M., Akerberg, J.: Security and privacy in the industrial internet of things: current standards and future challenges. IEEE Access **8**, 152351–152366 (2020)
34. Agrawal, N., Kumar, R.: Security perspective analysis of industrial cyber physical systems (I-CPS): a decade-wide survey. ISA Trans. (2022)
35. Dhirani, L.L., Armstrong, E., Newe, T.: Industrial IoT, cyber threats, and standards landscape: evaluation and roadmap. Sensors **21**(11) (2021) [Online]. Available: www.mdpi.com/1424-8220/21/11/3901

36. Al-Turjman, F., Alturjman, S.: Confidential smart-sensing framework in the IoT era. J. Supercomput. **74**(10), 5187–5198 (2018). https://doi.org/10.1007/s11227-018-2524-1
37. Casado-Vara, R., de la Prieta, F., Prieto, J., Corchado, J.M.: Blockchain framework for IoT data quality via edge computing. In: Proceedings of the 1st Workshop on Blockchain-Enabled Networked Sensor Systems, pp. 19–24 (2018)
38. Machado, C., Fröhlich, A.A.M.: IoT data integrity verification for cyber-physical systems using blockchain. In: 2018 IEEE 21st International Symposium on Real-Time Distributed Computing (ISORC), pp. 83–90. IEEE (2018)
39. Casado-Vara, R., Novais, P., Gil, A.B., Prieto, J., Corchado, J.M.: Distributed continuous-time fault estimation control for multiple devices in IoT networks. IEEE Access **7**, 11972–11984 (2019)
40. Yan, J., Liu, J., Tseng, F.-M.: An evaluation system based on the self-organizing system framework of smart cities: a case study of smart transportation systems in China. Technol. Forecast. Soc. Change **153**, 119371 (2020)
41. Algarni, A.: A survey and classification of security and privacy research in smart healthcare systems. IEEE Access **7**, 101879–101894 (2019)
42. Salman, T., Jain, R.: A survey of protocols and standards for internet of things. arXiv preprint arXiv:1903.11549 (2019)
43. Mezquita, Y., Gil-González, A.B., Martín del Rey, A., Prieto, J., Corchado, J.M.: Towards a blockchain-based peer-to-peer energy marketplace. Energies **15**(9), 3046 (2022)
44. Zhang, H., Liu, B., Wu, H.: Smart grid cyber-physical attack and defense: a review. IEEE Access **9**, 29641–29659 (2021)
45. Sengan, S., Subramaniyaswamy, V., Indragandhi, V., Velayutham, P., Ravi, L.: Detection of false data cyber-attacks for the assessment of security in smart grid using deep learning. Comput. Electr. Eng. **93**, 107211 (2021)
46. Zavala-Díaz, J., Reyes-Archundia, E., Olivares-Rojas, J.C., Chávez-Báez, M.V., Gutiérrez-Gnecchi, J.A., Méndez-Patiño, A.: Study of public key cryptography techniques for authentication in embedded devices for smart grids. In: 2021 IEEE International Autumn Meeting on Power, Electronics and Computing (ROPEC), vol. 5, pp. 1–5. IEEE (2021)
47. Pirayesh, H., Zeng, H.: Jamming attacks and anti-jamming strategies in wireless networks: a comprehensive survey. IEEE Commun. Surv. Tutorials (2022)
48. Ghayvat, H., Pandya, S.N., Bhattacharya, P., Zuhair, M., Rashid, M., Hakak, S., Dev, K.: CP-BDHCA: blockchain-based confidentiality-privacy preserving big data scheme for healthcare clouds and applications. IEEE J. Biomed. Health Inf. (2021)
49. Renuka, K., Kumari, S., Li, X.: Design of a secure three-factor authentication scheme for smart healthcare. J. Med. Syst. **43**(5), 1–12 (2019)
50. Xu, C., Yang, H.H., Wang, X., Quek, T.Q.: Optimizing information freshness in computing-enabled IoT networks. IEEE Internet Things J. **7**(2), 971–985 (2019)
51. Maimaris, A., Papageorgiou, G.: A review of intelligent transportation systems from a communications technology perspective. In: 2016 IEEE 19th International Conference on Intelligent Transportation Systems (ITSC), pp. 54–59. IEEE (2016)
52. Behzadan, V., Munir, A.: Models and framework for adversarial attacks on complex adaptive systems. arXiv preprint arXiv:1709.04137 (2017)
53. Azam, F., Kumar, S., Yadav, K., Priyadarshi, N., Padmanaban, S.: An outline of the security challenges in VANET. In: 2020 IEEE 7th Uttar Pradesh Section International Conference on Electrical, Electronics and Computer Engineering (UPCON), pp. 1–6. IEEE (2020)

54. Verma, S.B., Verma, S.B., et al.: Secure data transmission in BPEL (business process execution language) (2020)
55. Shen, X., Lu, Y., Zhang, Y., Liu, X., Zhang, L.: An innovative data integrity verification scheme in the internet of things assisted information exchange in transportation systems. Cluster Comput. 1–13 (2022)
56. Gupta, M., Abdelsalam, M., Khorsandroo, S., Mittal, S.: Security and privacy in smart farming: challenges and opportunities. IEEE Access **8**, 34564–34584 (2020)
57. Salamatian, S., Huleihel, W., Beirami, A., Cohen, A., Médard, M.: Why botnets work: distributed brute-force attacks need no synchronization. IEEE Trans. Inf. Forensics Secur. **14**(9), 2288–2299 (2019)
58. Ferrag, M.A., Shu, L., Djallel, H., Choo, K.-K.R.: Deep learning-based intrusion detection for distributed denial of service attack in agriculture 4.0. Electronics **10**(11) (2021) [Online]. Available: www.mdpi.com/2079-9292/10/11/1257
59. Sontowski, S., Gupta, M., Laya Chukkapalli, S.S., Abdelsalam, M., Mittal, S., Joshi, A., Sandhu, R.: Cyber attacks on smart farming infrastructure. In: 2020 IEEE 6th International Conference on Collaboration and Internet Computing (CIC), pp. 135–143 (2020)
60. Wang, Q., Chen, D., Zhang, N., Qin, Z., Qin, Z.: LACS: a lightweight label-based access control scheme in IoT-based 5g caching context. IEEE Access **5**, 4018–4027 (2017)
61. Li, Q., Zhang, X., Zheng, Q., Sandhu, R., Fu, X.: Live: lightweight integrity verification and content access control for named data networking. IEEE Trans. Inf. Forensics Secur. **10**(2), 308–320 (2015)
62. Gope, P., Amin, R., Hafizul Islam, S., Kumar, N., Bhalla, V.K.: Lightweight and privacy-preserving RFID authentication scheme for distributed IoT infrastructure with secure localization services for smart city environment. Future Gener. Comput. Syst. **83**, 629–637 (2018) [Online]. Available: www.sciencedirect.com/science/article/pii/S0167739X17313043
63. Sánchez, S.M.: Integral support predictive platform for industry 4.0. ADCAIJ: Adv. Distrib. Comput. Artif. Intell. J. **9**(4), 71–82 (2020)
64. Tuptuk, N., Hailes, S.: Security of smart manufacturing systems. J. Manuf. Syst. **47**, 93–106 (2018) [Online]. Available: www.sciencedirect.com/science/article/pii/S0278612518300463
65. Del Rey, Á.M., Batista, F.K., Dios, A. Q.: Malware propagation in wireless sensor networks: global models vs individual-based models. ADCAIJ: Adv. Distrib. Comput. Artif. Intell. J. **6**(3), 5–15 (2017)
66. ElMamy, S.B., Mrabet, H., Gharbi, H., Jemai, A., Trentesaux, D.: A survey on the usage of blockchain technology for cyber-threats in the context of industry 4.0. Sustainability **12**(21), 9179 (2020)
67. Lu, Y., Wang, D., Obaidat, M.S., Vijayakumar, P.: Edge-assisted intelligent device authentication in cyber-physical systems. IEEE Internet Things J. (2022)
68. Chhetri, S.R., Rashid, N., Faezi, S., Al Faruque, M.A.: Security trends and advances in manufacturing systems in the era of industry 4.0. In: 2017 IEEE/ACM International Conference on Computer-Aided Design (ICCAD), pp. 1039–1046 (2017)
69. Lupión, M., Redondo, J.L., Sanjuan, J.F., Ortigosa, P.M.: Deployment of an IoT platform for activity recognition at the ual's smart home. In: International Symposium on Ambient Intelligence, pp. 82–92. Springer, Berlin (2020)

Smart Wearable Systems for the Remote Monitoring of Venous and Diabetic Foot Ulcers: State of the Art

Julio Souza[1,2](✉), Sara Escadas[1,2], Daniel Rodrigues[1,2], and Alberto Freitas[1,2]

[1] Department of Community Medicine, Information and Health Decision Sciences (MEDCIDS), Faculty of Medicine, University of Porto, Porto, Portugal
juliobsouza@med.up.pt

[2] Center for Health Technology and Services Research (CINTESIS), Faculty of Medicine, University of Porto, Porto, Portugal

Abstract. We reviewed the state of the art of smart wearable systems (SWS) that employ data science techniques to continuously monitor the health status of venous and diabetic foot ulcers (DFU). Our main objective was to analyze the literature to identify data collection and data science methods currently employed, as well as key research challenges that need to be addressed regarding the development of SWS applied to these selected vascular diseases. A narrative review was conducted, analyzing publications retrieved from PubMed and SCOPUS. Publications were first screened based on whether they describe a SWS applicable to the two selected vascular diseases, and must necessarily include data processing (descriptive, predictive and/or prescriptive analytics), either at planning or at the implementation stage. A total of 44 publications were included in the full-text analysis after screening. Our results were structured and presented according to three main topics consisting of data collection, smart features, and limitations and challenges. Overall, research on SWS for the targeted diseases are already robust and well defined in terms of what parameters should be measured, sensory devices and algorithms/clinical rules for recommendation level functionalities. Nevertheless, these projects lack evaluation on usability, adherence, satisfaction, as well as clinical validation, as works are mostly at proof-of-concept stage, focusing on pilots/small-scale preliminary studies to assess the developed systems. Therefore, and according to future work perspectives, authors should focus on the refinement of current systems and further evaluation and clinical validation of existing projects.

Keywords: mHealth · Personalized healthcare · Vascular diseases · Remote monitoring

1 Introduction

The ageing of the world's population has created greater pressure over health systems due to the higher demand for care and assistance, driven by the growing prevalence

V. Julián et al. (Eds.): ISAmI 2022, LNNS 603, pp. 101–110, 2023.
https://doi.org/10.1007/978-3-031-22356-3_10

of chronic conditions, including cancers and cardiovascular diseases, multimorbidity patterns and disability [1]. In this light, health care has become one the main application areas requiring faster developments in the field of Smart Wearable Systems (SWS) [2]. These systems can potentially contribute with the implementation of personalized health systems, shifting from a reactive and hospital-centered model to a preventive and personalized-centered model, focusing mostly on wellbeing rather than disease, working towards smarter and sustainable health care systems [3]. SWS provide means to adapt daily environment to the individuals' physical and cognitive limitations, improving their overall health status and quality of life and preventing further deterioration in health. In the health sector, there have been a variety of SWS to support independent living for the elderly, postoperative rehabilitation, and for the analysis and improvement of individual health, technical or sportive abilities [4].

Overall, the architecture of a SWS should include a variety of implantable or wearable devices, including sensors and smart fabrics, actuators, power supplies, wireless communication networks, processing units, multimedia devices, software, user interfaces, algorithms for data processing, or even decision support functionalities [1]. These systems enable the measurement of physiological parameters (e.g., body and skin temperature, heart rate, respiration rate, blood oxygen saturation (SpO_2), electrocardiograms (ECG)) and forward these measurements via a wireless sensor network, either to a central node or directly to a healthcare setting [1]. Patients can wear the devices during routine and healthcare professionals are able to manage their condition and implement interventions whenever is needed according to the data transmitted via an information gateway (e.g., mobile phone) [5], sometimes outside of health facilities and in real-time, and for longer periods than are possible during the patients' visit or stay at the health services.

Chowdhury et al. [2] state that two main sub-sections characterize a complete wearable medical device: the hardware sub-section, which includes sensor selection and characterization, noise removal from the collected data, and communication with a decision-making (or data processing) subsystem. The other sub-section is the software, which involves data processing and decisions based upon the collected data. Artificial Intelligence (AI) methods can be used in this sub-section to represent, model and reason with medical knowledge and thereby demonstrate human-like capabilities for supporting diagnosis, prevention and guiding medical care [6]. In fact, the recent advances in the field of AI are what enhance the possibility of remote and continuous motoring, as well as diagnosis and recommendations with the data collected through the wearable devices [2].

Amongst the plethora of healthcare applications in the field of SWS, vascular diseases such as leg venous ulcers and diabetic foot ulcers (DFU) offer key opportunities in the sense that these conditions can lead to considerable loss of mobility and quality of life, often requiring constant follow-up and hospitalizations due to their severity [8]. Standard treatment plans are defined according to the patients' health condition and encompass pharmacological and physical activity interventions to control the disease progression and improve the overall condition. Nevertheless, follow-ups are only done sporadically due to the difficulty to monitor the patients and their activities outside of healthcare facilities, opening key opportunities and gains with the development of SWS. In this article, we conducted a narrative review of the literature related to SWS covering venous ulcers and DFU and provided an overview of the current state-of-the-art regarding data collection, smart features, and research challenges and limitations to be addressed.

2 Methods

2.1 Search Strategy

We conducted a literature search on the PubMed and SCOPUS repository for studies describing SWS focusing on the monitoring of the two afore mentioned vascular diseases, from inception. The last search was conducted in late March 2022. The search queries were built looking for search terms in titles, abstracts and author-identified keywords, to enhance query recall when compared to searching just based on the title. Queries essentially included broad terms to identify intended system, which consists of those providing remote and continuous monitoring and that included data processing ("smart monitoring") for decision support. Finally, broad terms for each one of the two vascular diseases were added to scope down the application areas, which are publications addressing venous ulcers and DFU. The final query is reported below. Duplicates between the two repositories were removed and irrelevant articles, namely applications that did not include a complete wearable or portable device and thereby did not constitute a SWS, were removed prior to the full-text analysis. ("smart monitoring" [tiab] OR "remote monitoring" [tiab] OR "continuous monitoring" [tiab]) AND (("venous ulcer" [tiab] OR "diabetic foot" [tiab] OR "DFU" [tiab])).

2.2 Eligibility Criteria

We considered articles written in English and published in peer-reviewed journals or conference proceedings. Based upon an initial screening of titles and abstracts, we excluded articles that did not meet any of the following criteria: (1) Publications presenting SWS applicable to any of the two selected vascular diseases, regardless of the development stage (e.g., protocols, system design, system testing and evaluation, clinical validation stage); (2) Publications should include the use of a complete wearable or portable device; (3) Publications should include data processing (descriptive, predictive or prescriptive analytics). Other review articles, publications with no abstract available, letters, editorials, comments and abstract-only publications were excluded from the review. The complete list of articles analyzed is included in the References section [8–53].

2.3 Analysis

Three researchers independently reviewed the included publications. Information extracted from publications included data collection; system objectives; outputs to the user; data science approaches to infer and predict the patient's state and/or to deliver coaching, recommendations or alerts; challenges and limitations reported; and authors' conclusions. Analyses were carried out by three relevant thematic areas: (a) data collection via wearable devices, including the main variables considered; (b) smart features obtained with the application of data science methods to infer and predict the patient's health status; including prescriptive analytics that enable patients and healthcare professionals with means to visualize, interpret and communicate the processed information; and (c) current research challenges and limitations to be addressed.

3 Results

3.1 State-of-the Art

A total of 44 studies were included for full-text analysis. Most of the publications focused exclusively on DFU (n = 33). A total of 8 studies targeted both, DFU and venous ulcers, whereas 3 publications developed an application for leg venous ulcer. In sub-Sects. 3.2, 3.3 and 3.4, we summarize the body of the literature according to data collection methods, smart features, and challenges and limitations, respectively.

3.2 Data Collection

Overall, in most studies, data from different types of sensors integrated to distinct wearable or portable devices were typically transmitted to mobile phones as the main information gateway, although some projects opted for transmitting the data to a web platform for data processing and coaching [20]. Other projects allowed data collection directly from the device, such as touch screen cameras [53] or continuously transmitted the results to the cloud for a more elaborate analysis by medical staff, eliminating the need for high computational and energy capabilities of the mobile devices [43]. Table 1 summarizes the variables collected by the different studies for each one of the targeted diseases, as well as the main wearable devices used.

3.3 Smart Features

Although several articles analyzed in this review do not mention nor explain which data science methods were used to process the data, a considerable number of publications recurred to predictive and prescriptive analytics techniques to analyze the data and further provide recommendations, smart coaching and alerts. Generally, these applications also have an interface in which the clinician can access and monitor the patient in real-time. Some works mention a follow-up and coaching provided by a healthcare professionals based upon the data measured by the monitoring sensors [24], without major data processing by AI algorithms.

Regarding leg venous ulcers applications, Wang et al. [46] proposed real-time feedback during the application of compression bandages and for ongoing in-vivo monitoring of the sub-bandage pressure. Yan et al. [47] developed an iPhone-based software to help users to do the prescribed exercises and to promote adherence to the prescribed physical activity and further encourage patient–provider communication, such as reminding the patient to do the exercise every day.

Diabetic foot was the condition for which most coaching functionalities have been implemented. The solution developed by Abbott et al. [8] provided offload instructions to

Table 1. Variables and wearable devices.

Disease	Wearable devices	Variables
Venous ulcers	Sensors integrated to compression apparatus [11, 19, 46, 52]; Sensor gauze [15]; Socks [27]; Insole [44] Sensing strips [26]; Slipper [47]	Temperature [11, 26]; Pressure [11, 26, 46, 52]; Wound metrics (e.g., area, depth, segmentation) [19, 53]; Wound classification (e.g., Wound Bed Preparation (WBP)) [53]; Oxygen saturation levels (SpO_2) [44]; Heart rate [44]; Perfusion index (PI) [44]; Physical activity parameters (e.g., accelerations of the lower limb movement) [27, 44, 47]; PH index [15, 18]; Moisture [11, 15]
Diabetic foot ulcers	Socks [16, 23, 27, 33, 36, 40, 45]; Insoles [8–10, 14, 28, 30, 31, 34, 39, 43, 44, 48, 51]; Footwear (shoes, boots) [29, 32, 35, 42, 49]; Electronic orthotics shoe [12]; Camera [17, 22, 53]; Foot mat [20, 21, 25, 41]; Sensing strips [26]; Elastic band [38]	Plantar pressure [10–14, 16, 23, 26, 28–30, 32–34, 36, 38, 39, 43, 44, 46, 48–51]; Humidity [10, 29, 42, 51]; Temperature [9–11, 13, 16, 17, 20, 21, 23, 25, 26, 29, 31–33, 39–42, 45, 48, 51]; PH index [15, 18]; Moisture [11, 15, 32]; DFU area [24, 37, 53]; Motion and physical activity (e.g., gait, dorsiflexion parameters) [12, 27, 34, 35, 44]; Oxygen saturation levels (SpO_2) [44, 51]; Heart rate [44]; Perfusion index (PI) [44]; DFU Healing associated metrics [22]

patients whenever aberrant pressures were identified. Ming et al. [31] proposed an algorithm that prompts a warning signal when temperature differences are higher than 1.5 °C between left and right foot corresponding sensor sites [31]. Killeen et al. [20] developed a system where the clinical staff is warned to call the patient to collect information for further triage whenever persistent localized temperature differences exceeding 1.75 °C between the left and right feet are detected. Torreblanca González et al. [45] employed a stepwise regression model to predict when temperature measurements exceed a certain threshold, so that a smart sock would send an alert to the telephone and the patient will know when to stop. Macdonald et al. [28] implemented alerts to the patient if pressures greater than 35 mmHg were sustained for 95–100% of the time in a 15 min sampling window on the sensor. These alerts were wirelessly transmitted from the individuals' footwear to a smartwatch through vibration or audible alarms, including the targeted site of the foot and offload instructions. Park et al. [35] developed a smart offloading boot combined with a smartwatch application and a cloud dashboard to remotely monitor patients' adherence and activity (e.g., balance, gait). Other projects applied to DFU, however, did not focus on temperature and pressure. Zoppo et al. [53] proposed a touch screen with digital camera with a sixteen infrared sensor array, in which wound pictures are analyzed by an IA algorithm (discrete time-cellular nonlinear network—DT-CNN) to identify the wound and provide relevant measurements. Another project focused on developing a mobile application that applies novel visual analytics to uploaded photos to detect the ulcer and calculate its size, letting patients and caregivers track DFU healing progress [37].

3.4 Open Research Challenges and Limitations

In what concerns the limitations of the reviewed studies, the main problems mentioned are: proof of concept stage or prototype stage, small trials with reduced number of participants, high cost and need of calibrating the sensors throughout the study. Most studies mentioned the need of further investigation, improvement of the devices or clinical validation. In a project dedicated to monitoring diabetic foot, Macdonald et al. [28] reported that participants' attitude and behavioral intention towards future adoption were negatively impacted by their experiences, although a focus group and in-depth interview findings indicate that this population remain optimistic about the role of technology in supporting foot monitoring. Furthermore, participants and podiatrists' comments suggest that evidence of device efficacy in preventing foot ulcerations would improve trust and would help to increase the likelihood of future adoption. Consults with target users regarding device design features were also recommended. Overall, recommendations and coaching level functionalities are still requiring further development and clinical validation.

4 Conclusions

In this paper, we conducted a narrative literature review of SWS applied to venous ulcers and DFU, summarizing aspects of data collection, smart features, and current research challenges and limitations related to the developments of these systems. Recommendation level functionalities for disease management to guide treatments or interventions were found in a considerable number of studies, and several works mentioned a follow-up and coaching provided by a healthcare professionals based upon the data measured by the monitoring sensors, without major data processing by AI algorithms. Coaching feedback, whenever developed, is usually carried out by a smartphone application, warning the patient when he/she should stop walking, offload or is at risk of ulcer development. Overall, research on SWS for the targeted diseases are already robust and well defined in terms of what parameters should be measured, sensory devices and algorithms/clinical rules for recommendation level functionalities. Nevertheless, these projects lack evaluation on usability, adherence, satisfaction, as well as clinical validation, as works are mostly at proof-of-concept stage, focusing on pilots/small-scale preliminary studies to assess the developed systems. Therefore, and according to future work perspectives, authors should focus on the refinement of current systems and further evaluation and clinical validation of existing projects.

Acknowledgements. This research work was developed under the project Inno4Health (ITEA-2019-19008-Inno4Health), funded by the European Regional Development Fund (ERDF) through the Operational Programme for Competitiveness and Internationalization (POCI) (ref. POCI-01-0247-FEDER-069523).

References

1. Chan, M., Estève, D., Fourniols, J.Y., Escriba, C., Campo, E.: Smart wearable systems: current status and future challenges. Artif. Intell. Med. **56**(3), 137–156 (2012)
2. Chowdhury, M.E., Khandakar, A., Qiblawey, Y., Reaz, M.B.I., Islam, M.T., Touati, F.: Machine learning in wearable biomedical systems. In: Marinho D.A., Neiva H.P., Johnson C.P., Mohamudally N. (eds.) Sports Science and Human Health—Different Approaches. IntechOpen (2020)
3. Chawla, M.N.: AI, IoT and wearable technology for smart healthcare–a review. Int. J. Green Energy **7**(1), 9–13 (2020)
4. Gatzoulis, L., Iakovidis, I.: Wearable and portable eHealth systems. Technological issues and opportunities for personalized care. EMBS, **26**(5), 51–56 (2007)
5. Patel, S., Park, H., Bonato, P., Chan, L., Rodgers, M.: A review of wearable sensors and systems with application in rehabilitation. J. NeuroEng. Rehabil. **9**, 21 (2012)
6. Parimbelli, E., et al.: A review of AI and data science support for cancer management. Artif. Intell. Med. **117**, 102111 (2021)
7. Kelechi, T.J., Johnson, J.J., Yates, S.: Chronic venous disease and venous leg ulcers: an evidence-based update. J. Vasc. Nurs. **33**(2), 36–46 (2015)
8. Abbott, C.A., et al.: Innovative intelligent insole system reduces diabetic foot ulcer recurrence at plantar sites: a prospective, randomised, proof-of-concept study. Lancet Digit Health **1**(6), e308–e318 (2019)
9. Beach, C., Cooper, G., Weightman, A., Hodson-Tole, E.F., Reeves, N.D., Casson, A.J.: Monitoring of dynamic plantar foot temperatures in diabetes with personalised 3D-printed wearables. Sensors (Basel) **21**(5), 1717 (2021)
10. Bencheikh, M.A., Boukhenous, S.: A wearable preventive insole for diabetic patients. In: International Conference on Computer Systems and Applications, pp. 1–2 (2018)
11. Chen, Y.-W., Toro, C., Tanaka, S., Howlett, R.J., Jain, L.C. (eds.): Innovation in medicine and healthcare 2015. SIST, vol. 45. Springer, Cham (2016). https://doi.org/10.1007/978-3-319-23024-5

12. Dabiri, F., Vahdatpour, A., Noshadi, H., Hagopian, H., Sarrafzadeh, M.: Electronic orthotics shoe: preventing ulceration in diabetic patients. In: Annual International Conference of the IEEE Engineering in Medicine and Biology Society, pp. 771–774 (2008)
13. De Pascali, C., et al.: Modeling, fabrication and integration of wearable smart sensors in a monitoring platform for diabetic patients. Sensors (Basel) **21**(5), 1847 (2021)
14. do Carmo Dos Reis, M., Soares, F.A., da Rocha, A.F., Carvalho, J.L.A., Rodrigues, S.S.R.: Insole with pressure control and tissue neoformation induction systems for diabetic foot. In: Annual International Conference of the IEEE Engineering in Medicine and Biology Society, pp. 5748–5751 (2010)
15. Du, Y.-C., Ciou, W.-S.: A sensor gauze with multi-channel moisture and pH monitoring for chronic wound care. IEEE Access **7**, 29185–29192 (2019)
16. El-Nahas, M., El-Shazly, S., El-Gamel, F., Motawea, M., Kyrillos, F., Idrees, H.: Relationship between skin temperature monitoring with smart socks and plantar pressure distribution: a pilot study. J. Wound Care **27**(8), 536–541 (2018)
17. Erdenechimeg, D., Enkhzul, D., Munkhtamir, O., Enkhbat, B.: Wireless monitoring method for diabetic foot temperature. In: International Conference Advance Communication, pp. 15–21 (2017)
18. Gao, Y., Nguyen, D.T., Yeo, T., Lim, S.B., Tan, W.X., Madden, L.E., Jin, L., Long, J.Y.K., Aloweni, F.A.B., Liew, Y.J.A., Tan, M.L.L., Ang, S.Y., Maniya, S.D., Abdelwahab, I., Loh, K.P., Chen, C.-H., Becker, D.L., Leavesley, D., Ho, J.S., Lim, C.T.: A flexible multiplexed immunosensor for point-of-care in situ wound monitoring. Sci. Adv. **7**(21), eabg9614 (2021)
19. Kekonen, A., et al.: Bioimpedance method for monitoring venous ulcers: clinical proof-of-concept study. Biosens. Bioelectron. **178**(112974), 112974 (2021)
20. Killeen, A.L., Walters, J.L.: Remote temperature monitoring in diabetic foot ulcer detection. Wounds **30**(4), E44–E48 (2018)
21. Killeen, A.L., Brock, K.M., Dancho, J.F., Walters, J.L.: Remote temperature monitoring in patients with visual impairment due to diabetes mellitus: a proposed improvement to current standard of care for prevention of diabetic foot ulcers. J. Diabetes Sci. Technol. **14**(1), 37–45 (2020)
22. Kong, L.Y., Ramirez-GarciaLuna, J.L., Fraser, R.D.J., Wang, S.C.: A 57-year-old man with type 1 diabetes mellitus and a chronic foot ulcer successfully managed with a remote patient-facing wound care smartphone application. Am. J. Case Rep. **22**, e933879 (2021)
23. Kulkarni, V.V., Seraj, F., Rathikumar, J.R.: An embedded wearable device for monitoring diabetic foot ulcer parameters. In: Pervasive Technology Related to Assistive Environment Conference, pp. 1–7 (2020)
24. Ladyzynski, P., et al.: Area of the diabetic ulcers estimated applying a foot scanner-based home telecare system and three reference methods. Diabetes Technol. Ther. **13**(11), 1101–1107 (2011)
25. Lavery, L.A., Petersen, B.J., Linders, D.R., Bloom, J.D., Rothenberg, G.M., Armstrong, D.G.: Unilateral remote temperature monitoring to predict future ulceration for the diabetic foot in remission. BMJ Open Diabetes Res. Care **7**(1), e000696 (2019)
26. Lim, R., Wai Choong, D.S., Cheng, M.-Y.: Development of integrated pressure and temperature sensing strips for monitoring venous leg ulcer application. In: Proceeding—Electronic Components and Technology Conference, pp. 1586–1591 (2020)
27. Lucangeli, L., D'Angelantonio, E., Camomilla, V., Pallotti, A.: SISTINE: Sensorized socks for telemonitoring of vascular disease patients. In: International Workshop on Metrology for Industry 4.0 & IoT, pp. 192–197 (2021)
28. Macdonald, E.M., Perrin, B.M., Cleeland, L., Kingsley, M.: Podiatrist-delivered health coaching to facilitate the use of a smart insole to support foot health monitoring in people with diabetes-related peripheral neuropathy. Sensors (Basel) **21**(12), 3984 (2021)

29. Maluf, K.S., Morley, R.E., Jr., Richter, E.J., Klaesner, J.W., Mueller, M.J.: Monitoring in-shoe plantar pressures, temperature, and humidity: reliability and validity of measures from a portable device. Arch. Phys. Med. Rehabil. **82**(8), 1119–1127 (2001)
30. Manohara Pai, M.M., Kolekar, S.V., Pai, R.M.: Development of smart sole based foot ulcer prediction system. In: International Conference on Intelligent Communication and Computational Techniques, pp. 87–90 (2019)
31. Ming, A., Walter, I., Alhajjar, A., Leuckert, M., Mertens, P.R.: Study protocol for a randomized controlled trial to test for preventive effects of diabetic foot ulceration by telemedicine that includes sensor-equipped insoles combined with photo documentation. Trials **20**(1), 521 (2019)
32. Moulaei, K., Malek, M., Sheikhtaheri, A.: A smart wearable device for monitoring and self-management of diabetic foot: a proof of concept study. Int. J. Med. Inform. **146**(104343), 104343 (2021)
33. Najafi, B., Mohseni, H., Grewal, G.S., Talal, T.K., Menzies, R.A., Armstrong, D.G.: An optical-fiber-based smart textile (smart socks) to manage biomechanical risk factors associated with diabetic foot amputation. J. Diabetes Sci. Technol. **11**(4), 668–677 (2017)
34. Najafi, B., Ron, E., Enriquez, A., Marin, I., Razjouyan, J., Armstrong, D.G.: Smarter sole survival: Will neuropathic patients at high risk for ulceration use a smart insole-based foot protection system? J. Diabetes Sci. Technol. **11**(4), 702–713 (2017)
35. Park, C., Mishra, R., Vigano, D., Macagno, M., Rossotti, S., D'Huyvetter, K., Garcia, J., Armstrong, D.G., Najafi, B.: Smart offloading boot system for remote patient monitoring: toward adherence reinforcement and proper physical activity prescription for diabetic foot ulcer patients. J. Diabetes Sci. Technol. 19322968211070850. Advance online publication (2022)
36. Perrier, A., et al.: Smart diabetic socks: embedded device for diabetic foot prevention. IRBM **35**(2), 72–76 (2014)
37. Ploderer, B., Brown, R., Seng, L., Lazzarini, P.A., van Netten, J.J.: Promoting self-care of diabetic foot ulcers through a mobile phone app: user-centered design and evaluation. JMIR Diabetes **3**(4), e10105 (2018)
38. Raviglione, A., Reif, R., Macagno, M., Vigano, D., Schram, J., Armstrong, D.: Real-time smart textile-based system to monitor pressure offloading of diabetic foot ulcers. J. Diabetes Sci. Technol. **11**(5), 894–898 (2017)
39. Rescio, G., Leone, A., Francioso, L., Siciliano, P.: Sensorized insole for diabetic foot monitoring. Eurosensors Conf. **2**(13), 860 (2018)
40. Reyzelman, A.M., et al.: Continuous temperature-monitoring socks for home use in patients with diabetes: observational study. J. Med. Internet Res. **20**(12), e12460 (2018)
41. Frykberg, R.G., Gordon, I.L., Reyzelman, A.M., Cazzell, S.M., Fitzgerald, R.H., Rothenberg, G.M., Bloom, J.D., Petersen, B.J., Linders, D.R., Nouvong, A., Najafi, B: Feasibility and efficacy of a smart mat technology to predict development of diabetic plantar ulcers. Diabetes Care **40**(7), 973–980 (2017)
42. Sandoval-Palomares, J., Yáñez-Mendiola, J., Gómez-Espinosa, A., López-Vela, J.M.: Portable System for monitoring the microclimate in the footwear-foot interface. Sensors (Basel) **16**(7), 1059 (2016)
43. Tavares, C., et al.: Gait Shear and plantar pressure monitoring: a non-invasive OFS based solution for e-health architectures. Sensors (Basel) **18**(5), 1334 (2018)
44. Texier, I., et al.: SWAN-iCare project: towards smart wearable and autonomous negative pressure device for wound monitoring and therapy. In: International Conference Wireless Mobile Communication Healthcare, 2014, pp. 357–360 (2015)
45. Torreblanca González, J., Gómez-Martín, B., Hernández Encinas, A., Martín-Vaquero, J., Queiruga-Dios, A., Martínez-Nova, A.: The use of infrared thermography to develop and

assess a wearable sock and monitor foot temperature in diabetic subjects. Sensors (Basel) **21**(5), 1821 (2021)
46. Wang, D.H.-C., Blenman, N., Maunder, S., Patton, V., Arkwright, J.: An optical fiber Bragg grating force sensor for monitoring sub-bandage pressure during compression therapy. Opt. Express **21**(17), 19799–19807 (2013)
47. Yan, R., Zhao, W., Sun, Q.: Research on a physical activity tracking system based upon three-axis accelerometer for patients with leg ulcers. Healthc. Technol. Lett. **6**(5), 147–152 (2019)
48. Yavuz, M., et al.: Temperature- and pressure-regulating insoles for prevention of diabetic foot ulcers. J. Foot Ankle Surg. **59**(4), 685–688 (2020)
49. Yong, K.F., Forero, J. P., Foong, S., & Nanayakkara, S.: FootNote: designing a cost effective plantar pressure monitoring system for diabetic foot ulcer prevention. In: International Conference Augmented Human, pp. 167–168 (2015)
50. You, Z., Zahid, A., Heidari, H., Imran, M., Abbasi, Q.H.: A compact wearable system for detection of plantar pressure for diabetic foot prevention. In: Asia Pacific Conference on Postgraduate Research in Microelectronics and Electronics, pp. 64–67 (2018)
51. Yuan, Z., Huang, J., Zhao, Z., Zahid, A., Heidari, H., Ghannam, R., Abbasi, Q.H.: A compact wearable system for detection and estimation of open wound status in diabetic patient. In: Asia Pacific Conference on Postgraduate Research in Microelectronics and Electronics, pp. 60–63 (2018)
52. Zhao, S., Liu, R., Fei, C., Guan, D.: Dynamic interface pressure monitoring system for the morphological pressure mapping of intermittent pneumatic compression therapy. Sensors (Basel) **19**(13), 2881 (2019)
53. Zoppo, G., et al.: AI technology for remote clinical assessment and monitoring. J Wound Care **29**(12), 692–706 (2020)

Electronic Plush Toy as a Non-pharmacological Tool in Therapies

Sergio Márquez-Sánchez[1,2](✉), Jorge Herrera-Santos[1], Sara Mora-Simon[3,4], and Sara Rodríguez-González[1]

[1] BISITE Research Group, University of Salamanca, Calle Espejo s/n. Edificio Multiusos I+D+i, 37007 Salamanca, Spain
smarquez@usal.es, jorgehsmp@usal.es, srg@usal.es

[2] Air Institute, IoT Digital Innovation Hub (Spain), 37188 Salamanca, Spain

[3] Department of Basic Psychology, Psychology and Behavioral Sciences Methodology, Faculty of Psychology, University of Salamanca, Avda.de la Merced 109-131, 37005 Salamanca, Spain
s_mora@usal.es

[4] Institute of Biomedical Research of Salamanca (IBSAL), Primary Care Research Unit of Salamanca (APISAL), Avda. Portugal 83-2o, 37005 Salamanca, Spain

Abstract. Due to the increase in life expectancy, the prevalence of chronic, cardiovascular and neurodegenerative diseases has increased. In this regard, the increase in cognitive impairment (CI), both in the mild phase, mild cognitive impairment (MCI), and in the more severe cases defined as dementia, is also noteworthy. It has been observed that between 5 and 15% of cases of MCI progress annually to dementia. It is therefore important to detect MCI early, not only because some cases are at high risk of progressing to dementia, but also because early detection of MCI allows to adopt strategies that improve the quality of life of those patients. Although there are numerous non-pharmacological therapies that have been shown to be effective, therapy with dolls or plush toys is noteworthy, as it has been shown to improve the well-being of people with dementia as well as other pathologies. The electronic plush toy is therefore able to provide solutions to different study scenarios and problems that we encounter in therapy, not only in the case of cognitive impairment. Among these needs it is also a good support for children and families against any kind of stress and anxiety caused by clinical treatments and hospitalisations. Through the therapeutic use of plush toys we are able to prepare patients for their clinical interventions and to be a form of temporary distraction from pain with integrated playful interactions and intervention methods.

Keywords: Non-pharmacological therapy · Innovative therapy · Doll therapy · Robot therapy

V. Julián et al. (Eds.): ISAmI 2022, LNNS 603, pp. 111–122, 2023.
https://doi.org/10.1007/978-3-031-22356-3_11

1 Introduction

Different preliminary studies show that users connect better, emotionally, with dolls or plush toys than with robotic or virtual characters [1]. All in all, this illustrates a great potential in the use of social robots in the form of plush toys to provide socio-emotional support during therapies. One of the innovations of this research is the combination of electronics and textiles, as currently there is a gap in the use of dolls and plush toys for therapeutic purposes and robots that are used as a non-pharmacological tool for therapy.

Robot therapies can nurture people and help them overcome their challenges and cope with different problems. They can help children and elderly people through emotionally stimulating interactions. Human beings rely on social relationships to give meaning to our existence, and these are the sources of the pleasures and disappointments in our lives. In addition, we can experience different psychic, psychological and physical symptoms caused by trauma, illness or clinical pathology that can be helped by the use of therapeutic devices.

Therapy in which dolls or plush toys are involved allows to generate a very realistic experience, easily becoming an integral part of people's lives and helping to resolve conflicts in any area of human relations, helping to visualise and find solutions. From a didactic and orderly perspective, plush toy therapy is a valuable tool for therapists as a basis for their work with patients. It can represent symbolically and tangibly the family, the organisational and structural system and its dynamics according to the individual situation.

Thanks to advances in robotics, the trend is moving from zootherapy towards robot therapy. The benefits of using pets in the rehabilitation or treatment of chronic diseases or psychological pathologies have long been widely accepted. Animals can have a positive impact on a patient's social, emotional, cognitive and even physical condition, such as "normalising heart rhythm". Robots have no real emotional capacity but are programmed. Behind such robots are teams of psychologists, doctors and therapists who "study emotions and assess what patients need", and then engineers programme them according to the guidelines they are given, using mathematical models.

In a world where virtual interactions and robots made of cold materials are becoming increasingly common, this proposal aims to analyse the possibilities offered by electronic plush toys in the clinical field. Specifically, their potential to help patients with problems such as Alzheimer's, dementia, autism, trauma, psychological problems as well as pathologies such as stress, anxiety and pain. Therefore, a very favourable scenario is established in this proposal, for an innovative therapeutic product, with which a large amount of scientific findings can be obtained. The product must be reliable, cheap and competent in terms of production.

Doll Therapy is a complementary methodology to pharmacological therapy, which consists of the use of therapeutic dolls in the different interventions carried out by specialist. Sessions with dolls stand out as one of the most useful among complementary therapies. Sometimes, the drugs needed to control certain pathologies adversely affect the users' faculties. Combining pharmacological therapy with doll therapy creates a synergy that slows down and reduces the progressive impairment of their faculties.

The aim of this paper is to design an experimental therapeutic electronic plush toy that adds value and enhances today's products and advances the research. Providing a set of interconnected intelligent tools, with the capacity to respond in real-time to the user's stimuli, developing a holistic system, personalised and adapted to both, the environment and to the physical, cognitive and sensory circumstances of the individuals in each case study.

2 State of Art

Due to the increase in life expectancy, the prevalence of chronic, cardiovascular and neurodegenerative diseases has increased. The increase in cognitive impairment (CI) is also noteworthy [2]. When we speak of CI we refer to a syndrome in which the patient may present impairment in one or more cognitive functions (attention/concentration, memory, executive functions, language, visuoconstructive skills, gnosias and praxias), and, depending on its severity, the impact on daily living (basic and/or instrumental) activities may vary. This impairment should be objectified by neuropsychological tests and not be explained by a psychiatric disorder [3,4]. According to the latter consensus, when the cognitive impairment has a minimal impact on instrumental daily living activities, the patient would be considered to have mild cognitive impairment (MCI), whereas if the impairment is so severe that it affects not only instrumental activities but also basic activities, the patient would be considered to have dementia [5].

It has been observed that between 5 and 15% of cases with MCI progress annually to dementia, while the remaining percentage of patients with MCI remain stable in this entity [6]. With this in mind, it is therefore important to detect MCI early, not only because some of the cases present a high risk of progression to dementia but also because early detection of this entity allows for strategies to improve the quality of life of these patients. When the presence of cognitive dysfunction (CD) is detected, it is important to begin the treatment of the patient, including non-pharmacological therapies. Currently, in the approach to different pathologies, non-pharmacological therapies are a useful, low-cost and highly versatile tool. The main objective of these therapies is to improve the quality of life of both the patient and the caregiver, as well as the clinical management of the pathology [7–11].

Numerous robots have been developed in various forms (cats, dogs, dinosaurs, bears) but the best known and most studied social robot is PARO, a seal designed in Japan by Shibata [12]. PARO is an interactive seal robot with light, touch, postural, temperature and sound sensors around its body, and can recognise voices when touched or spoken to [13–15]. Numerous studies have focused on the effects of PARO and other social robots on people with AD and other dementias, showing benefits in the quality of life of users and family members, as well as improved social interaction, emotional expression and reduced loneliness, and decreased SPCD such as agitation and anxiety [15–18].

Although there are numerous non-pharmacological therapies that have been shown to be effective, it is worth highlighting therapy with dolls or plush toys [1]. Doll therapy is based on Bowlby's attachment theory (1969) which suggests that human beings, from birth, need to develop a relationship with at least one primary caregiver for normal social and emotional development [19]. The aim of doll therapy is to use dolls, in a careful and controlled way, to improve the well-being of people with dementia [20]. There is little scientific evidence to date on the efficacy of doll therapy in different populations, however, existing studies show very beneficial effects, especially in reducing behavioural problems in dementia [15,21].

3 Methodology

The main objective of this work has been the creation of an electronic plush toy, based on the integration of different technologies, to improve the quality of life of people with different pathologies. The project has consisted of the following development phases.

In the first phase, the specification of requirements and the definition of the architecture have been carried out. This has involved the identification of commercial solutions and applicable research in the field of study, as well as all those pathologies that we can work on from a therapeutic field with this type of product. Accordingly, the functionalities to be incorporated in the proposed solution were defined, as well as the technical aspects required for its implementation, which materialised through the design of the architecture, the definition of both the components and the connectors and the correct identification of the flow of information through the plush toy and the mobile application [22–25].

In the second phase, the design and assembly of the hardware have been carried out. A PCB has been created where the different electronic components are integrated and connected. These components are responsible for making the plush toy interact, react and communicate in real-time. This phase also includes the activities related to the firmware programming for the acquisition and communication of the electronic components with the mobile app [26].

In the third phase, the mobile app has been developed to visualise the data collected by the hardware in real-time. For its development, all the features and functions discussed in previous points have been considered, such as the visualisation, per session, of data and of interactivity in real-time.

The fourth phase deals with the integration and validation of the system by interconnecting the electronic components and the mobile application. In this phase, it has been verified that all the developed functionalities comply with the requirements specified in the first phase of the project and that the implemented technologies offer their corresponding advantages. As for the validation of the system, tests have been carried out with real users to verify the correct performance of the system.

4 Design and Construction

The architecture and design of the plush toy is the differentiating element of this project (see Fig. 1). It is not a single device, as the plush toy is part of a series of elements that communicate with each other. Thanks to the technology they are to incorporate and the interconnection of the different tools, these plush toys are to be able to respond, in real-time, to the user's stimuli or the guidelines set by the therapist, so that we can adapt the behaviour of the doll to the physical, cognitive and sensory circumstances of the users. In terms of firmware programming, different principles and design criteria must be applied when developing the social and interactive electronic plush toy, among which the nature and characteristics of the case study, recipient and working environment must be taken into account.

4.1 HW Design

For data processing, an MCU with ARM architecture and multithread capacity has been used, which includes the RTC functionality enabling the periodic execution of events. The device has a wireless communications module that enables the communication between the device and a smartphone. To achieve data communication at the minimum consumption cost, the Bluetooth Low Energy protocol has been used.

Following the specified functional requirements, the proposed architecture is as shown in Fig. 1.

The developed product is distinctly different from other similar solutions on the market, mainly due to the plush toy's design and the communication with the mobile app. Among the functionalities that make it distinct is the programming which enables the toy to respond in real-time to the user's interactions or the guidelines set by the therapist. This is important as the behaviour of the plush toy can be adapted to the physical, cognitive and sensory circumstances of the users. The hardware design is modular so that it can be adapted to new dolls and include new sensors and functions. The firmware programming has taken into account the nature and characteristics of the case study, the recipient and the working environment.

For the device to operate correctly, external stimuli produced by the user are recognised. These shall be recognised by means of the following peripherals:

- IMU: an Inertial Measurement Unit has been used with which interruptions have been configured to recognise different stimuli such as soft movements or falls.
- Tactile sensor: conductive thread connected to capacitive pins on the board has been used to detect the interaction between the user and the plush toy.
- Microphone: to detect the possible agitation of the user.
- Battery charger: the device has a charging circuit.

For the doll to interact with the patient and react to certain stimuli, a series of reactions are generated to "bring the doll to life". These reactions are generated by means of the following elements:

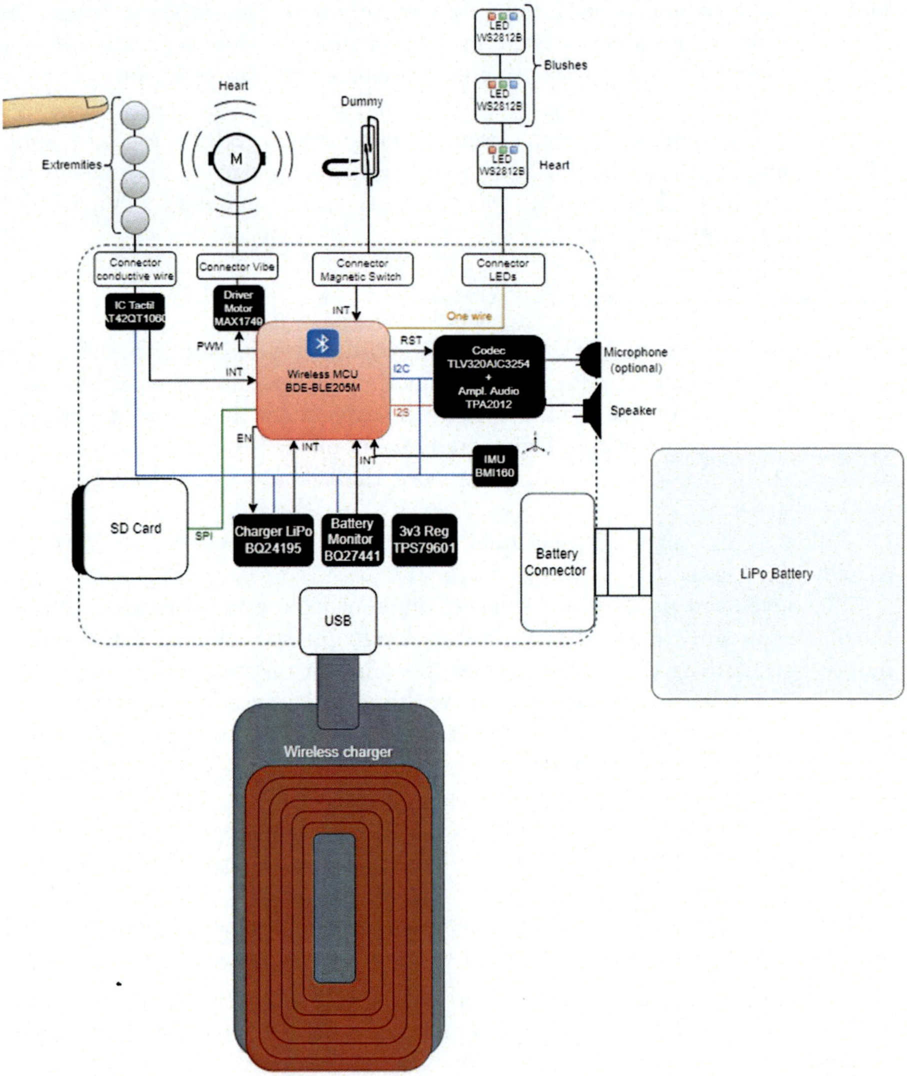

Fig. 1. Proposed architecture.

- Vibrator: a small vibrator has been installed to simulate the pulse, thus representing states of calmness (slow pulsation) or activity (fast pulsation).
- RGB LEDs: using LEDs located on the cheeks and heart, the area is illuminated according to the emotions of happiness or anger.
- Loudspeaker: for realistic sound reproduction.

All these functions are programmed to appear automatically in the plush toy, so that, after an interaction, for example, the plush toy's cheeks blush using LEDs or the heart beats faster. In addition, it is also possible to activate this behaviour from the mobile app, setting us apart from any other similar product by integrating several functions never seen before in this type of project. Figure 2 below shows the developed PCB.

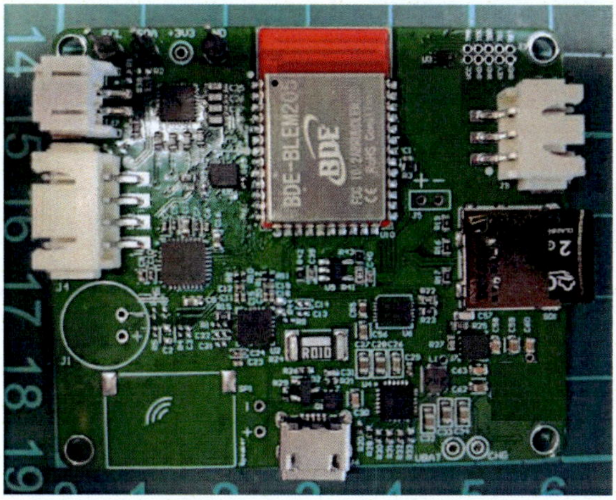

Fig. 2. Prototype PCB.

App As for the mobile app, the user interface is provided through a mobile app with support for iOS operating systems. Through the interface, it is possible to interact with the toy, as well as track sessions, view project data, assess progress and receive alerts. The recommendations of good visual design practices established by Material Design have been followed. The plush toy has an implemented state machine that establishes its operation. Thanks to this machine, the app can be used to change the states of the plush toy, thus acting as a slave of the App. It is worth mentioning that all screens related to user information are loaded locally as saved in the last update. Subsequently, as long as we have an internet connection, the data are loaded from the server (see Fig. 3).

Construction of the Robot The steps followed to make the plush toy were: first, the patterns on the fabrics (blue coral, Minky light blue, white cotton

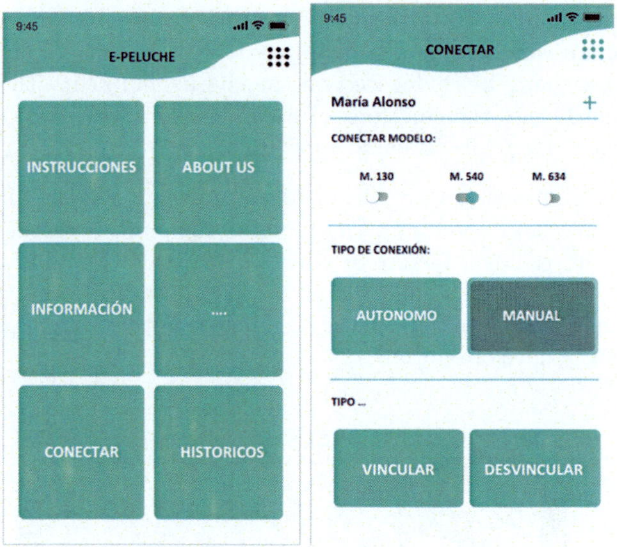

Fig. 3. Mobile app screens developed in Spanish language.

fabric) were marked. The expression of the plush toy was decided and the cotton fabric was painted with acrylic paint. Afterwards, the plush animal was given a body by gluing the friseline with an iron, securing it with stitching, cutting the outline with zigzag scissors, and sewing the front part of the body of the plush animal. Once this was done, the arms and legs were sewn onto the front body of the toy. The toy was turned right side out once sewn and filled with 100% wool plush.

As far as the integration of the electronics is concerned, the electronic components were sewn in such a way that they were effectively fixed to the fabric (see Fig. 4). In addition, for tactile interaction, conductive thread and conductive fabric made of stainless steel particles were used and integrated.

The finished plush toy achieved the result shown in Fig. 5.

5 Conclusions and Future Work

The aim of this work has been to create an electronic plush toy that allows to analyse the possibilities offered by electronic plush toys for use in the clinical field and help patients with problems such as Alzheimer's, dementia, autism, trauma, psychological problems as well as pathologies such as stress, anxiety and pain. The plush toy integrates a set of interconnected intelligent tools, with the capacity to respond, in real-time, to the user's stimuli, developing a holistic, personalised system adapted to the environment and to the physical, cognitive and sensory circumstances of the individuals in each case study.

Fig. 4. Integration of electronics in the plush toy.

Fig. 5. Finished plush toy with integrated electronics.

The aim of this study has been to evaluate the effectiveness of an intervention with electronic plush toys on the presence of psychological and behavioural symptoms in people with different pathologies, as well as their quality of life, affectivity, usability and emotional state. Its development allows for progress in the search for technological solutions that improve the quality of life of people with different pathologies. Participants who received a plush toy showed a decrease in the frequency of symptoms related to disinhibition, motor impairment, appetite and eating habits. In addition, a slight increase in the patient's caregiver-perceived quality of life and positive affect was observed.

As for the future, work is underway to develop new functionalities to address more issues and reach a wider market. The user will be able to interact verbally with their plush toy as it will listen and respond, sing, tell stories and play the voice of family members, or a song from their childhood. Users will also be able to interact with the plush toy, as it will respond to their touch or a change of location. In addition, the plush toy can be programmed to help the user manage some daily activities, such as physical activities to improve psychomotor skills, keep track of their time or help them remember basic actions such as taking their medication. Above all, however, this product will be designed to entertain, accompany, reassure, give affection and teach users.

Following the testing of the range of effects that different functionalities have on different patients, one of the main tasks to be carried out in the future is to exploit a variety of models and add-ons, making it possible for users to customise their plush toys according to their personal preferences or needs.

Acknowledgements. This work has been supported by the project "XAI - XAI - Sistemas Inteligentes Auto Explicativos creados con Módulos de Mezcla de Expertos", ID SA082P20, financed by Junta Castilla y León, Consejería de Educación, and FEDER funds.

References

1. Márquez-Sáncheza, S., Mora-Simonb, S., Herrera-Santosa, J., Roncerod, A.O., Corchadoa, J.M.: Intelligent Dolls and robots for the treatment of elderly people with dementia. Adv. Distrib. Comput. Art. Intell. J. **9**(1), 99–112 (2020)
2. Ferri, C.P., Prince, M., Brayne, C., Brodaty, H., Fratiglioni, L., Ganguli, M., Hall, K., Hasegawa, K., Hendrie, H., Huang, Y., Scazufca, M.: Global prevalence of dementia: a Delphi consensus study. Lancet **366**(9503), 2112–2117 (2005)
3. Albert, U., Tomassi, S., Maina, G., Tosato, S.: Prevalence of non-psychotic disorders in ultra-high risk individuals and transition to psychosis: a systematic review. Psychiatry Res. **270**, 1–12 (2018)
4. McKhann, G.M., Knopman, D.S., Chertkow, H., Hyman, B.T., Jack Jr, C.R., Kawas, C.H., Klunk, W.E., Koroshetz, W.J., Manly, J.J., Mayeux, R., Phelps, C.H.: The diagnosis of dementia due to Alzheimer's disease: recommendations from the National Institute on Aging-Alzheimer's Association workgroups on diagnostic guidelines for Alzheimer's disease. Alzheimer's Dementia **7**(3), 263–269
5. Khan, A., Zubair, S., Khan, S.: Comprehensive performance analysis of neurodegenerative disease incidence in the females of 60–96 year age group. ADCAIJ:

Adv. Distrib. Comput. Artif. Intell. J. **10**(2) (2021). https://doi.org/10.14201/ADCAIJ2021102183196
6. Sperling, R.A., Aisen, P.S., Beckett, L.A., Bennett, D.A., Craft, S., Fagan, A.M., Phelps, C.H.: Toward defining the preclinical stages of Alzheimer's disease: recommendations from the National Institute on Aging-Alzheimer's Association workgroups on diagnostic guidelines for Alzheimer's disease. Alzheimer's Dement. **7**(3), 280–292 (2011)
7. Olazarán, J., Reisberg, B., Clare, L., Cruz, I., Peña-Casanova, J., Del Ser, T., Muñiz, R.: Nonpharmacological therapies in Alzheimer's disease: a systematic review of efficacy. Dement. Geriatr. Cogn. Disord. **30**(2), 161–178 (2010)
8. Bondioli, M., Chessa, S., Narzisi, A., Pelagatti, S., Piotrowicz, D.: Capturing play activities of young children to detect autism red flags. Adv. Intell. Syst. Comput. **1006**, 71–79 (2020)
9. Smith, R.: Adaptivity as a service (AaaS): personalised assistive robotics for ambient assisted living. Adv. Intell. Syst. Comput. AISC **1239**, 239–242 (2021)
10. Smith, R., Ding, Y., Goussetis, G., Dragone, M.: A cots (UHF) RFID floor for device-free ambient assisted living monitoring. Adv. Intell. Syst. Comput. AISC **1239**, 127–136 (2021)
11. Stavropoulos, T., Meditskos, G., Papagiannopoulos, S., Kompatsiaris, I.: eHealth4MS: problem detection from wearable activity trackers to support the care of multiple sclerosis. Adv. Intell. Syst. Comput. AISC **1239**, 3–12 (2021)
12. Jung, M.M., Van der Leij, L., Kelders, S.M.: An exploration of the benefits of an animallike robot companion with more advanced touch interaction capabilities for dementia care. Front. ICT **4**, 16 (2017)
13. Iacono, I., Marti, P.: Narratives and emotions in seniors affected by dementia: A comparative study using a robot and a toy. In: 2016 25th IEEE International Symposium on Robot and Human Interactive Communication (RO-MAN), pp. 318–323. IEEE (2016)
14. Jøranson, N., Pedersen, I., Rokstad, A.M.M., Ihlebaek, C.: Effects on symptoms of agitation and depression in persons with dementia participating in robot-assisted activity: a cluster-randomized controlled trial. J. Am. Med. Directors Assoc. **16**(10), 867–873 (2015)
15. Moyle, W., Jones, C., Dwan, T., Ownsworth, T., Sung, B.: Using telepresence for social connection: views of older people with dementia, families, and health professionals from a mixed methods pilot study. Aging Mental Health **23**(12), 1643–1650 (2019)
16. Feng, Y., Yu, S., Van De Mortel, D., Barakova, E., Hu, J., Rauterberg, M.: LiveNature: ambient display and social robot-facilitated multi-sensory engagement for people with dementia. In: Proceedings of the 2019 on designing interactive systems conference, pp. 1321–1333
17. Jøranson, N., Pedersen, I., Rokstad, A.M.M., Ihlebaek, C.: Change in quality of life in older people with dementia participating in Paro-activity: a cluster-randomized controlled trial. J. Adv. Nurs. **72**(12), 3020–3033 (2016)
18. Jøranson, N., Pedersen, I., Rokstad, A.M.M., Aamodt, G., Olsen, C., Ihlebæk, C.: Group activity with Paro in nursing homes: systematic investigation of behaviors in participants. Int. Psychogeriatr. **28**(8), 1345–1354 (2016)
19. Bowlby, J.: Attachment and loss: volume I: attachment. In: Attachment and loss: Volume I: attachment, pp. 1–401. The Hogarth Press and the Institute of Psycho-Analysis, London (1969)
20. Verity, J.: Dolls in dementia care: bridging the divide. J. Dementia Care **14**(1), 25 (2006)

21. Mitchell, G., O'Donnell, H.: The therapeutic use of doll therapy in dementia. British J. Nurs. **22**(6), 329–334 (2013)
22. Jamal, A., Munshi, A., Aljojo, N., Qadah, T., Zainol, A.: Digital information needs for understanding cell divisions in the human body (2020)
23. Gupta, S., Meena, J., Gupta, O.: Neural network based epileptic EEG detection and classification. ADCAIJ: Adv. Distrib. Comput. Artif. Intell. J. **9**(2), 23–32 (2020)
24. Srivastav, R.K., Agrawal, D., Shrivastava, A.: A survey on vulnerabilities and performance evaluation criteria in blockchain technology. ADCAIJ: Adv. Distrib. Comput. Artif. Intell. J. **9**(2), 91–105 (2020)
25. Costa, Â., Novais, P., Corchado, J., Neves, J.: Increased performance and better patient attendance in an hospital with the use of smart agendas. Logic J. IGPL **20**(4), 689–698 (2012)
26. Miguel Tomé, S.: Towards a model-theoretic framework for describing the semantic aspects of cognitive processes. ADCAIJ: Adv. Distrib. Comput. Artif. Intell. J. 8(4), 83–96 (2020). https://doi.org/10.14201/ADCAIJ2019848396
27. Moyle, W., Jones, C., Murfield, J., Thalib, L., Beattie, E., Shum, D., Draper, B.: Using a therapeutic companion robot for dementia symptoms in long-term care: reflections from a cluster-RCT. Aging Mental Health **23**(3), 329–336 (2019)
28. Fatima, N. (2020) Enhancing performance of a deep neural network: a comparative analysis of optimization algorithms. ADCAIJ: Adv. Distrib. Comput. Artif. Intell. J. **9**(2), 79–90

Federated Learning with Exponentially Weighted Moving Average for Real-Time Emotion Classification

Arijit Nandi[1,2](✉), Fatos Xhafa[1], Laia Subirats[2,3], and Santi Fort[2]

[1] Department of CS, Universitat Politècnica de Catalunya, 08034 Barcelona, Spain
arijit.nandi@eurecat.org fatos@cs.upc.edu

[2] Eurecat, Centre Tecnològic de Catalunya, 08005 Barcelona, Spain
laia.subirats@eurecat.org santi.fort@eurecat.org

[3] ADaS Lab, Universitat Oberta de Catalunya, 08018 Barcelona, Spain

Abstract. Federated learning (FL) allows to develop a powerful shared prediction model while preserving the users' privacy by keeping the data local. In particular it is a useful framework to use resource-constrained edge computing devices, such as mobile and IoT devices (wearable sensors), as local clients. In FL, the local clients update their model by replacing it with the current global model, which may cause performance degradation due to the previous local model information is lost. We propose an exponentially weighted moving average (EWMA) for local model updates at the local client to address this issue and improve model performance. This work focuses mainly on improving our previously developed federated learning-based real-time emotion classification from multi-modal data streams Fed-ReMECS with a local model update by EWMA, called Fed-ReMECS-U. The experiment is carried out with the help of the widely used multi-modal benchmark DEAP dataset for emotion classification. Experiments showed that the accuracy of the proposed method has improved when compared to the former Fed-ReMECS.

Keywords: Federated learning · Real-time emotion classification · Multi-modal data streaming · Data-driven emotion systems

1 Introduction

Real-time multi-modal emotion state classification from wearable sensor-based devices is one of the hottest topic in the IoT era. Because of the availability of wearable devices, emotional health tracking (emotional state extraction) is possible to ensure a person's health, as physical health is inextricably linked to emotional health. Also, in such a scenario, a data stream is generated from wearable-sensor devices, data is decentralised, data access is limited to those

V. Julián et al. (Eds.): ISAmI 2022, LNNS 603, pp. 123–133, 2023.
https://doi.org/10.1007/978-3-031-22356-3_12

devices only, and usually high data generation rate needs to be processed to extract real-time emotional states. Furthermore, safeguarding users' sensitive information makes such data processing even more challenging. For this scenario, we developed and proposed Fed-ReMECS [7], a MQTT[1] protocol supported federated learning (FL) framework, capable of classifying emotions in real-time from multimodal physiological data streams while ensuring users' privacy in IoT environment. In Fed-ReMECS, when the users (local clients) receive the global model from global server their corresponding models are replaced with the new global model and they start the next round of FL. Replacing the corresponding local model by the received global model may cause performance degradation because the previous local model information is lost. To solve this problem and improve the model performance we propose an Exponentially Weighted Moving Average (EWMA) [9] at local client's end, that means when the local client receives the global model, it will not not replace the local model right away instead the local client will update its model by applying EWMA between the local model and recently received global model. The proposed work is an update on our existing Fed-ReMCES [7], it's called Fed-ReMECS-U.

The main contributions of our work are as follows:

- We have introduced EWMA at local client's end to update their corresponding local model.
- We have conducted a comparative evaluation of Fed-ReMECS-U against our previously proposed Fed-ReMECS to study performance and training stability using the most popular benchmark dataset called DEAP for real-time emotion state classification.

The rest of the paper is structured as follows: Sect. 2 introduces briefly the multi-modal data stream classification, federate learning and emotion state representation. The material and methods for the updated Fed-ReMECS-U are presented in Sect. 3. In Sect. 4, the experimental results and discussion are provided. Finally, the paper ends with the conclusion in Sect. 5.

2 Preliminaries

Here, a brief introduction to multi-modal stream classification, federated learning and exponentially weighted moving average model are provided.

2.1 Multi-modal Data Stream Classification

Multi-modal data stream refers to the continuous and unbounded data stream generated from multiple sensors. In multi-modal data streams classification scenario, the following assumptions are made:

- Data tuples arrive sequentially one at a time.

[1] https://mqtt.org/.

- The model first receives the unlabeled data tuples of the stream and then *immediately* after making the prediction the *true class labels* arrives.
- The base model is tested first and then trained (interleaved-test-then-train).
- The data tuple can be seen by the base model at most once.
- A temporary storage for data stream available, means no looped back over received data.

It should be noted that in a multi-modal data stream classification scenario, features are extracted from each modality (sensors) and then a feature fusion approach is applied so that we can fuse and fed in to the stream classifier for emotion classification. In literature the most popular feature fusion approach is feature concatenation because here all features get priority and it is easy to implement [10]. Based on the stream arrival, the classifier classifies emotional state and updates itself using progressive learning strategy [3]. Figure 1 depicts the overall approach of emotion classification from multi-modal data stream.

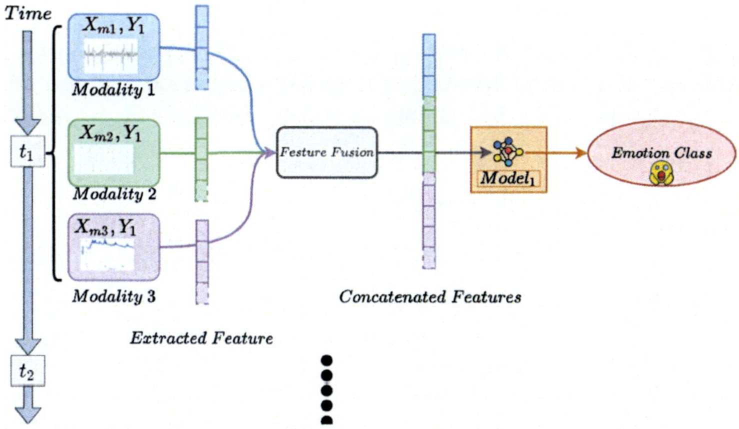

Fig. 1. Emotion classification from multi-modal data stream

As can be seen from Fig. 1, the time dimension (t) shows the data tuples $(t_1, t_2, t_3, \ldots)$ arrive at different times from each modality. Then, test-then-train and emotion classification takes place on arrived tuples.

2.2 Federated Learning

Federated Learning (FL [6,10]) is a promising distributed machine learning approach that enables mobile local clients to build a powerful global model by collaborating with a global server. The mobile devices share a local model developed from the sensitive data accessible to those devices without sharing the sensitive data with other parties. Clearly, in FL there are two parties involved in the whole approach, global server and local clients (end/edge users/devices) [6].

The main functions of local clients are: (1) access to the local data, develop the model and perform the corresponding learning task (2) sharing the developed local model to the global server (3) receive the global model and replace the local model with the global model.

The main functions of the global server are: (1) collect the shared local models from local clients; (2) create the global model by model averaging (called Federated Averaging) of all the collected local models; (3) broadcast the global model to all the local models participated in the Federated Averaging (FedAvg).

2.3 Exponentially Weighted Moving Average

An EWMA is also known as Auto-Regressive (AR) filtering. It is a type of moving average (MA) that gives the most recent data points more weight and significance. EWMA is frequently used to generate short-term forecasts of time series (e.g., sales and stocks). A simple first order EWMA is formulated as follows:

$$M_t = \beta M_t + (1-\beta)M_{t-1} \tag{1}$$

where M_t is the current measurements, M_{t-1} is the previous measurements, β determines the effect of the previous measurements on the current value. In general, the β value is set to 0.9 and its widely used.

3 Materials and Methods

Herewith, the proposed approach details, the dataset description, experimental study, experimental setup and finally the performance metric are provided.

3.1 Proposed Approach

Fed-ReMECS is developed by following traditional FL approach which is capable of classifying emotions from multi-modal physiological streams in IoT environment. In Fed-ReMECS, when the local client receives the global model after FedAvg, it replaces the local model immediately. That causes the performance degradation because the local model information is completely lost. Our main motivation in the proposed Fed-ReMECS-U is to utilize the previous the local model directly rather discarding it after receiving the global model in the beginning of each round. To utilize the previous local model information in the current round we have utilized exponentially weighted moving average (EWMA) to combine the previous local model and currently received global; then the combined model is considered as the current local model for the current round. This EWMA happens in every round at local clients after receiving the current global model from global server. To apply the EWMA in FL at local clients end, the Eq. (1) needs a little modification as follows:

$$w_t^l = \beta \cdot w_t^g + (1-\beta) \cdot w_{t-1}^l \tag{2}$$

where $w_t^{(l)}$ and $w_t^{(g)}$ are the local model weight and global model weights at time t, respectively. $w_{t-1}^{(l)}$ is the local model weights from previous round or at $t-1$. Here β is the important factor.

3.2 Dataset Description

The proposed approach's effectiveness is assessed using the most popular and widely used benchmark dataset DEAP [5] for emotion recognition from multi-modal physiological signals. The following is a brief description of the DEAP[2] [5] (Database for Emotion Analysis using Physiological Signals) dataset. In the DEAP dataset, the channel no. 37 is for Electrodermal Activity (EDA) signal and channel no. 38 is for Respiratory Belt (RB) signal. For our experiment, we have considered EDA and RB signals for the multi-modal physiological data stream (see Table 1 for a brief description).

Table 1. Brief description of DEAP dataset

Description	Details
No. participants	32 (16 male-16 female)
Stimuli	40 music videos of length 60s each
Physiological signals	EEG, EDA, RB
No. channels	Total of 47 channels (32 channels for EEG, 12 channels of peripheral, 3 unused)
Saved data file system	.dat file of total 32 participants
Data representation	3D matrix representation $40 \times 40 \times 8064$, which represents video/trial $\times$ channel $\times$ data
Label representation	2D matrix representation (40×4)
Available emotion labels	valence, arousal, dominance and liking

3.3 Experimental Study

The experimental study of our proposed Fed-ReMECS-U approach for real-time emotion classification using multi-modal physiological data streams (for example, EDA+RB data streams for DEAP dataset) is presented here. The following steps describes the whole experimental study:

– *Data set consideration and data rearrangement*: The pre-processed multi-modal DEAP data is used to create the multi-modal data stream. The experiment is carried out with the DEAP dataset, which contains EDA and RB signal

[2] DEAP dataset link: https://www.eecs.qmul.ac.uk/mmv/datasets/deap/.

streams. A data rearrangement is performed to convert the data to 1D matrix form for easier EDA and RB data because the DEAP is stored in 3D matrix form, representation as follows:

$$[participant, video, data, valence\ class, arousal\ class]$$

In the experiment using DEAP dataset, the EDA and RB multi-modal data streams are used to classify discrete emotion states based on valence-arousal measurements. While streaming from DEAP dataset, an automatic mapping of the valence (V) and arousal (A) scores to 0–1 are done. Both the V-A mapping and the next stage discrete emotion conversion are performed based on our Fed-ReMECS experiment [7]. Table 2 presents the discrete emotion labels (where L-Low, M-Middle and H-High):

Table 2. Discrete emotion mapping with the help of valence–arousal status

Valence–arousal level	Discrete emotions
LV-LA	sad (0)
LV-MA	Miserable (1)
LV-HA	Angry (2)
MV-LA	Tired (3)
MV-MA	Neutral (4)
MV-HA	Tense (5)
HV-LA	Calm (6)
HV-MA	Happy (7)
HV-HA	Excited (8)

– *Stream reading*: We used a sliding window protocol to stream data for each participant in the multi-modal data (DEAP) stream. The sliding window size is set to 10s because it has previously been used in emotion literature [1,4] and its effectiveness has been validated. The multi-modal data stream rate for the DEAP dataset is approximately 3 Mb/10s. The Fed-ReMECS-U is built with the MQTT protocol.

– *Feature extraction and fusion:* In this experiment, the wavelet feature extraction technique is used to extract features from multi-modal signal streams (EDA and RB signals from the DEAP dataset). The base function for feature extraction is the wavelet Daubechies 4 (Db4). Decomposition of EDA and RB into three levels is performed in our experiment. Following that, a feature fusion approach (concatenation approach) is used to combine the extracted features from EDA and RB modality. The fused features are then sent to the client-side emotion classifier.

– *Emotion classifier*: To classify the discrete emotion labels (in Table 2), we have employed a three-layer Feed Forward Neural Network (FFNN) as the base classifier to do the classification in real-time from multi-modal data streams (EDA and RB). It is already been established in our previous work [8] that the FFNN is best classifier for real-time emotion classification from multi-modal physiological data streams. According to the Table 2, we have classified 9 discrete emotions.

– *Local model test-train*: Stochastic Gradient Descent (SGD) is used to train the FFNN model in online fashion. The interleaved test-then-train [3] method is used to evaluate all considered base classifiers. It tests the model before it is trained, and then the performance metrics are updated using the received data tuple. As a result, the base classifier is always tested on new data tuples. The base classifier performs poorly at first, but as it sees more data tuples from multi-modal data streams, it gains stability and improves performance.

– *Local model sending and global model receiving*: The exact time it takes to send the local model and receive the global model depends entirely on the problem. According to the DEAP dataset experimental design, we analyze the sending and receiving time in our experiment. In DEAP dataset, each participant watches a 60-second video at a time, so the local model is built at that time, and after finishing each 60-second video, the local model is sent to the MQTT broker, and then to the global server for global model creation. Once the global model is created, global server sends the global model to the MQTT broker, then MQTT Broker forwards the global model to all the clients participated in the federated learning, finally all the clients update their local model with the received global model.

– *Global model creation*: The global server is responsible of creating the global model after performing Federated Averaging (FedAvg) on all the received local models. The FedAvg formula is as follows (in Eq. (3)):

$$w_t^g = \frac{1}{N_{total}} \sum_{i=1}^{N_{total}} w_{t-1,i}^l \tag{3}$$

where w_t^g is the global model created at time t, N_{total} is the total number of local model received at global server, $w_{t-1,i}^l$ is the local model received from all clients at time t. Here we assume full participation from all clients in FedAvg.

– *Local model update*: When each client receives the global model, its local model is updated with the global one; then, each client starts for the next federated learning iteration. Each local model weight is updated using Eq. (2). Note that in the Fed-ReMECS-U, the categorical cross-entropy loss function is used.

3.4 Experimental Setup

Herewith the machine setup, software development, experimental environment and parameter setup are as follows:

Machine configuration: Ubuntu 18.04 and 20.04 LTS 64 bit OS, processor core-i7-7700HQ with RAM 24 Gb–2400MHz and 4Gb-Nvidia GTX-1050.

Software development: The Fed-ReMECS-U is implemented in Python 3.7 with the help of TensorFlow 2.0 and Keras in the backend. For the Subscriber and Publisher paho-mqtt library and for the MQTT broker Eclipse Mosquitto broker is used.

Parameter setup: The learning rate for SGD is 0.05 for the whole emotion classification. For the hidden layer and output layer *sigmoid* and *softmax* is used, respectively. The *epoch* is set to 1 implies that model training happens once for each incoming data tuple.

Source-code: The source code and implementation can be found in *Github*.[3]

3.5 Performance Metric

For evaluating the classifiers performance, accuracy (Acc) [2] according to formulae: $Acc = \frac{\sum_{i=1}^{|l|} \frac{TP_i + TN_i}{TP_i + FN_i + FP_i + TN_i}}{|l|}$, where $|l|$ is the number of classes. True positives (TP_i), True negatives (TN_i), False positives (FP_i) and False negatives (FN_i).

4 Results, Analysis and Discussion

In this section we summarize the experimental results of our proposed Fed-ReMECS-U and make a comparison with our previously developed Fed-ReMECS [7] approach under different number of clients running in parallel. A total of six different number of client scenario (5, 10, 15, 20, 25 and 32) are considered for the experiment. The average accuracy comparison of the global model in case of Fed-ReMECS and Fed-ReMECS-U is shown in Table 3 and the performance change comparison in terms of accuracy is shown in Fig. 2.

Table 3. Global model's average accuracy comparison in Fed-ReMECS versus Fed-ReMECS-U

No. of clients	Fed-ReMECS [7] Accuracy	Fed-ReMECS-U accuracy
5	0.8384 (±0.29)	0.9875 (±0.06)
10	0.8423 (±0.30)	0.9812 (±0.10)
15	0.8205 (±0.33)	0.8416 (±0.29)
20	0.8384 (±0.28)	0.8515 (±0.27)
25	0.8263 (±0.26)	0.8474 (±0.26)
32	0.8192 (±0.21)	0.850 (±0.25)

The Fed-ReMECS-U follows the same models as with the Fed-ReMECS, where the real-time emotion classification is done by each clients' local model because of the data accessibility restrictions, therefore, there is no way the global model has access to the data. That is why the global model performance is calculated by taking the average of the local model's performance for the very first data tuple. Now from the average accuracy comparison Table 3, we can see that our proposed Fed-ReMECS-U has our performed our previously proposed

[3] The source code can be found in GitHub at: https://github.com/officialarijit/Fed-ReMECS-U.

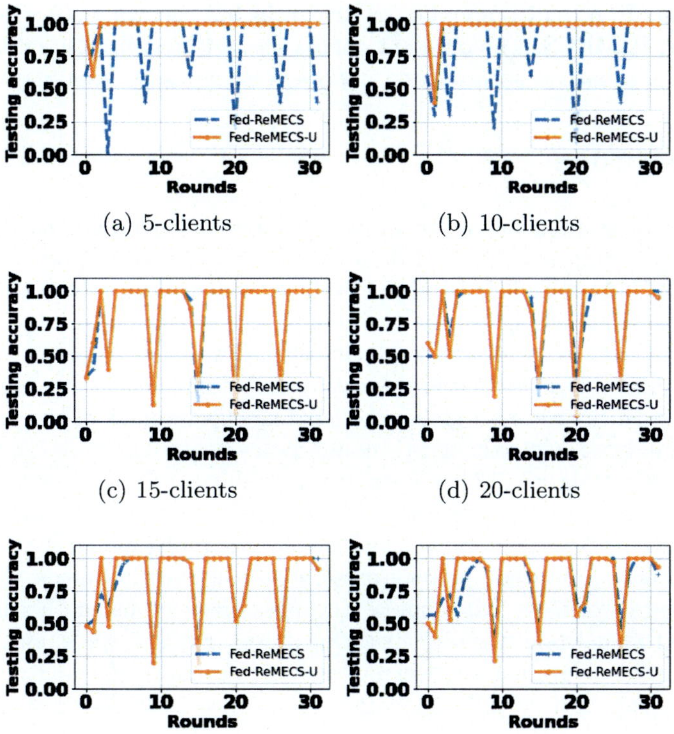

(a) 5-clients (b) 10-clients

(c) 15-clients (d) 20-clients

(e) 25-clients (f) 32-clients

Fig. 2. Global model's accuracy comparison in Fed-ReMECS versus Fed-ReMECS-U for different number of clients

Fed-ReMCES in real-time emotion classification from physiological data stream in IoT environment.

It can also noticed that the overall performance improvement of Fed-ReMECS-U than the Fed-ReMECS is 10.41% for 5-clients, 10.58% for 10-clients, 2.11% for 15-clients, 1.31% for 20-clients, 2.12% for 25-clients and finally 3.08% for 32-clients. From this comparison we can easily say that the proposed Fed-ReMECS-U is better than Fed-ReMECS in terms of accuracy. Now, from Fig. 2, the global model's performance (accuracy) changes in different rounds for different number of clients, we can see that the performance of Fed-ReMECS-U has hold up very good than Fed-ReMECS. In some rounds the global model's performance has dropped is because of the fact that the global model has no access to the local data that is why the accuracy is dependent on the average performance of all the local models, so one local model's performance drop can cause a drop in global model's performance.

Finally, based on the average comparison and overall performance comparison we can conclude that the proposed Fed-ReMECS-U with the EWMA incorporation at the local client end has improved the emotion classification accuracy in real-time from multi-modal data stream in IoT environment when compared to the Fed-ReMECS. The different clients scenario testing also ensures that the proposed approach is scalable and it preserves the users' local data privacy.

5 Conclusion

The paper presented an improved federated learning-based real-time emotion classification from multi-modal data streams with a local model update strategy by EWMA, called Fed-ReMECS-U. In Fed-ReMECS-U, an exponentially weighted moving average is incorporated in the local clients end to update their local model but not by directly replacing the local model by global model; rather, the local clients use the previous local model's information and combine (using EWMA) with the received global model. The effectiveness of the proposed approach is validated using the multi-modal data streams (EDA and RB) generated from the DEAP dataset. The experimental result confirmed that Fed-ReMECS-U (local model update with the EWMA) has better performance in terms of accuracy for real-time emotion classification from multi-modal data streams with high scalibility and users' privacy preserved in IoT environment.

In future work, we will explore the possibility of using this model into a goal-based model (e.g. for vital track of users). Finally, we would like to make an experimental evaluation using a realistic scenario of using a considerable number of end devices and a global server such as in an Augmented Workspace context.

Acknowledgement. Work partially funded by ACCIÓ under the project TutorIA.

References

1. Ayata, D., Yaslan, Y., Kamaşak, M.: Emotion recognition via random forest and galvanic skin response: comparison of time based feature sets, window sizes and wavelet approaches. In: Medical Technologies National Congress, pp. 1–4 (2016)
2. Ayata, D., Yaslan, Y., Kamasak, Mustafa, E.: Emotion recognition from multimodal physiological signals for emotion aware healthcare systems. J. Med. Biolog. Eng. 149–157 (2020)
3. Bifet, A., Holmes, G., Kirkby, R., Pfahringer, B.: Moa: Massive online analysis. J. Mach. Learn. Res. **11**, 1601–1604 (2010)
4. Candra, H., Yuwono, M., Chai, R., Handojoseno, A., Elamvazuthi, I., Nguyen, H.T., Su, S.: Investigation of window size in classification of eeg-emotion signal with wavelet entropy and support vector machine. In: 37th Annual International Conference of the IEEE EMBS, pp. 7250–7253 (2015)
5. Koelstra, S., Muhl, C., Soleymani, M., Lee, J.S., Yazdani, A., Ebrahimi, T., Pun, T., Nijholt, A., Patras, I.: Deap: A database for emotion analysis; using physiological signals. IEEE Trans. Affect. Comput. **3**(1), 18–31 (2012)
6. Konečný, J., McMahan, H.B., Yu, F.X., Richtarik, P., Suresh, A.T., Bacon, D.: Federated learning: Strategies for improving communication efficiency. In: NIPS Workshop on Private Multi-Party Machine Learning (2016). arxiv.org/abs/1610.05492

7. Nandi, A., Xhafa, F.: A federated learning method for real-time emotion state classification from multi-modal streaming. Methods **204**, 340–347 (2022)
8. Nandi, A., Xhafa, F., Subirats, L., Fort, S.: Real-time multimodal emotion classification system in e-learning context. In: Proceeding of the 22nd Engineering Applications of Neural Networks Conference, pp. 423–435 (2021)
9. Perry, M.B.: The Exponentially Weighted Moving Average. Wiley (2011)
10. Zhang, Y., Jiang, C., Yue, B., Wan, J., Guizani, M.: Information fusion for edge intelligence: a survey. Inf Fusion **81**, 171–186 (2022)

User-Adapted Semantic Description Generation Using Natural Language Models

Javier Sevilla Salcedo(✉), Laura Martín Galván, Jose C. Castillo, Álvaro Castro-González, and Miguel A. Salichs

Robotics Lab, Universidad Carlos III de Madrid, Getafe, Spain
javier.sevilla@uc3m.es, 100444738@alumnos.uc3m.es, jocastil@ing.uc3m.es, acgonzal@ing.uc3m.es, salichs@ing.uc3m.es

Abstract. Nowadays, conversational agents are making their way into our lives in many fields. The agent's speech is an important element in human-computer interaction, and to appear natural and friendly it should avoid predefined texts. With this premise, together with the significant growth in natural language generation models, this work explores the capabilities of natural language models to lead to a more fluent human-computer interaction and open up a range of new opportunities and applications. The present work proposes a system for the generation of descriptions tailored to the user's profile from different random topics using Natural Language models. After implementing the different applications, integrating natural language generation in conversational agents has proven to be highly effective.

Keywords: Natural language generation · Natural language processing · Conversational agent · Human-computer interaction · Deep learning

1 Introduction

Intelligent conversational agents get increasingly integrated into public-facing tasks such as customer service, including troubleshooting, information-giving, or assistive chatbots used in multiple fields [8]. For the interaction to be good and smooth, we need the agent to present itself naturally and intelligently to the user [2]. One of the main barriers is the public's rejection of interacting with a computer due to unnatural texts and, as a result, a not very fluent conversation, which leads to mistrust and uneasiness. [3]. In order to move forward in this line, we must focus on what nurtures a conversation so that it is natural and fluid, such as moving away from predefined and repetitive texts, that lack naturalness [7]. Another aspect would be to provide agents with creativity to generate topics

V. Julián et al. (Eds.): ISAmI 2022, LNNS 603, pp. 134–144, 2023.
https://doi.org/10.1007/978-3-031-22356-3_13

of conversation or even opinion [4]. Finally, human beings adapt their utterances according to the characteristics of the person we are talking to (e.g. their age, background, and level of familiarity).

This research work builds on this philosophy; hence, working with Natural Language Processing (NLP) models stems from the desire to provide conversational agents with a new layer of user involvement and a natural feel [11]. The way it expresses itself is crucial for smooth interaction. The user's perception of the agent can be affected by the responsiveness of its language. We have used natural language models for these purposes, which provide us with a new approach to the generation of non-pre-written texts. Moreover, thanks to advances in NLP, the current models allow numerous applications beyond the generation of simple texts.

Given both the exponential growth of the Deep Learning and the models' capabilities, companies and industries' interest in integrating such technology in their products grows simultaneously. In this way, we have been able to check the maturity of current models, prices, advantages and disadvantages. The integration of NLP into conversational agents is an essential step toward a seamless Human-Computer Interaction, thus bringing agents closer to people. To the same end, numerous works are exploring NLP in this field. In 2016, Hammed presented a work exploring the idea of using NLP as a tool to enable social robots to learn from people through interactions [6]. The ultimate goal was to achieve trust and comfort between the robot and the user, seeking a long and stable relationship. To this end, during the interactions with the user, they performed stay analysis and extracted personal information about the user, including personal interests, to model each user's profile and thus adapt and personalise the activities and conversations to the person.

For a conversational agent that wants to adapt its behaviour to the user, an essential part is the knowledge of human emotional states. Graterol et al. proposed a framework that allows robots to detect emotions through NLP techniques and collect them in a semantic repository based on ontologies [5]. The main idea is that the robot starts the conversation in a neutral tone. As the conversation develops, it will adjust its tone according to the emotions detected. They use several transformer-based models together with a multilayer perceptron network to detected emotions. These emotions get stored in *EMONTO*, an extensible ontology that represents emotions semantically.

Woo et al. proposed a conversational system for robotic companions [15]. The system has three main modules. The first uses the relationship between how the user pronounces and expresses the sentence with the robot's perception to extract emotional information-inspired by humans. The second one is a sentence construction module based on predefined rules, thanks to which the robot can create sentences using only the user's sentence as a reference. Finally, there is a database of pre-written sentences when the robot cannot understand the user or the answer is archaic. This way, a smooth and stable conversation can be maintained.

Aiming to a fluent and spontaneous feel during a conversation the main objective of this work is the design and implementation of an application capable of generating user-adapted semantic descriptions using natural language models. The defined pipeline must be designed in a modular way in order to be able to use each module independently.

The rest of the document takes the following structure, Sect. 2 develops the current state of the NLP model of text generation used for the development of the project. Section 3 describes the software design of the desired application and discusses the different developments and modules developed for this work, defining the methodology and progress. Finally, the general conclusions of the project are presented in Sect. 4.

2 Natural Language Generation Model

In this work we have used Generative Pre-trained Transformer 3 (GPT-3), a Transformer-based autoregressive language model capable of producing human-like text [1]. The philosophy of this model is the predictive generation of text from an input sequence entered by the user. The language models of GPT-3 are defined as "few-shot learners". With "Few-shot"/"One-shot"/"Zero-shot", they refer to the number of examples of the task to be performed given to the model. The model was trained with an large corpus composed by about 45TB of text data from different datasets, which allowed it to learn the language domain in a very refined way showing outstanding performance on different language processing fields. The model is even able to learn new mnemonic rules without the need to be retrained, just by adding new examples of the desired behaviour to the model's input when performing inference. Model applications have already been demonstrated performing sentiment analysis, generating programming code, or translating text, among others.

2.1 Transformers in Generative Models

Transformer is a deep learning model based on the mechanism of self-attention. Self-Attention evaluates the input data by differentially weighting the importance of each part of the input data. Even though Attention mechanism was introduced with RNN structures, transformers are built on these attention technologies by its own, highlighting the fact that attention mechanisms alone can match the performance of RNNs with attention [13]. The model presents an encoder-decoder architecture. The encoder layer made of encoding modules that process sequentially the input one module after another. On the other hand, the decoder layer is made of decoding modules that process the encoder's output. Each encoder layer generates encodings containing information about the parts of the inputs relevant to each other. Each decoder layer decodes the generated encodings, using the encodings and taking the encoded contextual information to construct an output sequence [12]. To this aim, each encoder and decoder module uses the attention mechanism. At the language model level, more and

Listing 2.1. N-shot learning examples. Top: Zero-shot learning; Middle: One-shot learning; Bottom: Few-shot learning

```
- Zero-shot -
input_Prompt:
Translate English to French:     #task description
cheese =>                        #prompt
```

```
- One-shot -
input_Prompt:
Translate English to French:     #task description
sea otter => loutre de mer       #example
cheese =>                        #prompt
```

```
- Few-shot -
input_Prompt:
Translate English to French:     #task description
sea otter => loutre de mer       #example
peppermint => menthe poivree
plush girafe => girafe peluche   #example
cheese =>                        #prompt
```

more networks with transformers have appeared and continue to emerge, for example with GPT-3, the model structure is composed solely by Transformers decoder blocks to achieve this. They also added modularity since modifying the number of decoder blocks in the model can directly obtain lighter models for less demanding tasks. GPT-3 integrates 96 decoder blocks in its full version.

2.2 From Fine-Tuning to Few-Shot Learning

Over the years, models attempted to move away from task-specific training through transfer learning. Fine-tuning was later introduced to allow performing specific tasks based on pre-trained models. When we talk about fine-tuning, we refer to a technique in Deep Learning in which the weights of a pre-trained model get readjusted by training without losing their initial settings. The main model layers are frozen so that it adapts better to the tasks with which it is trained [9]. One of the advantages is that fine-tuning does not involve training the entire model but rather an update of its gradient, which is a significantly faster process. Later, more advanced models, such as GPT-3, went a step further and evolved from fine-tuning to N-shot learning [10]. In contrast to fine-tuning, when talking about N-shot learning or in-context learning, it adds examples within the model's input prompt to better understand the expected inference behaviour without the need to be fine-tuned or specifically trained. Listing 2.1 shows an example of different types of n-shot learning. One of the advantages of n-shot learning is the inclusion of encodings that help the model understand the task it has to perform. For example, in the example shown in Listing 2.1 we find the encoding "=>". This pattern indicates the model to look at the word before the encoding and add its translation after it.

3 User-Adapted Semantic Description Generation

This work proposes the application of Natural Language Generation techniques to generate user-adapted semantic descriptions. For the design process of the application, we proposed a modular development. In this way, the three modules work independently, allowing code reuse in further applications beyond the goal of this work. Figure 1 shows a diagram of the developed pipeline for the project where the *Random Topic Generator* module generates a topic which is later fed to *Semantic Description Generator* module, which creates a description from the given topic. Finally, *User-Adapted Text Paraphraser* module takes the generated description and adapts it following the introduced profile.

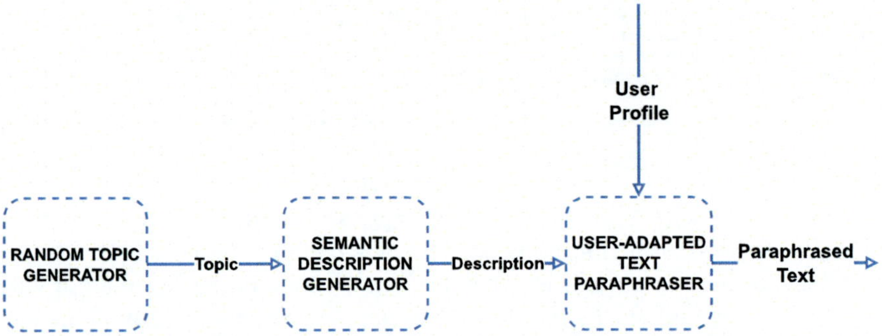

Fig. 1. User-adapted random semantic description generator diagram.

3.1 Random Topic Generation

This module is able to randomly generate a topic to have a tool that introduces randomness to the generation of words in natural language that the agent can use. The idea was to make inferences from the model and obtain a different theme with each execution. As mentioned in Sect. 2, one of the significant advantages of using language models such as GPT-3 is the use of n-Shot Learning to adjust the model to your needs without the need for application-specific fine-tuning training. With this in mind, several input prompts were designed and tested for the model to generate the random topics. Listing 3.1 (lines *1* to *10*) shows how, using Few-shot learning, the model is given several examples with the desired behaviour. In addition, using the encoding "***subject: -topic-***" the model understands the expected behaviour in a refined way and therefore it generates a few-worded topic only. As can be seen in Listing 3.1, the input prompt details what we are looking for, and we added eight examples so that the model can understand the pattern to follow. In this case, the topic generated has been "Trees".

Listing 3.1. Random Topic Generator tuned.

```
input_Prompt: The following is a list of random subjects for generation knowledge about the
      world:
subject: Pencils
subject: Radiology
subject: Fishing
subject: Napoleon
subject: Dolphins
subject: Keys
subject: Pigeons
subject: Newspapers
subject:
model_Output: Trees
```

Listing 3.2. Semantic Description Generator tuned.

```
1 input_Prompt: I have been doing deep research about ladders, let me give you brief relevant
      semantic information about it:
2 model_Output: Ladder is a tool used to climb up or down. It can be made of wood, metal,
      plastic and other materials. Ladders are also called as steps or rungs in the ladder
      system. They are usually placed on walls and floors for easy access to different
      levels of buildings such as apartments, offices etc.
```

3.2 Theme-Based Semantic Description Generation

This module is designed as a tool to generate a semantic description given an input topic. Through numerous iterations, the prompt used was "*I have been doing deep research about -any topic-, let me give you brief relevant semantic information about it:*", shown in Listing 3.2, which generated much information in a precise way without much extra text around it.

The adjusted model gives promising results as it fits the initial purpose correctly even with Zero-shot Learning (see Sect. 2). As for the implementation, like with the previous module, it is developed as a stand-alone component. The input prompt has been fixed to the one shown in line *1* of Listing 3.2, with the difference of the input topic (*ladders*, in the example). This module outputs the description associated to the given topic.

3.3 User-Adapted Text Paraphrasing

Conversational Agents interact with very different and diverse profiles of people; without further going, they can go from talking to a child to an adult or an older adult. Therefore, when talking about specific or complex topics, the interaction could be enhanced with a mechanism capable of adapting the text that we want to transmit and paraphrasing it to the user. To this end, after previous experience working with generative models, we have searched for a natural language structure capable of generating the paraphrase using Zero-Shot Learning with GPT-3.

Listing 3.3. User-adapted text paraphrasing examples.

```
1 input_Prompt: My granddad, who is really old, asked me what this passage means:
2 """
3 Gravity, also called gravitation, in mechanics, the universal force of attraction acting
      between all matter. It is by far the weakest known force in nature and thus plays no
      role in determining the internal properties of everyday matter. On the other hand,
      through its long reach and universal action, it controls the trajectories of bodies in
       the solar system and elsewhere in the universe and the structures and evolution of
      stars, galaxies, and the whole cosmos. On Earth all bodies have a weight, or downward
      force of gravity, proportional to their mass, which Earth's mass exerts on them.
      Gravity is measured by the acceleration that it gives to freely falling objects. At
      Earth's surface the acceleration of gravity is about 9.8 metres (32 feet) per second
      per second. Thus, for every second an object is in free fall, Its speed increases by
      about 9.8 metres per second. At the surface of the Moon the acceleration of a freely
      falling body is about 1.6 metres per second per second.
4 The works of Isaac Newton and Albert Einstein dominate the development of gravitational
      theory. Newton's classical theory of gravitational force held sway from his Principia,
       published in 1687, until Einstein's work in the early 20th century. Newton's theory
      is sufficient even today for all but the most precise applications. Einstein's theory
      of general relativity predicts only minute quantitative differences from the Newtonian
       theory except in a few special cases. The major significance of Einstein's theory is
      its radical conceptual departure from classical theory and its implications for
      further growth in physical thought.
5 """
6 I rephrased it for him, in plain language an elder can understand:
7 """
8 model_Output: Gravity is a force that exists between two objects. It causes the objects to
      have an attractive force on each other. The more massive the object is, the stronger
      this attractive force becomes.
```

```
1 input_Prompt: My son, who is a child, asked me what this passage means:
2 """
3 - Text about Gravity -
4 """
5 I rephrased it for him, in plain language a child can understand:
6 """
7 model_Output: Gravity is the force that makes things fall down. It is the weakest force in
      nature, but it is very powerful because it affects everything in the universe. Gravity
       is measured by how fast something falls. The faster it falls, the more gravity is
      pulling it down. Gravity is what makes the Earth's surface curved.
```

Listing 3.3 shows two examples of inference in which we copy the definition and history of gravity from an encyclopaedia[1] and ask the model to adapt it in the first example for an **older person** and in the second for a **child**. Line *1* of the listing shows the input prompt used for this application. As seen in the listing, we used the encoding (""") within the prompt; this indicates a change between the narrator and the quote in the text to paraphrase. Before the definition, we indicate the model that a person has asked us to tell him/her what the text means. Within the input prompt, we give a brief description of this person, indicating the profile (an older person, a child...), so that the model can adapt the paraphrase to the given profile.

As with the applications described above, this application has been implemented as a stand-alone function in a modular way. In this case, the input prompt has been kept fixed to the main structure shown in Listings 3.3; however, mak-

[1] https://www.britannica.com/science/gravity-physics.

ing use of the user profile entered, we modify the first and last sentence of the prompt, while the text entered for paraphrasing is introduced into the input prompt following the defined structure, as shown in lines *3* of Listings 3.3. The output of the module will be the paraphrased text. As can be seen in the Listings, both examples show adapted descriptions; while the elder output explains gravitation as an attraction force among objects using less technical and lighter language. On the other hand, the child output simplifies even further the description leaving behind the "attraction" concept while focusing instead on "falling" and "pull down" concepts which are more relatable to a child.

3.4 Case Study

Following the processes presented in the previous sections, this case study shows how the user-adapted text paraphrase module has been joined to the Random Topic Generator and Semantic Description Generator modules. Figure 1 shows the pipeline of the connection between the modules where, starting from a random topic, the semantic description generator outputs a random description, later paraphrased to an elder and a child user profiles.

Listing 3.4 shows an example adapting the text for the elderly while Listing 3.5 shows the same example with the adaptation for children. Even though no input prompt is shown for simplicity, its contents follows the ones introduced above; however, all the outputs shown in brown in the listings have been generated and output by the respective modules.

Listing 3.4. Elder-adapted random semantic description generation.

```
random_Topic_Generator_Output: Wine

semantic_Description_Generator_Output: Wine is a product of fermentation. Fermentation
        happens in the presence of yeast, which produces alcohol and carbon dioxide gas. The
        process can be done by any type of plant or animal (e.g., bacteria). It is also
        possible to make wine from grapes grown on other plants such as corn, wheat or rice;
        however this requires special equipment and techniques that are not available for home
         winemaking. Wine production involves three main steps: Winemaking - This step
        involves making wine from grape juice using a variety of different methods including
        fermentation with yeasts, vinification (the final stage), malolactic fermentation and/
        or secondary aging in oak barrels for up to several years after harvest until it
        reaches its desired quality level called "bottle age".

user_Adapted_Text_Paraphrasing_Output: Wine is a fermented drink made by yeast. The wine
        gets its alcohol and carbon dioxide gas from the juice of grapes that are grown on
        special plants called vines.
```

Listing 3.5. Child-adapted random semantic description generation.

```
random_Topic_Generator_Output: Wine

semantic_Description_Generator_Output: Wine is a product of fermentation. Fermentation
        happens in the presence of yeast, which produces alcohol and carbon dioxide gas. The
        process can be done by any type of plant or animal (e.g., bacteria). It is also
        possible to make wine from grapes grown on other plants such as corn, wheat or rice;
        however this requires special equipment and techniques that are not available for home
         winemaking. Wine production involves three main steps: Winemaking - This step
        involves making wine from grape juice using a variety of different methods including
        fermentation with yeasts, vinification (the final stage), malolactic fermentation and/
        or secondary aging in oak barrels for up to several years after harvest until it
        reaches its desired quality level called "bottle age".
```

```
user_Adapted_Text_Paraphrasing_Output: To make wine, you need to first get grape juice. You
      can do this by picking grapes off a vine, or you can get juice from a store. Then, you
      need to add yeast to the juice. This is what makes the wine. The yeast will eat the
      sugar in the juice and turn it into alcohol and carbon dioxide gas. After a while, the
      wine will be done and you can drink it!
```

Each module runs an independent inference with its own individually tuned parameters for optimal performance. Table 1 lists the parameters used in each of the modules. Within the listing of parameters, we find *Engine*, which indicates the type of model used; *Temperature* controls the degree of randomness of the response. *Response length* limits the maximum number of tokens to be generated; *Top P* controls to some extent the randomness and creativity of the response. *Frequency penalty* penalises the repetition of tokens in the output, while *Presence penalty* penalises the generation of new tokens already present in the input text. Finally, *Stop sequences* are encodings to stop generation.

Table 1. Inference parameters used.

Parameter	Random topic generation	Semantic description generation	User-adapted text paraphrasing
Engine	Babbage	Babbage	Davinci
Temperature	0.64	0	0.6
Response length	54	500	200
Top P	1	1	1
Frequency penalty	2	1.92	0.4
Presence penalty	2	0	0.2
Stop sequences	\n,subject:	\n	”””

4 Conclusions

After working hand in hand with generative language models, its integration in conversational agents, as mentioned several times in the report, presents a vast and novel scope of application. The ability to adapt to the user's profile and the possibility of autonomously selecting conversational topics may enhance the perceived intelligence and naturalness of the agent, thus improving interaction. Nevertheless, there are still some challenges to overcome in full integration. One of the main limitations derived from the large size of these models is the computational capacity required for training and inference and its related costs. Additionally, being a large decoder-based model its use leads to a higher latency in inference, which when implemented with conversational agents can affect their immediacy, thus the naturalness and fluidity of the interaction.

In general, the integration of generative language models in this scope gives us flexibility and provides creativity to some extent, without losing naturalness. The correct design of the prompt in each module allowed us to correctly match the expected performance in the preliminary results. Although the initial objectives have been met, despite having made several templates adjusted to the profiles shown in the examples, a dynamic adaptation should be made so as not to restrict the possible user profiles [14].

Acknowledgement. This work was partially supported by project Robots sociales para estimulación física, cognitiva y afectiva de mayores (ROSES) RTI2018-096338-B-I00 funded by Agencia Estatal de Investigación (AEI), Ministerio de Ciencia, Innovación y Universidades. This publication is part of the R&D&I project PLEC2021-007819 funded by MCIN/AEI/10.13039/501100011033 and by the European Union NextGenerationEU/PRTR. This work has also been supported by the Madrid Government (Comunidad de Madrid-Spain) under the Multiannual Agreement with UC3M ("Fostering Young Doctors Research", SMM4HRI-CM-UC3M), and in the context of the V PRICIT (Research and Technological Innovation Regional Programme).

References

1. Brown, T.B., Mann, B., Ryder, N., Subbiah, M., et al.: Language models are few-shot learners (2020)
2. Cassell, J., Sullivan, J., Churchill, E., Prevost, S.: Embodied Conversational Agents. MIT Press (2000)
3. Clark, L., Pantidi, N., Cooney, O., Doyle, P., Garaialde, D., Edwards, J., Spillane, B., Gilmartin, E., Murad, C., Munteanu, C., Wade, V., Cowan, B.R.: What makes a good conversation? Challenges in designing truly conversational agents. In: Proceedings of the 2019 CHI Conference on Human Factors in Computing Systems, pp. 1–12. CHI '19 (2019)
4. Clavel, C., Callejas, Z.: Sentiment analysis: from opinion mining to human-agent interaction. IEEE Trans. Affect. Comput. **7**(1), 74–93 (2016)
5. Graterol, W., Diaz-Amado, J., Cardinale, Y., Dongo, I., Lopes-Silva, E., Santos-Libarino, C.: Emotion detection for social robots based on NLP transformers and an emotion ontology (2021)
6. Hameed, I.A.: Using natural language processing (NLP) for designing socially intelligent robots. In: 2016 Joint IEEE International Conference on Development and Learning and Epigenetic Robotics (ICDL-EpiRob) (2016)
7. Klingspor, V., Demiris, J., Kaiser, M.: Human-robot-communication and machine learning, p. 41
8. Laranjo, L., Dunn, A.G., Tong, H.L., Kocaballi, A.B., Chen, J., Bashir, R., Surian, D., Gallego, B., Magrabi, F., Lau, A.Y.S., Coiera, E.: Conversational agents in healthcare: a systematic review. J. Am. Med. Inf. Assoc. **25**(9), 1248–1258 (2018)
9. Radford, A., Narasimhan, K., Salimans, T., Sutskever, I.: Improving language understanding by generative pre-training, p. 12
10. Raffel, C., Shazeer, N., Roberts, A., Lee, K., Narang, S., Matena, M., Zhou, Y., Li, W., Liu, P.J.: Exploring the limits of transfer learning with a unified text-to-text transformer. arXiv:1910.10683 [cs, stat] (2020)
11. Rheu, M., Shin, J.Y., Peng, W., Huh-Yoo, J.: Systematic review: trust-building factors and implications for conversational agent design. Int. J. Human-Comput. Int. **37**(1), 81–96 (2021)
12. Sutskever, I., Vinyals, O., Le, Q.V.: Sequence to Sequence Learning with Neural Networks. In: Advances in Neural Information Processing Systems, vol. 27. Curran Associates, Inc. (2014)

13. Vaswani, A., Shazeer, N., Parmar, N., Uszkoreit, J., Jones, L., Gomez, A.N., Kaiser, L., Polosukhin, I.: Attention is all you need (2017)
14. Weiss, K., Khoshgoftaar, T.M., Wang, D.D.: A survey of transfer learning. J. Big Data **3**(1), 1–40 (2016). https://doi.org/10.1186/s40537-016-0043-6
15. Woo, J., Botzheim, J., Kubota, N.: Conversation system for natural communication with robot partner. In: 2014 10th France-Japan/8th Europe-Asia Congress on Mechatronics (MECATRONICS2014- Tokyo) (2014)

An Integration of Packet Routing and Data Processing in Sensor Networks

Ichiro Satoh(✉)

National Institute of Informatics, 2-1-2 Hitotsubashi, Chiyoda-ku, Tokyo 101-8430, Japan
ichiro@nii.ac.jp

Abstract. Modern sensor networks are not used only to collect data from a large number of sensor nodes, but also to process the data for their applications. Although networking for data, e.g., the routing of data packets, and processing data have been handled separately so far, data processing for several applications in sensor networks process depends on networking. It can improve performance to change the routes of packets according to the requirements of data processing for applications, such as collecting data from multiple sensor nodes. This paper proposes an approach to defining itineraries for packets among nodes and data processing in a unified fashion. The framework has several advantages. It can adapt its networking to the requirement of each packet, even when there are multiple packets whose requirements are different. Unlike existing protocols, each packet can also travel one or more nodes under its own control. The framework is constructed based on a process calculus for specifying routings and function calls at nodes. The paper presents a prototype implementation of the framework.

Keywords: Sensor network · Programmable packet · Routing specification

1 Introduction

Sensor networks have been used widely for various applications, e.g., monitoring environments in fields and detecting abnormality in factories. A typical sensor network consists of multiple sensor nodes and one or more intermediate nodes. Each sensor node has sensing systems to monitor its targets and periodically transmits the data measured from its sensing systems to other nodes in a peer-to-peer manner or to its intermediate nodes through its own wired or wireless network equipment. Intermediate nodes are not only responsible for collecting data from sensor nodes, but also for processing the collected data.

Nowadays, sensor nodes themselves have become smarter and smarter in the sense that such sensor nodes have enough computational resources to execute

V. Julián et al. (Eds.): ISAmI 2022, LNNS 603, pp. 145–154, 2023.
https://doi.org/10.1007/978-3-031-22356-3_14

lightweight data processing programs, for example noise reduction and simply statistical programs. As a result, this gives two changes to sensor networks. The first is for sensor networks to process the data measured from their sensor nodes within them as a self-contained data monitoring and processing system. The second is to make sensor nodes to be used for multiple applications rather than single ones in sensor networks in the same time. A sensor network needs to handle multiple packets according to their network requirements, where each application has its own network requirements, e.g., required routing of packets to transmit data for the application.

However, when conventional sensor networks have been used only to collect data measured by sensors at nodes, their communication protocols have been designed with only data transfer as design principle for networking. In networks other than sensor networks, data transfer through communication has been also designed and operated independently of data processing. Although such independence has several advantages, but it causes a mismatch between data processing requirements and communication requirements, resulting in unnecessary costs and complicated processing when communication troubles. Since most sensor networks are not necessarily wide bandwidth nor reliable, as data processing has been performed inside a sensor network, networking for transmitting data is required to be adapted to data processing.

This paper proposes a framework for enabling us to define and control the routing and data processing of each packet according to the packet's requirements. It seems to be similar to *active networking* approaches studied around the year 2000. However, there is several essential differences between our approach and active networking. Whereas packets in active networking have single destinations, packets in the framework have one or more destinations. Active networking aims at transmitting packets through application-specific routings, but their packets themselves are independent from their applications. Unlike active networking, our framework enables packets to invoke application-specific functions at their visiting nodes, where such functions define to gather data on nodes and to process data in the payloads of packets.

We need to verify whether the transmission of packets in sensor networks can satisfy the requirements of their applications. Therefore, we propose a specification language for describing the itineraries of packets and the references of functions invoked by packets. Although the language is defined based a theoretical foundation for formally analyzing routing of packets, the framework provides an interpreter to execute the language and then forwards packets and invoke data processing functions. The size of the description must be small since it is to be embedded in a packet. To this, it may be better to have a variety of instructions for this purpose, but sensor networks have been used widely for various applications, e.g., monitoring environments in fields and detecting abnormality in factories.

2 Basic Approach

This framework provides packet routing optimized for sensor networks and integrates data transmission with application-specific data processing, unlike conventional communication protocols.

2.1 Example Scenario

Before discussing the framework proposed in this paper, we describe our basic example scenario. Figure 1 shows a small sensor network consisting of four sensor nodes and one Intermediate node. The Intermediate node tries to average the data the data measured by each of the sensor nodes. The left of Fig. 1 corresponds to a sensor network with a conventional networking. An intermediate node sends query messages to all the sensor nodes. When each sensor node receives a query message, it returns the data measured from its sensing system to the intermediate node. Next, the intermediate node calculate an average of the collected data. The right of Fig. 1 illustrates to calculate an average of data in these sensor nodes with our proposed framework. Next, we intend to have the following differences between conventional approaches and our framework when calculating average of the data. On the other hand, the Intermediate node in a conventional sensor network needs to send query messages to all the sensor nodes and then to receive data from them. Next, it calculates the average of the collect data outside the network. On the other hand, in the framework, a packet can travel among all the sensor nodes. When arriving at a sensor node, the packet invokes a function to calculate a provisional average for the data collected so far.

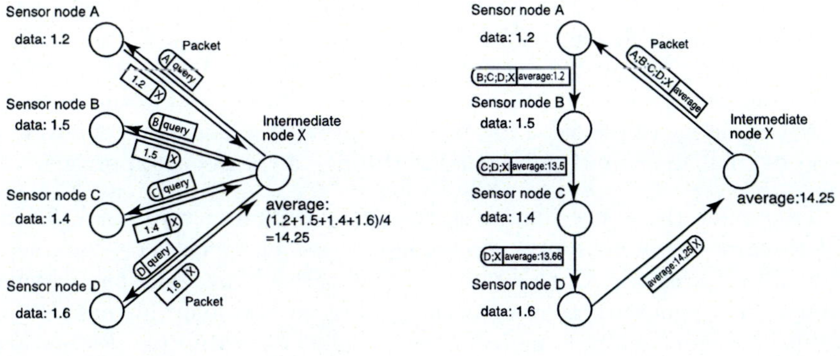

Fig. 1. The left of a conventional sensor network and the right of our proposed sensor network

2.2 Design Principle

To realize the above example scenario, the framework is required to provide the following functions.

- **Adaptation**: Packets in sensor networks are imposed with their own requirements regarding their routing among nodes, for example, packets should not be forwarded by certain nodes for reason of security. However, a mechanism for routing in each node forwards such packets to intermediate nodes, which are on a shortest route between them and the packets destinations, independently of the requirements of packets. The framework should adapt routing mechanisms in nodes to the routings of packets while the packets are forwarded at the nodes.
- **Multiple destinations**: We often need to collect the data measured by multiple sensor nodes. Rather than inquiring from the management node to each sensor node to obtain data, it is more efficient to have a packet sent from the management node travel to multiple nodes to collect data, because of reducing the number of communications. Unlike existing communication protocol, the framework allows each packet to have one or more destinations.
- **Arrival order**: When a packet travel among one or more destinations, the arrival order of the packet affects the efficiency and security in transmitting the packet. On the other hand, some packets have no constrains on their arrival order. The framework allows each packet to specify constraints on the arrival order of the packet.
- **Data processing**: Routing of packets among nodes are strictly bound with applications for which the packets are used. As mentioned previously, many sensor nodes have capabilities to execute application-specific data processing nowadays. Furthermore, the framework enables packets to simply process data by invoking data processing functions on the data in the node and the data carried in the packet's payloads. These functions are provided from nodes as built-in ones and designed to process the data obtained from the packets' current nodes and the data stored in the payloads of packets and then store their results in the payload of the packet.
- **Soundness**: We should reason about the routing of packets because such routings seriously affect the correctness of activities, monitoring sensors and processing measurement data, in sensor networks. Therefore, the framework needs to analyze routings of packets through a theoretical approach.

To provide the above functions, this framework seems to break down several assumptions of conventional communication protocols. For example, most existing protocols assume that their unicast packets have only one destination, whereas our framework allows unicast packets to have multiple destinations to collect data from multiple nodes in sensor networks. Multiple destinations for each packet are not given as a set of these destinations but also itineraries among them. Since protocols assume that communication protocols are neutral to applications, communication and computation are clearly separated in the sense that communication was independent of the application, including the content of the computation. Such a separation may be rational and beneficial for reason of reusability and portability, but data processing within sensor networks is not often separated from communication for the processing. In this framework functions for processing data provided by their arriving nodes can be invoked within routing information. In this paper we assume that the names and arguments of such functions provided in the destination node are known to the packet at the stage when its route is defined.

Each node interprets the itineraries of its receiving packets. forwards the packet to the next destination as a result of interpreting the itinerary. The itinerary of each packet's routing is contained within the packet and such a packet is transmitted by nodes according to its itinerary. However, the size of each packet should be limited. Each packet should be transmitted as a UDP packet, where the size of each UDP packet is limited within 65535 bytes.

2.3 Packet Description

The description of each packet's, e.g., packet itinerary and function calls, is stored within that packet itself. Due to the size limitation of a packet, such as description must be compact. Therefore, this paper proposes a language specialized for describing itineraries and calling functions provided at nodes. The description of such a packet is also performed by the node that receives the packet, and even if the node's computing performance has increased, the execution of the description needs to be fast and light-weight. When receiving a packet, each node will invoke functions for data processing according to the description of the packet and then forward that the packet to the next destination. The description of each packet is interpreted and executed at its current node. Therefore, the language is designed to quickly execute descriptions written in it. Before executing descriptions, language-level construction in the descriptions can be directly converted to byte code and stored in this format in the sense that descriptions written in the language and byte code in a one-to-one mapping style. To define correspondence between such language-level construction and their byte-code, we define the operational semantics of the language, where each operation defined in a theoretical approach corresponds to its byte code.

The current implementation, the description is executed at nodes based on *reduction semantics* in the sense that the route description is reduced at each time it arrives at the destination node, i.e., the route description that has already been executed is deleted, so that the node can always execute it from the beginning of the description. This is used to reduce the cost of identifying the execution point, but also to reduce the amount of data being communicated by shortening the route description in order to reduce packet communication costs.

3 Routing Specification Language

This section outlines our proposed specification language, which aims at specifying routings of packets on sensor networks and being interpreted at each node. Since specification for routing is embedded in the header part of each packet, it must be able to specify routing within a limited space. Routing specification needs to be *safe* in the sense that packets are delivered to their specified nodes according to their itinerary orders. Routings specified in the languages can invoke application-specific processing, e.g., gathering data at sensor nodes and processing data at specified nodes. Routings specified in the languages can be verified and analyzed based on a theoretical manner to trace their itineraries in a theoretical manner Nodes for data stream processing are heterogenous, so that programs written in the language need to be executed on different underlying systems.

As mentioned previously, our language is defined as a process calculus for specifying routings of packets because flows should be verified to satisfy their given conditions.

Definition 1 The set $\mathcal{E}$ of expressions of the language, ranged over by $E, E_1, E_2, \ldots$ is defined recursively by the following abstract syntax:

$$E ::= 0 \quad | \quad \langle \ell, f \rangle \quad | \quad E_1 \; ; E_2 \quad | \quad E_1 + E_2 \quad | \quad E_1 \,\%\, E_2$$

where we assume $\mathcal{F}$ is a set of functions on data, ranged over by f and $\mathcal{L}$ is the set of location names, ranged over by $\ell, \ell_1, \ell_2, \ldots$. We often omit 0, i.e., $E \; ; 0$ is abbreviated to E.

We describe the semantics of $\mathcal{E}$ intuitively. 0 represents a terminated itinerary. $\langle \ell, f \rangle$ represents a packet to be forwarded to a node named ℓ and then execute a function called f. $E_1 ; E_2$ means that if a packet is forwarded to nodes according to the E_1 itinerary terminates, then is according to the E_2 itinerary. $E_1 + E_2$ represents a packet to be forwarded according to either E_1 or E_2, where the selection is based on the reachability of their next destinations. $E_1 \% E_2$ means that an agent can follow either E_1 before E_2 or E_2 before E_1 in its itinerary.

Like a process calculus, the semantics of the language is defined as a labeled transition system and each transition is given as a reduction between expressions in $\mathcal{E}$.

Definition 2 The language is a labeled transition system $\langle\, \mathcal{E}, \mathcal{L} \cup \{\tau\} \{ \stackrel{\alpha}{\longrightarrow} \subseteq \mathcal{E} \times \mathcal{E} \,|\, \alpha \in (\mathcal{L} \times \mathcal{F}) \cup \{\tau\} \} \,\rangle$ defines as the induction rules given below:

$$\textbf{(Forward:)}\ \frac{-}{\langle ell, f \rangle \stackrel{\ell}{\longrightarrow} 0} \qquad \textbf{(Sequence:)}\ \frac{E_1 \stackrel{\langle \ell, f \rangle}{\longrightarrow} E_1'}{E_1 \; ; E_2 \stackrel{\langle \ell, f \rangle}{\longrightarrow} E_1' \; ; E_2}$$

$$\textbf{(Selection 1:)}\ \frac{E_1 \stackrel{\langle \ell, f \rangle}{\longrightarrow} E_1'}{E_1 + E_2 \stackrel{\langle \ell, f \rangle}{\longrightarrow} E_1'} \qquad \textbf{(Selection 2:)}\ \frac{E_2 \stackrel{\langle \ell, f \rangle}{\longrightarrow} E_2'}{E_1 + E_2 \stackrel{\langle \ell, f \rangle}{\longrightarrow} E_2'}$$

$$\textbf{(Non-order 1:)}\ \frac{-}{E_1 \,\%\, E_2 \stackrel{\tau}{\longrightarrow} E_1 \; ; E_2} \qquad \textbf{(Non-order 2:)}\ \frac{-}{E_1 \,\%\, E_2 \stackrel{\tau}{\longrightarrow} E_2 \; ; E_1}$$

where $0 \; ; E$ is treated as being syntactically equal to E. E^* is recursively defined as $0 \,\#\, (E \; ; \; E^*)$. We often abbreviate $E_0 \stackrel{\tau}{\longrightarrow} \cdots \stackrel{\tau}{\longrightarrow} E_n$ to $E_0 (\stackrel{\tau}{\longrightarrow})^n E_n$. □

In Definition 2, each $\langle \ell, f \rangle$-transition means that a packet is forwarded to node ℓ and then a data processing function, named as f, is executed at the node. For example $E \stackrel{\langle \ell, f \rangle}{\longrightarrow} E'$ means that the packet is forwarded to a node named ℓ and executes the f function and then it behaves as E'. Also, if there are two possible transitions $E \stackrel{\langle \ell_1, f_1 \rangle}{\longrightarrow} E_1$ and $E \stackrel{\langle \ell_2, f_2 \rangle}{\longrightarrow} E_2$ in a packet, the processing by the packet chooses one of the destinations, ℓ_1 and ℓ_2. The τ-transition, on the other hand, corresponds to a non-deterministic choice in a packet's itinerary. For example, the first means that a packet is forwarded to $\texttt{node}_\texttt{a}$ and then collects data from $\texttt{node}_\texttt{a}$. Next, it is forwarded to $\texttt{node}_\texttt{b}$ and then collects data from $\texttt{node}_\texttt{b}$. Finally, it is forwarded to $\texttt{node}_\texttt{c}$ and then process an average of data obtained from $\texttt{node}_\texttt{a}$ and $\texttt{node}_\texttt{b}$.

$$
\begin{aligned}
&(\langle \texttt{node}_\texttt{a}, \texttt{collect}\rangle \ ; \langle \texttt{node}_\texttt{b}, \texttt{collect}\rangle \ ; \langle \texttt{node}_\texttt{c}, \texttt{average}\rangle \\
&\qquad\qquad \xrightarrow{\langle \texttt{node}_\texttt{a}, \texttt{average}\rangle} \langle \texttt{node}_\texttt{b}, \texttt{collect}\rangle \ ; \langle \texttt{node}_\texttt{c}, \texttt{average}\rangle \\
&\qquad\qquad \xrightarrow{\langle \texttt{node}_\texttt{b}, \texttt{collect}\rangle} \langle \texttt{node}_\texttt{c}, \texttt{average}\rangle \\
&\qquad\qquad \xrightarrow{\langle \texttt{node}_\texttt{c}, \texttt{average}\rangle} 0
\end{aligned}
$$

As an additional note on the semantics of the language, it is possible to invoke a function that processes data, but only when it is forwarded, so there must be a forwardable destination and the destination must have the function available. The function may be a combination of multiple functions.

4 Design and Implementation

The framework is constructed as middleware systems designed of running at sensor and intermediate nodes (Fig. 2). They transmit packets according to their itineraries and invoke functions provided from nodes. In the current implementation each packet is embodied within a UDP packet and each system is implemented with Java language.

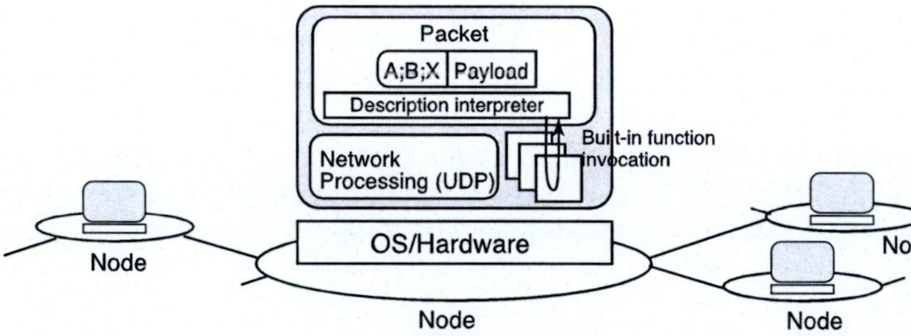

Fig. 2. System architecture

4.1 Interpretation of Description Embedded in Packets

When each system receives a packet, it interprets the routing descriptions embedded in the packet based on the operative semantics defined in Definition 2, up to the next destination forwarding decision. For example, if the beginning of the description is $\langle \ell, f\rangle$, the interpreter attempts to invoke function f, where f is the interface of a function, e.g., name and arguments, provided at the node as explained later. If it is E_1 ; E_2, description E_2 is executed after finishing the execution of E_1. In this interpreter, the description is subject to interpretation from left to right and the right part, where the first SEQ appears, is executed before the rest description is executed. If it is $E_1 + E_2$, one of either sub-description E_1 or E_2 is selected on the reachability to forward the packet to the first destination in each of the two sub-descriptions. The interpreter first checks the reachability of the first destination in the sub-descriptions. The current implementation provides three kinds of criteria for determining the reachability of the first destination. First, the node has communicated with the destination so far. Second, the destination must be included in the node's routing information. Third, the node's network management can confirm reachability, e.g., ping packet or routing protocol. As **Non-order** in the Definition 2 $E_1 \,\%\, E_2$ means the execution of multiple routes in any order To execute $E_1 \,\%\, E_2$ the interpreter selects one of either E_1 ; E_2 or E_2 ; E_1 according to the reachability of E_1 and E_2.

4.2 Invocation for Data Processing Function

When executing $\langle \ell, f \rangle$ in the description of a packet, an interpreter at the current node forwards the packet to the node named as ℓ and then an interpreter at the destination node invokes function f provided from the node. We assume that the name of the function is given as a string like other programming languages, where function f may be a composite function consisting of one or more sub-functions. Two types of arguments are supported. The first is the data stored in the node and The second is the data stored in the payload of each packet. To specify such data, the framework introduces key-value-store (KVS) database into nodes and the payloads of packets. The database consists of one or more pairs of keys and values, where each is given as a string, and value is in any data type, such as numeric, string, binary, and so on. Such values are indicated by their key. The results of the function are also stored in the keyed storage area in the payload of the packet when at least one key is given at the time of the function call. Therefore, in this framework the payload of each packet is treated as a data storage area in key-value-store format.

4.3 Packet Receiving and Forwarding

The current implementation delegates receiving and forwarding of packets to the underlying systems, i.e., OS. It can basically work independent of OSs because it is available with common programming interfaces for UDP in addition to Java VM. Although this framework was originally intended to be implemented as an OS-level communication protocol, the current implementation is built as a middleware system running on the underlying system, i.e., OS. As a result, not only data but also descriptions are stored in the payload portion of UDP packets, which slightly reduces the data capacity of the packets, but this is not a major limitation in practical use since packets are not sent and received frequently due to the characteristics of sensor networks. Functions can be invoked from packets can be implemented as Java programs.

4.4 Current Status

The current implementation of the framework was still in an early stage so that it was not optimized for performance. We evaluated that of several basic operations in a distributed system consisting of eight networked embedded computers as sensor nodes connected through Fast (100 Mbps) Ethernet via an Ethernet switch. Each embedded computer was a Raspberry Pi, where its processor was Broadcom BCM2835 (ARMv6-architecture core with floating point) running at 700 MHz and it has 512 MB memory, a Fast Ethernet port, and SD card storage (16 GB SDHC), with Raspbian, which was a Linux optimized to Raspberry Pi. Nevertheless, we describe the basic performance of the implementation. The cost of reading the route description in a packet and decide the next destination is within 10 ms. Although the cost of invoking a function depends on the function itself, when the function is simple, e.g., calculating averages of data, the cost within about a few milli seconds. The cost of transmitting packets in this framework is almost equal to the cost of a packet in UDP. Although the size of description stored in a packet cannot be exactly estimated, for example, the size of description $\langle \texttt{node}_\texttt{a}, \texttt{collect} \rangle$; $\langle \texttt{node}_\texttt{b}, \texttt{collect} \rangle$; $\langle \texttt{node}_\texttt{c}, \texttt{average} \rangle$ with KVS database for its arguments is about 300 bytes.[1]

5 Related Work

Data aggregation in sensor networks has been well studied in this decade. Like ours, many approaches to processing data inside sensor networks have been explored. Most of them focused on two purposes: data aggregation and query processing. The former focused on algorithms for aggregating data from sensor nodes under some constraints, e.g., energy [1,2]. The latter aimed at energy-efficient mechanism for querying at multiple sensor nodes [3,4]. Most of existing works on the former assumed that peculiar network topologies, e.g., tree-based [1] and cluster-based ones [5]. TinyDB provided a query mechanism for sensor networks but was not optimized to networks.

To customize distributed computing, particularly network processing between computers, several researchers have explored active networking technologies [6]. However, these existing technologies have focused on configurations for low-level network processing, e.g., routing and QoS protocols, so that they are not suitable for applications in sensor networks. They could be classified into two approaches: programmable infrastructures and programmable packets. Most researchers had explored the former rather than the latter. For example, SensorWare [7] enabled program codes, which run on nodes to be dynamically deployed for customizing query operations for sensor networks. PLAN [8] and

[1] The size depends on the size of data given as the arguments in the KVS.

Smart Packets [9] attempted to insert program codes, written their own custom languages, into packets to customize the routings of the packets. However, they intend to customize low-level handling for packets, whereas our framework customize the arrival order of packets. Existing programmable packet approach could not customize data processing.

6 Conclusion

This paper proposed a framework for transmitting and processing data measured from sensor nodes in sensor networks. The framework enabled each packet to have its own itinerary and to invoke functions for processing data in a unified specification. It was useful in sensor networks whose applications have different requirements, e.g., routings of packets for the applications. It could also reduce the number of communications in sensor networks because at most one packet can transmit data to multiple destinations. It was defined based on a novel process calculus for specifying routings and function calls, so that it enables us to theoretically reason about the routings and processing of packets for sensor networks. We presented a prototype implementation of this framework. As it did not aim at performance of network processing, it processed data among sensor nodes in a reasonable cost.

References

1. Jamieson, K., Moss, D., Gnawali, O., Fonseca, R., Levis, P.: Collection tree protocol. In: Proceedings of the 7th ACM Conference on Embedded Networked Sensor Systems, pp. 1–14 (2009)
2. Rajagopalan, R., Varshney, P.K.: Data-aggregation techniques in sensor networks: a survey. IEEE Commun. Surveys Tutor. **8**(4), 48–63 (2006)
3. Kamra, A., Misra, V., Rubenstein, D.: Counttorrent: ubiquitous access to query aggregates in dynamic and mobile sensor networks. In: Proceedings of the 5th International Conference on Embedded Networked Sensor Systems, SenSys '07, pp. 43–57. Association for Computing Machinery (2007)
4. Yao, Y., Gehrke, J.: The cougar approach to in-network query processing in sensor networks. ACM SIGMOD Rec. **31**(3), 9–18 (2002)
5. Krishna, R.K., Bhakare, K.R., Bhakare, S.: An energy-efficient grid based clustering topology for a wireless sensor network. Int. J. Comput. Appl. **39**(14), 24–28 (2012)
6. Tennenhouse, D.L., Smith, J.M., Sincoskie, W.D., Wetherall, D.J., Minden, G.J.: A survey of active network research. IEEE Commun. Magaz. **35**(1), 80–86 (1997)
7. Han, C.-C., Boulis A., Srivastava, M.B.: Design and implementation of a framework for efficient and programmable sensor networks. In: Proceedings of 1st International Conference on Mobile Systems, Applications, and Services (MobiSys 2003). ACM, pp. 187–200
8. Moore, J.T., Gunter, C.A., Hicks, M., Kakkar, P., Nettles, S.: Plan: a packet language for active networks. In: Proceedings of Functional Programming (ICFP), vol. 34. ACM, pp. 86–93
9. Schwartz, B., Jackson, A.W., Strayer, W.T., Zhou, W., Rockwell, R.D., Partridge, C.: Smart packets for active networks. In: Proceedings of 1999 IEEE Second Conference on Open Architectures and Network Programming. Proceedings. OPENARCH '99, pp. 90–97 (1999)

Quantum AI: Achievements and Challenges in the Interplay of Quantum Computing and Artificial Intelligence

Iñaki Fernández Pérez[1(✉)], Fernando de la Prieta[1], Sara Rodríguez-González[1], Juan M. Corchado[1,2], and Javier Prieto[1,2]

[1] BISITE Research Group, University of Salamanca, Salamanca, Spain
ifernandez@usal.es,
fer@usal.es, srg@usal.es, corchado@usal.es, javierp@usal.es

[2] AIR Institute, IoT Digital Innovation Hub, Salamanca, Spain
https://bisite.usal.es , https://www.innovationhub.es

Abstract. In recent times, Quantum Computing (QC) is receiving growing attention, thanks to the enormous advances in the construction of operational quantum computers, quantum materials and quantum cryptography. Given the advances in the physical construction and the scaling up of quantum computers, it is now necessary to foster the creation of quantum algorithms and methods that are adapted to such computers and that make the most of their intrinsic computational and communication capabilities. In the era of Big Data, some of the most computationally demanding tasks fall into the field of Artificial Intelligence (AI), including tasks that are currently computationally intractable due to physical limitations. The intrinsic parallelism, computational efficiency, and representational power offered by QC make for an excellent alternative to binary computers, holding the promise of enhanced AI models. This novel Quantum Artificial Intelligence (QAI) concept will result in the detection of patterns that classical AI algorithms are unable to identify, and in the time reduction of several orders of magnitude. In this review, we describe the scientific progresses in the confluence of AI and QC. We start by presenting both areas, basic concepts and the timeline of most significant advances in the history of AI and QC, to then focus on existing research made in the bidirectional approaches of QC benefiting from AI and AI benefiting from QC. Finally, we describe future avenues of research for the QAI incipient scientific area, and conclude.

Keywords: Quantum Artificial Intelligence · Artificial Intelligence · Quantum Computing

V. Julián et al. (Eds.): ISAmI 2022, LNNS 603, pp. 155–166, 2023.
https://doi.org/10.1007/978-3-031-22356-3_15

1 Introduction

Since the beginning of the 20th century, the study of quantum theory has given rise to quantum-based technologies, which are currently having an impact on areas such as cryptography, superconductors and quantum computing (QC). The capabilities of these technologies are highly promising, offering gains in performance and potential solutions to problems that were intractable with other types of technologies [8, 23, 39]. Precisely, the 80s-90s saw major breakthroughs in computation based on quantum effects, particularly in terms of (1) hardware, among which, most significant is the very recent achievement of quantum supremacy (demonstration of a quantum device solving problems beyond the reach of state-of-the-art binary computers) thanks to larger quantum computers (Nov. 2021, IBM Eagle [13]); and (2) large-scale initiatives in quantum computing technologies, such as the EC-funded Quantum Flagship launched in 2018 to advance towards a "Quantum Web" of quantum devices interconnected through quantum communication networks sharing quantum resources [58]. Consequently, the grounds are perfectly set for a major advance in algorithmics that leverage quantum computational capabilities in hard problems (including those tackled by Artificial Intelligence (AI) methods [3, 56]).

However, quantum supremacy does not necessarily mean a supremacy in intelligence. In 1945, John Von Neumann described his architecture [70] and in 1948 Alan Turing published his celebrated essay Intelligent Machinery [69]. However, even though the term 'Artificial Intelligence' was coined at the famous Dartmouth conference in the summer of 1956 [48] and there were some advances in the area in the decades after, it was not until the late 90*s* when AI actually turned effective in solving real-life problems (*e.g.* Deep Blue victory against Kasparov [36]). AI has now become crucial in virtually all aspects of society [4, 11, 12, 14, 15, 20–22, 26, 27, 32, 42, 47, 49, 53, 54, 64, 66–68, 71]. However, this is not the case for the more recent Quantum-based AI, or for AI-based Quantum methods. Therefore, it is necessary that the AI scientific community is ready in the short-term to avoid this gap when the generalized use of quantum computers becomes a reality (Fig. 1).

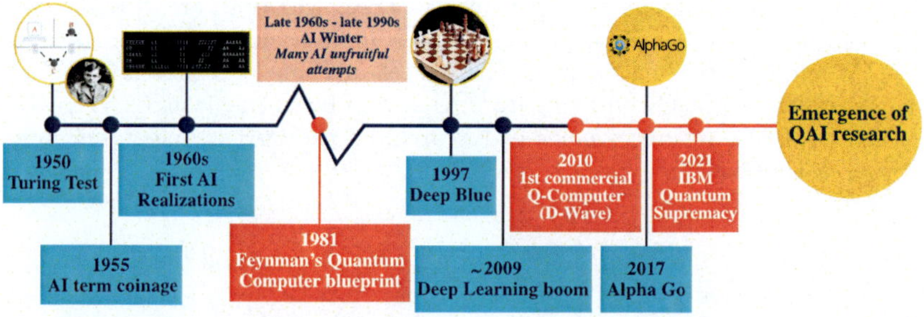

Fig. 1. Timeline of advances in Classical machines and classical AI versus quantum machines and quantum AI.

2 Quantum Computing and Artificial Intelligence

Since the 80*s* and the 90s, quantum computing and quantum algorithmics have had some very significant developments, including Feynman seminal blueprint for a quantum computer, Shor's quantum number factorization and discrete log [65], as well as quantum cryptokey distribution [29], with relevance in cryptography and cryptanalysis [1], Grover's quantum unordered search to find entries in data structures [31], or quantum pattern matching [59], *e.g.*, to find occurrences of substrings in a string. In the field of Quantum Machine Learning (QML), there have been some attempts related to the confluence of Quantum Computing and AI, such as the quantum annealing algorithm to solve optimization problems [24], quantum constraint satisfaction to address 3-SAT problems [10], quantum adiabatic algorithms to tackle NP-complete combinatorial optimization problems [2], quantum Principal Component Analysis (PCA) to find the main information-carrying components from a dataset [43], the quantum k-NN algorithm to reduce complexity in clustering problems [46], or quantum training of Boltzmann machines and neural networks to outperform theoretical performance of their binary counterparts [5,19]. These attempts have opened the way for more structured and focused research in an incipient field of Quantum AI.

Since the seminal inception of a blueprint by Feynman in 1981 [23], and the first successful physical (small-scale) quantum computer by D-Wave in 2010 [39], different physical realizations of quantum computers have been proposed (*e.g.*, quantum gate arrays, one-way quantum computers, adiabatic quantum computers, and topological quantum computers) until the achievement of the quantum supremacy in 2019 (Google Inc., 53 qubits [6]), 2020 (USTC China, 76 qubits [72]) and 2021 (IBM Eagle, 127 qubits [13]). This milestone in quantum computing (QC), aligned with relevant breakthroughs in other disciplines (*e.g.* major progress in advanced AI methods like Deep Learning [38,41]; large computational power and data availability [15]; or High Performance Computing [33]) have established the basis for research into an upcoming AI in which quantum computers can learn from quantum information. This *Quantum AI* (QAI) would leverage progress in a number of revolutionary technologies, such as medical Magnetic Resonance Imaging (MRI), superconductors, quantum computing and communication, quantum cryptokey distribution, and quantum measuring. This discipline (usually referred as quantum Machine Learning [8]) is in the intersection of AI and QC. Most of the initial approaches have been addressed to classical AI to reduce its computational complexity and speedup learning processes [46]. Other works have been oriented to QC to develop alternative learning algorithms and enable better solutions. Among others, the proposed solutions include quantum clustering [34], quantum autoencoders [40], or quantum reinforcement learning [18]. However, these solutions have been proposed independently, lacking a unifying framework that enables a seamless transition from classical AI to quantum AI and allows their cooperation during this transition.

As classical AI is supported by information storage (data) and manipulation (algorithms) [41,44] quantum AI is supported by quantum information storage (quantum data) and manipulation (quantum algorithms) [18,43]. Quantum data take the form of quantum binary digits (qubits) [16], while quantum algorithmics correspond to operations on qubits (*e.g.* quantum logic gates [51]). While qubits can be embodied in different physical systems (*e.g.* trapped ions, spin qubits, resonant cavities, or semiconducting qubits), the quantum effects of matter in all of them are fundamental for QC in providing unique advantages. These properties include: quantum superposition, qubit states simultaneously being in a combination of $|0\rangle$ and $|1\rangle$ values [25]; decoherence, loss of information of a quantum system to the environment [61]; collapse, irreversible reduction of a superposed quantum state to a single state [57]; entanglement, linkage of 2+ quantum particles with each other, with interdependent states even when separated by large distances [35]; or quantum tunneling, capability of quantum wave functions to propagate through potential barriers [30]. Therefore, QAI must be sustained by 3 elements: data, algorithms, and a computational framework. This unifying paradigm might allow scientists to develop QAI algorithms that manipulate data in the quantum sphere, not only speeding up machine learning processes, but also benefiting from QC properties. Pioneer attempts at QML have born fruit in the form of proposals of algorithms, targeted at the solution of specific problems. However, a technological framework structuring the incipient QAI field of research to facilitate the development of new QAI algorithms is currently lacking, and the problem of translating binary data into quantum states and vice versa (quantization/dequantization) is still open. Furthermore, in a near future, classical AI and QAI will coexist, and the AI scientific community will need a conceptual ecosystem in which both approaches will cooperate and develop synergies, facilitating the transition. To this aim, the new field of QAI will face the following fundamental problems [63], as described in Table 1.

In the next section, we analyze the state-of-the-art related to QAI, from the perspective of data, algorithmics, and computational framework dimensions.

3 Quantum Artificial Intelligence

3.1 Data Dimension

As said before, QC is supported by quantum information storage and manipulation, in the form of quantum binary digits (qubits) [16], and operations on qubits, *e.g.*, quantum logic gates [51]. Qubits represent information differently from classical two-state bits: in a qubit, both $|0\rangle$ and $|1\rangle$ values are simultaneously present with a probability due to superposition, until the state is collapsed by observing or measuring it. Information representation capability of qubits increases exponentially compared to binary representation [50]: one qubit can simultaneously represent two values, and N qubits can simultaneously represent up to $2N$ values, while N bits can only represent one in $2N$ values. As such, in Quantum AI, superposition in qubits and simultaneous operation on individual points of a superposed qubit allow to perform massively parallel computations [60], with the corresponding potential for hard problems currently unfeasible or requiring prohibitive amounts of time and resources [37]. However, classical AI

Table 1. Problems to be addressed by Quantum Artificial Intelligence [62].

Scientific problem	Description
Input problem	The cost of reading input data can potentially surpass the computational gain of quantum algorithms, limiting the obtained speed-up
Output problem	To translate the output of quantum algorithms as a binary string, it is necessary to learn an exponential number of bits, which makes some applications of QAI unfeasible
Noise problem	Random fluctuations in quantum states propagate very fast and can lead to the complete loss of information, which is at the root of the large effort in quantum error correction. This is particularly relevant for QAI because it entails series of unitary operations on large data volumes: quantum errors could invalidate entire analyses
Benchmarking problem	Some incipient benchmarks exist for the assessment of quantum algorithmic performance in specific cases; however, there is a lack of a solid benchmarking framework for the evaluation of QAI performance in comparison to classical AI
Costing problem	While theoretical bounds suggest that QAI algorithms will offer big advantages in solving large problems, it is currently not possible to estimate the actual number of quantum gates required to create a quantum circuit for a given algorithm

algorithms rely on datasets that store classical data, while the process of loading large amounts of data into a quantum system may be prohibitive (input problem [8]). In addition, quantum algorithms also suffer from the output problem [8], when reading out the data after being processed on a quantum device. Precisely, one open problem for QAI concerns the interfacing between classical and quantum information and how to encode data in the quantum sphere and vice versa. Further, classical data are represented in a different geometry compared to quantum embeddings, requiring special attention when closing the gap between representations [37]. Recently, a geometry test to assess the adequacy of quantum embeddings for given classical kernel embeddings has been proposed [45], which will be of a greatest importance in the development of the new QAI scientific area. This new area will investigate soft, mathematical methods to translate classical data to the quantum sphere and back, as opposed to hardware-based approaches, such as qRAMs that encode data in quantum mechanical form [28], and would benefit as well from defining quantum ontologies to add semantic richness to data used by QAI algorithms.

3.2 Algorithmic Dimension

The most usual quantum computational model, making use of qubits, are quantum circuits [51], in which quantum gates (which represent unitary operations on qubits) are combined. It includes measurement operations to produce a sequence of quantum operations for a desired result, *i.e.*, a way of representing quantum algorithms. An open problem of quantum circuits is the current lack of a universal quantum gate set that could express any possible operation on a quantum computer, with one approach to sidestep it being the approximation of quantum operations with a restricted gate set, for which efficient and powerful approximation methods are needed. Overall, the main framework for algorithmic description and design consists in creating quantum circuits by combining elementary operations, which is typically used in combination with algebraic matricial tools that allow for model checking and the verification of properties in the algorithms. Until now, QAI algorithmic solutions for concrete problems have been proposed; however, an integrative framework for the creation of new algorithms, that would facilitate the development of this scientific area, is still lacking. In QAI, enhanced qubit representational power coupled with parallel operations on them can provide exceptional grounds for a paradigm shift in computation based on quantum effects. This computation is exponentially more powerful than binary computation. It will allow to address problems with AI that are now intractable, due to high data dimension or pattern complexity. Since its beginnings, quantum algorithmics has had a small number of very significant developments. These include: Shor quantum factorization and discrete log [65]; Grover unordered search [31]; or quantum pattern matching [59]. Since the 00s-20s, quantum AI algorithmics has also seen progress, including the quantum annealing algorithm [24] for optimization; quantum constraint satisfaction [10] including 3-SAT; quantum PCA [43], to find main information-carrying components in a dataset; or quantum algorithms to train Boltzmann machines and neural networks [5,19]. They all offer significant theoretical performance gains compared to classical algorithms. Further, quantum effects are able to generate data with atypical patterns [55], and classical AI (*e.g.* deep neural networks) can frequently both recognize statistical data patterns, and produce data with similar patterns: recognizing patterns that they produce [52]. As such, it is possible that, if quantum computers can produce statistical patterns difficult to produce with classical computers, they might also recognize patterns that are difficult to recognize with classical AI [8]. These pioneering attempts into QAI are very significant in opening the way for more structured research in this field, in which current efforts are being now done. Figure 2 shows a graphical representation of a classical and a quantum Neural Network.

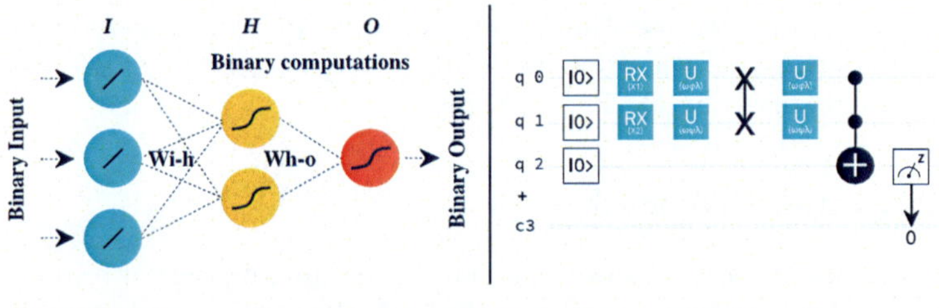

Fig. 2. Simple classical neural net (left) and quantum neural net circuit (right).

3.3 Computational Framework

The intrinsically enhanced data representational power of qubits, coupled with parallel quantum operations on them, provides exceptional grounds for computation based on quantum effects. Such QC, orders of magnitude more powerful than binary computation, allows to address problems using AI that are currently computationally intractable, due to the high dimension and volume of data to be processed or the complex patterns to be recognized or predicted by the AI algorithms. To produce this benefit, it is necessary to investigate a QAI computational framework that facilitates the development and assessment of AI algorithms in the quantum sphere and in the transition to it from the classical world (see *e.g.* [9]). The existing frameworks have usually relied on AI-dedicated quantum processing devices. In deep quantum learning, commercial quantum annealers are used to implement quantum learners. These dedicated devices are easier to construct and to scale up than general-purpose quantum computers [7]. QAI could massively benefit from developing a computational framework for the upcoming general-purpose quantum computers that, at the same time, cooperates with the existing von Neumann machines and classical AI.

4 Future Avenues of Research in QAI

QC and AI are both transformational technologies, and AI is likely to require QC to achieve significant progress. Many ground-breaking scientific results have been delivered recently in both areas, advancing technological exploitation in these two disciplines, mainly separately. The conjunction of QC and AI is the real disruption nowadays: recent large-scale initiatives exist to confront this challenge (*e.g.* the EU Quantum Flagship), and in the next few years, quantum technologies will enable doing things that simply cannot be done today: it could be possible to perform complex computational tasks, like modelling large-scale biomolecular and chemical reactions, that the most powerful supercomputers cannot currently manage, send sensitive information safely to anywhere, and diagnose diseases more quickly and accurately by looking inside cells [17]. To make QAI operational in the near future, the problems mentioned in Table 1 need to be addressed, and advances are needed in the three aforementioned dimensions, *i.e.* data, algorithmics and computational framework. Table 2 summarizes the state of the art according to these dimensions.

Consequently, given the current state of the art and needs for QAI to be solidly posed as a novel technological area, facilitating and maximizing its future impact, research is needed in:

1. **Data**: *e.g.* creating efficient and generic quantization/dequantization processes to standardize the *binary-to-quantum* and *quantum-to-binary* translation.
2. **Algorithmics**: *e.g.* creating specific QAI quantum interoperable representations, possibly enriched with semantic annotations, and a systematic methodology to create new and better QAI algorithms that do not necessarily require a classical AI counterpart.

Table 2. Current state of the art for Quantum AI in the Data, Algorithmic and Computational Framework dimensions.

Data	Ad hoc techniques for binary-to-quantum (B2Q) and quantum-to-binary (Q2B) data translation exist, specific for each problem
Algorithmics	Quantum ML and AI algorithms developed for specific problems or as the quantum version of a specific classical algorithm
Computational Framework	A structured high-level computational framework for QAI algorithms does not exist, basing new algorithm creation of low-level quantum circuit specification and algebraic matricial operations and tools for verification

3. **Computational Framework**: *e.g.* creating a computational framework specific to QAI allowing for the coexistence and cooperation of classical and quantum AI processes, in the aim of facilitating the transition from one to the other in the near future.

5 Conclusion

Here, we have presented a detailed review of the progresses in the crossroads of Quantum Computing and Artificial Intelligence. We have put the emphasis in the reciprocal benefits of QC for AI and of AI for QC, to result in the creation of a new scientific area integrating both disciplines, *i.e.* Quantum Artificial Intelligence. We have listed the main fundamental problems that are faced in QAI, as well as lines of future research that are necessary to fully unveil the potential of QAI. This novel scientific area presents itself as an extremely promising technology that is already starting to bear fruits, and which we believe that will have an enormous impact in the very near future, thanks to the incoming availability of medium-scale quantum processors.

Acknowledgements. This work has been supported by the project "XAI - XAI - Sistemas Inteligentes Auto Explicativos creados con Módulos de Mezcla de Expertos", ID SA082P20, financed by Junta Castilla y León, Consejería de Educación, and FEDER funds.

References

1. Ahmed, H., Fosong, A.: Towards quantum-secure authentication and key agreement via abstract multi-agent interaction. In: Practical Applications of Agents and Multi-Agent Systems, pp. 14–26. Springer, Heidelberg (2021)
2. Albash, T., Lidar, D.A.: Adiabatic quantum computation. Rev. Modern Phys. **90**(1), 015002 (2018) (number: 1 Publisher: APS)
3. Alonso, R.S., Prieto, J., García, O., Corchado, J.M.: Collaborative learning via social computing. Front. IT Electronic Eng. **20**(2), 265–282 (2019) (number: 2 Publisher: Springer)

4. Alonso, R.S., Sittón-Candanedo, I., Casado-Vara, R., Prieto, J., Corchado, J.M.: Deep reinforcement learning for the management of software-defined networks and network function virtualization in an edge-IoT architecture. Sustainability **12**(14), 5706 (2020) (number: 14 Publisher: Multidisciplinary Digital Publishing Institute)
5. Amin, M.H., Andriyash, E., Rolfe, J., Kulchytskyy, B., Melko, R.: Quantum boltzmann machine. Phys. Rev. X **8**(2), 021050 (2018) (number: 2 Publisher: APS)
6. Arute et al.: Quantum supremacy using a programmable superconducting processor. Nature **574**(7779), 505–510 (2019) (Nature Publishing Group)
7. Benedetti, R., Biswas, P.: Estimation of effective temperatures in quantum annealers for sampling applications: a case study with possible applications in deep learning. Phys. Rev. A **94**(2), 022308 (2016) (number 2, APS)
8. Biamonte, W., Pancotti, R., Wiebe, L.: Quantum machine learning. Nature **549**(7671), 195–202 (2017) (Nature Publishing Group)
9. Broughton et al.: Tensorflow Quantum: A Software Framework for Quantum Machine Learning. arXiv:2003.02989 (2020)
10. Campbell, E., Khurana, A., Montanaro, A.: Applying quantum algorithms to constraint satisfaction problems. Quantum **3**, 167 (2019) (publisher: Verein zur Förderung des Open Access Publizierens in den Quantenwissenschaften)
11. Cardoso, F.P.: Automated planning and BDI agents: a case study. In: International Conference on Practical Applications of Agents and Multi-Agent Systems, pp. 52–63. Springer, Heidelberg (2021)
12. Carneiro, D., Silva, F., Guimarães, M., Sousa, D., Novais, P.: Explainable intelligent environments. In: International Symposium on Ambient Intelligence, pp. 34–43. Springer, Heidelberg (2020)
13. Chow, J., Dial, O., Gambetta, J.: IBM Quantum Breaks the 100-qubit Processor Barrier (2021)
14. Costa, A., Novais, P., Corchado, J.M., Neves, J.: Increased performance and better patient attendance in an hospital with the use of smart agendas. Logic J. IGPL **20**(4), 689–698 (2012) (Oxford University Press)
15. Davenport, T.H., Barth, P., Bean, R.: How 'big data' is different. In: MIT Sloan Management Review (2012) (Publisher: MIT Sloan Management Review)
16. Debnath et al.: Demonstration of a small programmable quantum computer with atomic qubits. Nature **536**(7614), 63–66 (2016)
17. Djezzar et al.: Quorum sensing digital simulations for the emergence of scalable and cooperative artificial networks. IJAIML **9**(1), 13–34 (2019)
18. Dong, C., Li, T.: Quantum reinforcement learning. IEEE Trans. Syst. Man Cybern. Part B (Cybernetics) **38**(5), 1207–1220 (2008)
19. Ezhov, A.A., Ventura, D.: Quantum neural networks. In: Future Directions for Intelligent Systems and Information Sciences, pp. 213–235. Springer, Heidelberg (2000)
20. Faia, P., Abrishambaf, F., Vale, C.: Case based reasoning with expert system and swarm intelligence to determine energy reduction in buildings energy management. Energy Build. **155**, 269–281 (2017) (Elsevier)
21. Fatima et al.: Enhancing Performance of a Deep Neural Network: A Comparative Analysis of Optimization Algorithms (2020) (ediciones Universidad de Salamanca)
22. Fernández Pérez, I., Boumaza, A., Charpillet, F.: Decentralized innovation marking for neural controllers in embodied evolution. In: Annual Conference on Genetic and Evolutionary Computation, pp. 161–168 (2015)
23. Feynman, R.P., et al.: Simulating physics with computers. Int. J. Theor. Phys. **21**(6/7) (1982) (Number: 6/7)

24. Finnila et al.: Quantum annealing: a new method for minimizing multidimensional functions. Chem. Phys. Lett. **219**(5–6), 343–348 (1994) (Elsevier)
25. Friedman, J.R., Patel, V., Chen, W., Tolpygo, S., Lukens, J.E.: Quantum superposition of distinct macroscopic states. Nature **406**(6791), 43–46 (2000)
26. García, O., Alonso, R.S., Prieto, J., Corchado, J.M.: Energy efficiency in public buildings through context-aware social computing. Sensors **17**(4), 826 (2017) (number: 4 Publisher: Multidisciplinary Digital Publishing Institute)
27. Gazafroudi, C., Keane, S.: Decentralised flexibility management for EVs. IET Renew. Power Gener. **13**(6), 952–960 (2019) (iET)
28. Giovannetti, V., Lloyd, S., Maccone, L.: Quantum random access memory. Phys. Rev. Lett. **100**(16), 160501 (2008) (number: 16 Publisher: APS)
29. Gisin, N., Ribordy, G., Tittel, W., Zbinden, H.: Quantum cryptography. Rev. Modern Phys. **74**(1), 145 (2002) (number: 1 Publisher: APS)
30. Grifoni, M., Hänggi, P.: Driven quantum tunneling. Phys. Rep. **304**(5-6), 229–354 (1998) (number: 5-6 Publisher: Elsevier)
31. Grover, L.K.: A fast quantum mechanical algorithm for database search. In: 28th Annual ACM Symposium on Theory of Computing, pp. 212–219 (1996)
32. Gupta, M., et al.: Neural Network Based Epileptic EEG Detection and Classification (2020). Ediciones Universidad de Salamanca (España)
33. Hager, G., Wellein, G.: Introduction to High Performance Computing for Scientists and Engineers. CRC Press (2010)
34. Horn, D., Gottlieb, A.: Algorithm for data clustering in pattern recognition problems based on quantum mechanics. Phys. Rev. Lett. **88**(1), 018702 (2001)
35. Horodecki, R., Horodecki, P., Horodecki, M., Horodecki, K.: Quantum entanglement. Rev. Modern Phys. **81**(2), 865 (2009) (number: 2 Publisher: APS)
36. Hsu, F.H.: IBM's deep blue chess grandmaster chips. IEEE Micro **19**(2), 70–81 (1999) (number: 2 Publisher: IEEE)
37. Huang, B.: Mohseni, Babbush, Boixo, Neven, McClean: power of data in quantum machine learning. Nat. Commun. **12**(1), 2631 (2021)
38. Jasim, Y.A.: High-performance deep learning to detection and tracking tomato plant leaf predict disease and expert systems. ADCAIJ: Adv. Distributed Comput. Artif. Intell. J. **10**(2) (2021)
39. Jones: The quantum company: D-Wave pioneering a way of making quantum computers but also courting controversy. Nature **498**(7454), 286–289 (2013)
40. Lamata et al.: Quantum autoencoders via quantum adders with genetic algorithms. Quantum Sci. Technol. **4**(1), 014007 (2018)
41. LeCun, Y., Bengio, Y., Hinton, G.: Deep learning. Nature **521**(7553), 436–444 (2015) (number: 7553 Publisher: Nature Publishing Group)
42. Li, S., Liu, C.: Approximate Gaussian conjugacy: parametric recursive filtering under nonlinearity, multimodality, uncertainty, and constraint, and beyond. Front. IT Electronic Eng. **18**(12), 1913–1939 (2017) (Springer)
43. Lloyd, S., Mohseni, M., Rebentrost, P.: Quantum principal component analysis. Nat. Phys. **10**(9), 631–633 (2014) (number 9, Nature Publishing Group)
44. Márquez, S., Mora, S., Herrera, J., Roncero, A., Corchado, J.M.: Intelligent dolls and robots for the treatment of elderly people with dementia. Adv. Distributed Comput. Artif. Intell. J. **9**(1), 99–112 (2020)
45. McClean, H.: Quantum Machine Learning and the Power of Data (2021). https://ai.googleblog.com/2021/06/quantum-machine-learning-and-power-of.html
46. Mezquita, A., Casado-Vara, P., Corchado: A review of k-NN algorithm based on classical and quantum machine learning. In: International Symposium on Distributed Computing and AI (DCAI), pp. 189–198. Springer, Heidelberg (2020)

47. Mishra, D.: Brain Inspired Computing Approach for the Optimization of the Thin Film Thickness of Polystyrene on the Glass Substrates. arXiv:2107.12156 (2021)
48. Moor, J.: The Dartmouth College artificial intelligence conference: the next fifty years. AI Mag. **27**(4), 87–87 (2006) (number: 4)
49. Mugunthan, R., Kagal: BlockFLow: decentralized, privacy-preserving, and accountable federated machine learning. In: International Congress on Blockchain and Applications, pp. 233–242. Springer, Heidelberg (2021)
50. Mäkelä, H., Messina, A.: N-qubit states as points on the Bloch sphere. Physica Scripta **2010**(T140), 014054 (2010) (number: T140 Publisher: IOP Publishing)
51. Möttönen, M., et al.: Quantum circuits for general multiqubit gates. Phys. Rev. Lett. **93**(13), 130502 (2004) (number 13, APS)
52. Najafabadi, M.M., Villanustre, F., Khoshgoftaar, T.M., Seliya, N., Wald, R., Muharemagic, E.: Deep learning applications and challenges in big data analytics. J. Big Data **2**(1), 1–21 (2015) (number: 1 Publisher: Springer)
53. Nguyen, T.T., Hatua, A., Sung, A.H.: Blockchain approach to solve collective decision making problems for swarm robotics. In: International Congress on Blockchain and Applications, pp. 118–125. Springer, Heidelberg (2019)
54. Patil, N., Rigoli, K., Richardson: Dynamical perceptual-motor primitives for better deep reinforcement learning agents. In: Practical Applications of Agents and Multi-Agent Systems, pp. 176–187. Springer, Heidelberg (2021)
55. Perdrix, S.: Quantum patterns and types for entanglement and separability. Electronic Notes Theor. Comput. Sci. **170**, 125–138 (2007) (Elsevier)
56. Prieto, M.: Win: context-aided inertial navigation via belief condensation. IEEE Trans. Signal Process. **64**(12), 3250–3261 (2016)
57. Rempe, G., Walther, H., Klein, N.: Observation of quantum collapse and revival in a one-atom maser. Phys. Rev. Lett. **58**(4), 353 (1987) (number: 4, APS)
58. Riedel, K., Zoller, M., Calarco: Europe's quantum flagship initiative. Quantum Sci. Technol. **4**(2), 020501 (2019) (number 2, IOP Publishing)
59. Sasaki, M., Carlini, A., Jozsa, R.: Quantum template matching. Phys. Rev. A **64**(2), 022317 (2001) (number: 2 Publisher: APS)
60. Schliemann, J., Khaetskii, A.V., Loss, D.: Spin decay and quantum parallelism. Phys. Rev. B **66**(24), 245303 (2002) (number: 24 Publisher: APS)
61. Schlosshauer: Quantum decoherence. Phys. Rep. **831**, 1–57 (2019) (Elsevier)
62. Schuld, M., Petruccione, F.: Supervised Learning with Quantum Computers, vol. 17. Springer, Heidelberg (2018)
63. Schuld, M., Sinayskiy, I., Petruccione, F.: An introduction to quantum machine learning. Contemporary Phys. **56**(2), 172–185 (2015) (number 1, Taylor & Francis)
64. Shoeibi, K., Corchado: Artificial intelligence as a way of overcoming visual disorders: damages related to visual cortex, optic nerves and eyes. In: Distributed Computing and Artificial Intelligence, pp. 183–187. Springer, Heidelberg (2019)
65. Shor: Polynomial-time algorithms for prime factorization and discrete logarithms on a quantum computer. SIAM J. Comput. **26**(5), 1484–1509 (1997)
66. Silva et al.: Classification of chest diseases using deep learning. In: Distributed Computing and Artificial Intelligence, pp. 152–158. Springer, Heidelebrg (2020)
67. Sinanc, D., Demirezen, U., Sağıroğlu, S., et al.: Explainable Credit Card Fraud Detection with Image Conversion (2021) (ediciones Universidad de Salamanca)
68. Tadepalli, T.: COVID-19 early symptom prediction using blockchain and machine learning. In: International Congress on Blockchain and Applications, pp. 243–251. Springer, Heidelberg (2021)
69. Turing, A.M.: Intelligent Machinery (1948)

70. Von Neumann, J.: First draft of a report on the EDVAC. IEEE Ann. Hist. Comput. **15**(4), 27–75 (1993) (number: 4 Publisher: IEEE)
71. Yigitcanlar, K., Regona, R., Rowan, R., Desouza, C., Mehmood, L.: Artificial intelligence technologies and related urban planning and development concepts: how are they perceived and utilized in Australia? J. Open Innov.: Technol. Market Complexity **6**(4), 187 (2020) (mDPI)
72. Zhong et al.: Quantum computational advantage using photons. Science **370**(6523), 1460–1463 (2020)

An Architecture for a Coaching System to Support Type 2 Diabetic Patients

Diogo Martinho[1(✉)], Vítor Crista[1], Andreia Pinto[2,3], José Diniz[2,3], Alberto Freitas[2,3], João Carneiro[1], and Goreti Marreiros[1]

[1] Research Group On Intelligent Engineering and Computing for Advanced Innovation and Development (GECAD), Institute of Engineering, Polytechnic of Porto, Porto, Portugal
{diepm,vvrpc,jrcmgt,mgt}@isep.ipp.pt

[2] Department of Community Medicine, Information and Health Decision Sciences (MEDCIDS), Faculty of Medicine, University of Porto, Porto, Portugal
{andreiapinto,up201404234,alberto}@med.up.pt

[3] Center for Health Technology and Services Research (CINTESIS), Faculty of Medicine, University of Porto, Porto, Portugal

Abstract. Diabetes mellitus type 2 (DMT2) is an increasing worrying disease that has become a major public health challenge over the last decades. It can have serious health consequences and lead to long-term health complications that greatly impact the quality of life of the diabetic patient. Therefore, it becomes vital to define supportive mechanisms that can decrease the effect of the different risk factors and at the same time provide knowledge on how to deal with DMT2 and rely on new mobile technologies to help manage the disease. The work here presented had the main goal to propose the architecture for a coaching system that supports the self-management of diabetes mellitus type 2 disease which is being developed in the context of the FoodFriend project, with the goal of stimulating the innovation in the continuous monitoring of the patients' health. The coaching system here proposed aims to support both health managers and diabetic patients through the processing of food-intake and health data and provide actionable feedback that can improve the health condition of the patient. This includes the presentation of personalized recommendations and challenges adapted to patient current health condition and their motivation to change health behaviors. By doing so, we believe it will be possible to improve the patients' self-management of their disease as well as decrease the risk of disease-related complications associated to DMT2.

Keywords: mHealth · Personalized healthcare · Diabetes mellitus type 2 · Coaching system

1 Introduction

Diabetes is one of the most impactful and growing disease worldwide. In fact, latest epidemiological reports[1] [1] show that there are already over 500 million people with

[1] https://www.who.int/news-room/fact-sheets/detail/diabetes.

V. Julián et al. (Eds.): ISAmI 2022, LNNS 603, pp. 167–178, 2023.
https://doi.org/10.1007/978-3-031-22356-3_16

diabetes with this becoming a major cause of blindness, kidney failure, heart attacks, stroke, and lower limb amputation among other health complications. In 2019, diabetes observed a record of 1.5 million deaths thus being the ninth leading cause of death. In Portugal, over the last few years, the estimated prevalence of diabetes was of 13.6% for people between 20 and 79 years old [2]. Additionally, Type 2 diabetes (resulting from the ineffective use of insulin by the human body) has been observed in more than 95% of people with diabetes. This type of diabetes is often associated to the lack of healthy eating habits and the adoption of sedentary lifestyles that contribute to metabolic conditions and behavior disorders such as overweight and obesity.[2,3] The poor management of diabetes can lead to several complications, such as the development of micro- and macrovascular diseases, diabetic neuropathy and nephropathy, diabetic foot, higher probability of infections, and, in the most extreme cases, death [3] Social consequences brought by the high prevalence of diabetes include more expensive healthcare and employment worries with decreased productivity and attendance [4]. With the growing social and health footprint of this disease, it becomes clear that new intelligent and sustainable ways to manage diabetes are essential and new and improved ways to dimmish the effect of risk factors associated with the disease should be promoted. By doing so we are contributing to healthier lifestyles comprising healthier diets, physical activity, and better glycemic control which are essential to correctly manage diabetes [5].

Nowadays we are surrounded by smart environments combined with mobile technologies, which have also transformed the digitalization of healthcare and have become a key asset in our daily lives [6–8]. Smartphone and cloud-based solutions are being used by more and more people every day [9], even those in older age groups [10–12]. According to our latest studies we have learn that there is a growing interest in the topic of computer systems associated with diabetes disease. There are already several persuasive solutions such as mobile apps developed targeting the disease and there is some evidence of a decrease in some clinical outcomes associated with higher cardiovascular risk and poor disease management, such as HbA1C and weight, among others. Such kind of apps often rely on coaching mechanisms to motivate and persuade the user to adopt healthier lifestyles [13, 14]. In the context of diabetes management some of the most recent coaching systems proposed in the literature include the use of recommendation strategies [15, 16]. For instance, authors in [17] presented a virtual coaching conversational assistant to provide psychosocial support and motivate adults with diabetes to adopt and follow healthy coping strategies. These strategies are in accordance with the American Association of Diabetes Educators guidelines and are adapted to the users' needs. The assistant relies on different machine learning and natural language techniques to processed text inputs and provide adequate motivational feedback based on the user current emotional state and state of change (for this the authors defined different actions based on the states of changes provided with the Transtheoretical Model of Change). Authors in [18] presented an app called SweetGoals which sends messages about self-management goals, goal adherence, and encouragement about adherence in the upcoming week to help patients with type 1 diabetes manage their disease. This app also presented a virtual coaching conversational assistant to communicate with the user and provide scripted

[2] https://www.cdc.gov/chronicdisease/resources/publications/factsheets/nutrition.htm.
[3] https://www.nhs.uk/conditions/obesity/causes/

feedback regarding current health goals. These goals are mainly of two types related with glucose monitoring and mealtime and they process these goals according to user data inputted directly with the app or obtained using a glucometer. Additionally, the app also considers an incentive component with earnings to reward the users that can achieved established goals correctly. Authors in [19] presented a Clinical Decision Support System for Diabetic patients which processed anthropometric parameters such as blood pressures, biochemical data, and current medication information to provide tailored feedback to the user. For this the authors developed a Rule engine containing clinical practice recommendations of both American and Korean Diabetes Associations and provided actionable feedback regarding both diet and exercise improvements based on the data collected.

Many of the existing applications are targeted to type 1 diabetes, which has specific risk factors and different prevention and therapeutic indications compared to type 2 diabetes. Therefore, it is important to develop intelligent solutions adapted not only to type 2 diabetes but that can enhance user-driven inputs, making them more accessible, and at the same time provide personalized feedback (using behavioral, sentiment analysis [20, 21] and even real-time monitoring [22] strategies) based on the data collected. By doing so it will be possible to guide the user toward effective behavioral improvements to better manage his/her disease. This work concerns a project named FoodFriend,[4] which aims to develop a mobile application with the objective of helping type 2 diabetic patients to change their behavior and, in this way, adopt a healthier lifestyle, through better eating habits and more active practice of physical exercise. For this, behavioral change models will be used as well as coaching and gamification strategies, in order to promote continuous interaction with the app.

Thus, this work has the particular aim of presenting the architecture of the coaching system that will help diabetic patients to adopt better choices in order to monitor their disease.

2 Proposed Architecture

Here we describe the architecture considered for the development of the coaching framework within the diabetes use case of the FoodFriend project (see Fig. 1). To note that the architecture here presented is the initial approach to model the several components of the system still currently being developed. However, the goal is to define a microservice-oriented architecture with several components to process data retrieved from mobile and web applications with specific web services to do so.

At the front-end level two main applications are being developed. The application is the FoodFriend mobile app which will be used by the patient himself/herself and collect data on different aspects part of his/her daily life related to food-intake insights and behaviors. In return this application will present feedback about the patient's current health condition and personalized recommendations and challenges regarding activities that can be performed and/or behaviors that can be adopted for the patient to improve his/her current health condition. Figure 2 presents an example of the interfaces already developed to share feedback with the patient.

[4] https://itea4.org/project/food-friend.html.

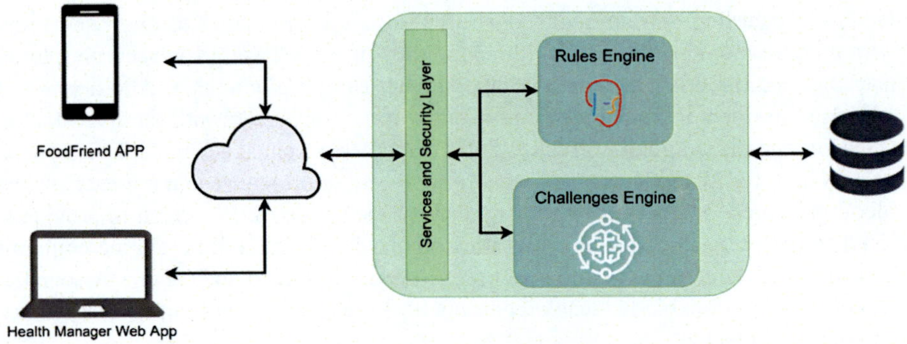

Fig. 1. Coaching architecture.

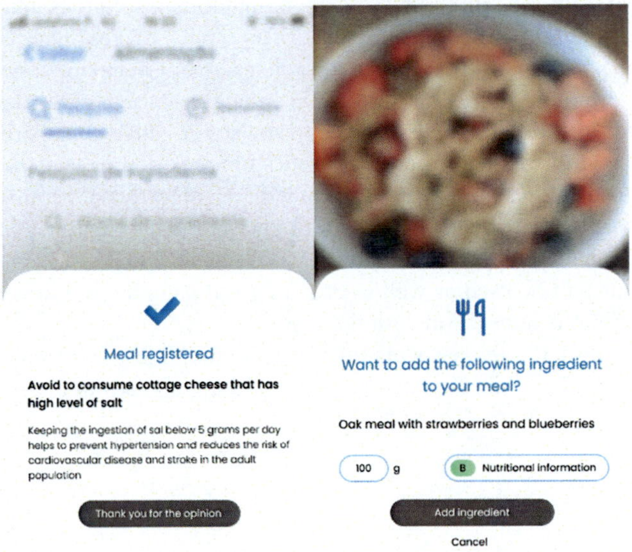

Fig. 2. Insertion of food and example of feedback to the user regarding food-intake.

The second application is a knowledge management web app which will be used by the health manager to provide/define new knowledge and actionable coaching inputs (such as the definition of additional food-intake recommendations and challenges). At the back-end level an API is being developed to support the necessary services to process all the information obtained from the developed applications. As such, the main features of the API include a set of restful services developed within JAX-RS to access the different processing components of the coaching framework: a rules engine component which will support patients the self-management of their disease through the presentation of personalized recommendations regarding their current health condition and behaviors to adopt in order to improve it; a challenges engine to generate actionable checkpoints and goals to be achieved by the user also regarding their current health condition and behaviors that should be improved.

2.1 Rules Engine

Family history, advanced age, inadequate diet, and sedentary lifestyle contributing to excess weight are risk factors for the onset of type 2 diabetes. Even though there is a genetic predisposition for the disease, its onset is widely considered to be associated with behavioral modifiable risk factors [23]. However, despite the challenges in establishing and maintaining these healthy habits, these changes are always beneficial and often sufficient to control their condition properly [24]. Accordingly, lifestyle management, particularly through adequate and eating and physical activity, as well as proper adhesion and compliance to the prescribed drug therapies, when there is any, are fundamental pillars of good management of type 2 diabetes [25].

Thus, one of the main focuses of the coaching system to be implemented will be on the feedback related to the patients' food choices, according to the user input. This feedback will be created through a rules engine implemented using JBoss Drools Framework and will be provided when users enter a meal or food on the mobile app. This framework was chosen given the possibility to model inferential knowledge very close to the one used in the clinical context through the diagnosis of diseases, thus being a suitable approach. Besides that, the specific environment also allows the definition of rules with grammar close to natural language which facilitates the comprehension within the medical team working in the definition of nutritional rules and other aspects related with the diabetic patient's health.

According to the literature for type 2 diabetes management, there are some parameters that have consensus due to their importance in controlling the disease, namely, weight management, caloric balance, and attention to some characteristics of certain foods, such as their glycemic index or nutritional complexity. However, there is still some uncertainty in the optimal distribution of macronutrients in the diet [26]. Therefore, our rules were generated in accordance with the World Health Organization (WHO) guidelines to promote a healthy diet.[5] This is also in accordance with current national guidelines for clinical practice [27, 28]. Portuguese guidelines only mention healthy diets, without any specification. On the other hand, WHO guidelines define a healthy diet, complementing Portuguese guidelines. The implementation of the KBS was developed by researchers in the areas of programming and software, but the definition of the values that compose the system was developed by a nutritionist and a doctor leaning on the previously mentioned guidelines. As a result, the developed rule engine currently considers ten different rules implemented regarding the consumption of several nutrients. All the rules developed consider the following aspects comprising a healthy diet:

- Salt consumption needs to be lower than 5 g per day.
- Fat consumption should not exceed 30% of total energy intake.
- Saturated fats consumption should be less than 10% of total energy intake.
- Carbohydrates consumption should be less than 50% of total energy intake.

[5] https://www.who.int/news-room/fact-sheets/detail/healthy-diet.

- Dietary fibers consumption should be at least 25g.
- Ideally, a person should have a consumption of 0 g of trans fat per day.
- Protein consumption should be less than 20% of total energy intake.

The system will calculate the macronutrients of all foods that are registered in the app and will make a daily calculation—namely, carbohydrates, proteins, fats and fiber—as well as some micronutrients. During meal and food registration, each component is identified, and its composition obtained from a database. As the user fills their nutritional log throughout the day, the key macro- and micronutrients will be registered and summed cumulatively, indicating whether they have exceeded daily recommended limits, for example, for saturated fats, salt, and carbohydrates. The same is available for nutrients consumed insufficiently, such as diminished fiber intake due to limited consumption of fruit and vegetables. Hence, users will be nudged into increasing their integration in their diets.

The system will also recognize and suggest some alternatives to the food opted for by the specific user, offering for instance, a recommendation to substitute their previous option for nutritionally richer and more complex carbohydrates with lower salt content (e.g. "brown" whole-wheat bread instead of "white" bread).

Additionally, a score associated with each food will be also calculated and indicated. Thus, a color is generated according to the score calculation, which can be from green to red, with the color green being considered the healthiest and the color red the least recommended and should not be consumed frequently [29].

The coaching system will signalize and generate personalized feedback messages and suggestions regarding food consumption improvements that take into account the user's motivation for behavior change. In an initial approach, we have defined different profiles based on the Transtheoretical model of behavior change (TTM), a commonly used framework to design tailored health interventions [30–32]. TTM mainly explains that behavior change involves progress through five stages of change that are differentiated by the individuals' willingness to change their current behavior and perception on health status. To advance over the different stages, individuals apply stage-specific cognitive and behavioral processes. Early stages of change (i.e., pre-contemplation and contemplation) correspond to those in which people are often unaware of the need to change or are ambivalent toward changing their behavior. The following stage correspond to which people become more convinced about the need to change their behavior (preparation stage) and start changing it (action and maintenance stages. It is essential to design feedback messages that can match individuals' processes of change since it will guide them to advance to the next stage of change. As a result, the initial approach undertaken in the development of the proposed coaching system was to gather stages of change with similar challenges into three different profiles:

- **Profile A**—corresponds to users in the pre-contemplation and contemplation stages;
- **Profile B**—corresponds to users in the preparation stage;
- **Profile C**—corresponds to users in the action and maintenance stages.

After defining these profiles, we specified feedback messages based on the most appropriate processes of change to address the specific challenges of each profile.

(see Table 1 for an example).

Table 1. Example with Feedback messages for each considered profile.

Profile	WHO Rule: Keep salt intake to less than 5 g per day	
	Process of change	Example of feedback message
Profile A: People are unaware of the need to change or are Ambivalent toward	Consciousness-raising: increase personal understanding of the need to change or the benefits of adopting healthy behavior	Did you know that excessive salt consumption is one of the main causes of hypertension and cardiovascular disease? Avoid consuming foods with too much salt like [example of a food consumed by the user]
Profile B: People want to change their behavior in the next month, and they take small steps toward behavior change believing in its positive outcomes	Self-liberation: create commitment to change behavior based on the belief that achievement of healthy behavior is possible	Have you tried to progressively reduce the consumption of foods with too much salt like [example of a food consumed by the user]? After a while, your taste buds will get used to it. Give it a try!
Profile C: Behavior change has occurred in the short-term and people intend on keeping with healthy behavior	Counterconditioning: promote consistency in the behavior by replacing unhealthy ways of acting and thinking with desirable ways	To keep your salt intake under control, try to replace foods with too much salt like [example of a food consumed by the user] with healthier foods like [example of alternatives to the unhealthy food consumed by the user]

All these interactions are currently contemplating only direct feedback and only processing food-intake over a single day (see Fig. 2). However, in the second phase of development (ongoing work), the coaching system will analyze user data over a larger time frame and identify key aspects of food-intake that can be improved with the generation of automatic and personalized challenges for the upcoming time periods (days, weeks, etc.) as is explained in the following section of this work.

2.2 Challenges Engine

Challenges are specific to each user and will be challenging depending on how close the user is to correctly follow a healthy diet. Additionally, these challenges also contribute to the gamification process that will be applied within the Food Friend app (translating correctly achieved challenges to rewards and experience the user can obtain).

A challenge engine is currently being developed consisting of two distinct yet correlated components. Since each user is unique and has different routines, it is a priority to find a way to interact with them in a most efficient way. So, a machine learning hybrid model was used, consisting of building the different clusters (user profiles) with the help of a hierarchical clustering method. Once these profiles have been defined, a decision tree algorithm (C5.0) was applied to describe each cluster. A created an initial model consuming a dataset composed of thirty-three users,[6] with data of their daily steps, taken for the duration of 3 months. After this phase, it was possible to find out different profiles based on temporal features (day of the week and time of the day).

The second component consists of proposing challenges regarding physical activity. To do so, we are evaluating 30 days of physical data of a specific user retrieved from GoogleFit app, to frame the user into a category of activity. We followed the specifications provided by the WHO [33] that has various categories regarding the level of physical activity depending on the daily steps taken. The process to propose a challenge consists of gathering 30 of user data, calculating the daily average steps, adjusting the user into a category (sedentary, low active, active, and highly active), and proposing a challenge with three different difficulties that vary considering a percentage of the average daily step value. After this, we will be able to identify the type of user and propose adaptative challenges that will be gradually incremented/decremented based on physical activity inputs the user provides each day.

2.3 Knowledge Management System

A Web Application will be provided to health managers, to allow sharing of knowledge within the context of diabetes management. Therefore, the main features of this application will be related to the insertion of knowledge regarding rules and challenges that can be used to provide personalized feedback to the patient according to his/her health condition, Additionally the health manager may also insert clinical guidelines used in the treatment of patients and consult reports describing all inserted knowledge in the system.

Figure 3 presents a mock-up example of the interface that will be developed within the health manager web app for the consult of statistical data (reports) on the inserted knowledge. All information inserted by the health manager will be sent to the coaching API and then be used by both the Rules and Challenges Engine when processing patient data to provide personalized feedback accordingly.

3 Conclusions and Future Work

Diabetes disease is growing worldwide and is one of the top leading causes of death observed over the last years. As a result of poor healthy eating and lifestyles, the predominance of this disease can only be tackled by defining effective measures that can diminish the impact of the risk factors associated with the disease. While doing so we are contributing not only to the well-being but improved and healthier lifestyles of diabetic patients. In this context it has been noticed the key role new technologies can provide to offer intelligent and personalized solutions to diabetic patients to better manage their health condition and control the disease correctly.

[6] https://www.kaggle.com/datasets/arashnic/fitbit.

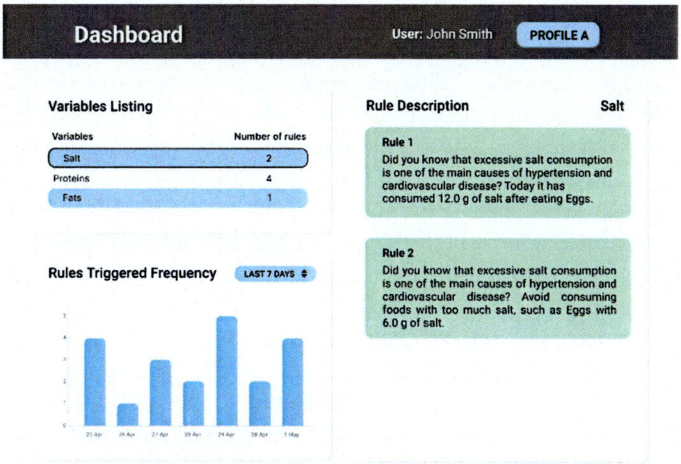

Fig. 3. Knowledge management system dashboard.

In this work it is presented an architecture for a coaching system to support the self-management of diabetes disease (more specifically the diabetes mellitus type 2 disease) that is currently being developed under the Portuguese use case of the FoodFriend project. The Portuguese use case aims to develop a mobile application with the objective of helping type 2 diabetic patients to change their behavior and, in this way, adopt a healthier lifestyle, through better eating habits and more active practice of physical exercise. The proposed architecture consists of two main applications at the front-end level to be used by the patient himself/herself and the health manager to insert domain knowledge and an API at the back-end level to process patient data and provide personalized feedback based on the existing knowledge in terms of recommendations and challenges. We are currently testing the first version of the proposed coaching system together with the mobile app for patients with type 2 diabetes. These patients will be recruited from a hospital in the North of Portugal, and they must follow specific inclusion criteria. They must be aged at least 45 years old, have a diagnosis of diabetes type 2, and be smartphone users. The patients will all access the same version of the app and will have a set of tasks that they must complete without any help, to verify whether the app is intuitive and self-explanatory. At the end of a certain period, the opinion regarding the use of developed will be collected through questionnaires and those features will be improved based on the information provided and to allow us to consider additional features to be developed in the future. To identify the usability and the satisfaction of the app, the System Usability Scale (SUS) and the User Satisfaction Evaluation Questionnaire (USEQ), will be also applied respectively.

As the next steps, a set of rules regarding food-intake data have already been established and implemented using a rules engine and further specification of more food-intake related rules will be next defined based on micronutrients and other aspects that comprise a healthy diet. Additionally, and as future work, we intend to develop the challenges engine that has been described. Our first approach and as explained previously, considers different profiles using a dataset that is very small and lacks the diversification of features such as age, gender, and culture, thus the final profiles output is too generic. Therefore, and in a first phase of development, we are going to retrieve GoogleFit data regarding physical activity to define personalized and adaptative challenges to each user based on their physical activity profile and motivation for behavior change, together with demographic data retrieved from the user profile page. In a second phase, we will apply the same process in the definition of dietary challenges based on the food-intake data acquired from real users in the meantime. We are currently testing the first version of the mobile app with real users in a study to assess the usability and the satisfaction of the app, applying the System Usability Scale (SUS) and the User Satisfaction Evaluation Questionnaire (USEQ), respectively. Additionally, we also intend to collect the opinion of the users regarding the features that should be improved and whether the information provided is clear as well as the opinion on the usefulness of the features to be developed in the near future. After this phase, we will have patient food-intake data to define profiles in a similar way as done to physical activity. These challenges will be the basis for the gamification features we intend to include in the mobile app that will in turn result in user engagement and improve user experience (with acquisitions of points, rewards, and unlockable content). By considering all these features we are defining an intelligent and personalized solution capable of motivating the diabetic patient to stay connected to the application and be motivated to improve health related behaviors to better manage his/her health condition, thus contributing to an improved well-being a better self-management of the disease.

Acknowledgments. This research work was developed under the project Food Friend –"Autonomous and easy-to-use tool for monitoring of personal food intake and personalised feedback" (ITEA 18032), co-financed by the North Regional Operational Program (NORTE 2020) under the Portugal 2020 and the European Regional Development Fund (ERDF), with the reference NORTE-01-0247-FEDER-047381 and by National Funds through FCT (Fundação para a Ciência e a Tecnologia) under the project UI/DB/00760/2020 and the PhD grant number UI/BD/151485/2021.

References

1. Roglic, G.: WHO Global report on diabetes: a summary. Int. J. Noncommunicable Diseases **1**, 3 (2016)
2. de Diabetologia, S.P.: Diabetes: Factos e números–O Ano de 2015—Relatório Anual do Observatório Nacional da Diabetes. Lisboa: SPD (2016)
3. Knight, J., Andrade, M., Nigam, Y.: Diabetes management. long term complications due to poor control. Nursing Times **113**, 45–48 (2017)
4. Hill, J., Nielsen, M., Fox, M.H.: Understanding the social factors that contribute to diabetes: a means to informing health care and social policies for the chronically ill. The Permanente J. **17**, 67 (2013)
5. LeRoith, D., Smith, D.O.: Monitoring glycemic control: the cornerstone ofdiabetes care. Clin. Ther. **27**, 1489–1499 (2005)
6. Scarry, A., Rice, J., O'Connor, E.M., Tierney, A.C.: Does the use of mobile applications or mobile health technology improve diet quality in adults? a protocol for a systematic literature review. HRB Open Research **5**, 1 (2022)

7. Alshutayria, A., Aljojo, N., Alharbia, B., Banjarb, A., Alshehria, A., Alargoubia, M., Barradha, O., Helabia, R.: An Interactive Mobile Application to Request the Help of the Nearest First Aider by the Injured. (2021)
8. Nakahara, T., Fukuyama, K., Hamada, M., Matsui, K., Nakatoh, Y., Kato, Y.O., Rivas, A., Corchado, J.M.: Mobile device-based speech enhancement system using lip-reading. In: International Symposium on Distributed Computing and Artificial Intelligence, pp. 159–167. Springer (2020)
9. Rathod, Y.A.: An access control and authorization model with open stack cloud for smart grid. ADCAIJ: Adv. Distrib. Comput. Artif. Intell. J. **9**, 69–87 (2020)
10. Portenhauser, A.A., et al.: Mobile apps for older adults: systematic search and evaluation within online stores. JMIR Aging **4**, e23313 (2021)
11. Janols, R., Guerrero, E., Lindgren, H.: A pilot study on personalised coaching to increase older adults' physical and social activities. In: International Symposium on Ambient Intelligence, pp. 140–148. Springer (2017)
12. Lindgren, H., Guerrero, E., Janols, R.: Personalised persuasive coaching to increase older adults' physical and social activities: a motivational model. In: International Conference on Practical Applications of Agents and Multi-Agent Systems, pp. 170–182. Springer (2017)
13. Gershkowitz, B.D., Hillert, C.J., Crotty, B.H.: Digital coaching strategies to facilitate behavioral change in type 2 diabetes: a systematic review. J. Clin. Endocrinol. Metab. **106**, e1513–e1520 (2021)
14. Márquez-Sáncheza, S., Mora-Simonb, S., Herrera-Santosa, J., Roncerod, A.O., Corchadoa, J.M.: Intelligent dolls and robots for the treatment of elderly people with dementia. Adv. Distrib. Comput. Articial Intell. J. **9**, 99–112 (2020)
15. Baptista, S., Wadley, G., Bird, D., Oldenburg, B., Speight, J., Group, M.D.C.R.: User experiences with a type 2 diabetes coaching app: qualitative study. JMIR Diabetes **5**, e16692 (2020)
16. Gong, E., Baptista, S., Russell, A., Scuffham, P., Riddell, M., Speight, J., Bird, D., Williams, E., Lotfaliany, M., Oldenburg, B.: My diabetes coach, a mobile app⇓based interactive conversational agent to support type 2 diabetes self-management: randomized effectiveness-implementation trial. J. Med. Internet Res. **22**, (2020)
17. Bassi, G., Giuliano, C., Perinelli, A., Forti, S., Gabrielli, S., Salcuni, S.: A virtual coach (Motibot) for supporting healthy coping strategies among adults with diabetes: proof-of-concept study. JMIR Hum. Factors **9**, e32211 (2022)
18. Stanger, C., et al.: A digital health intervention (SweetGoals) for young adults with type 1 diabetes: protocol for a factorial randomized trial. JMIR Res. Protocols **10**, e27109 (2021)
19. Lim, S., et al.: Multifactorial intervention in diabetes care using real-time monitoring and tailored feedback in type 2 diabetes. Acta Diabetol. **53**(2), 189–198 (2015). https://doi.org/10.1007/s00592-015-0754-8
20. Khan, R., Siddiqui, S., Rastogi, A.: In: Crime detection using sentiment analysis (2021)
21. Basarslan, M.S., Kayaalp, F.: Sentiment Analysis with Machine Learning Methods on Social Media. (2020)
22. Farias, G.P., Pereira, R.F., Hilgert, L.W., Meneguzzi, F.R., Vieira, R., Bordini, R.H.: Predicting plan failure by monitoring action sequences and duration. Adv. Distrib. Comput. Artif. Intell. J. (2017)
23. Galicia-Garcia, U., et al.: Pathophysiology of type 2 diabetes mellitus. Int. J. Mol. Sci. **21**, 6275 (2020)
24. Van den Arend, I., Stolk, R., Krans, H., Grobbee, D., Schrijvers, A.: Management of type 2 diabetes: a challenge for patient and physician. Patient Educ. Couns. **40**, 187–194 (2000)
25. Asif, M.: The prevention and control the type-2 diabetes by changing lifestyle and dietary pattern. J. Educ. Health Promotion **3** (2014)

26. Forouhi, N.G., Misra, A., Mohan, V., Taylor, R., Yancy, W.: Dietary and nutritional approaches for prevention and management of type 2 diabetes. Bmj **361**, (2018)
27. Saúde, D.G.d.: Processo assistencial integrado diabetes mellitus tipo 2. Departamento da Qualidade na Saúde (2013)
28. Carvalheiro, M., Diniz, A., Candeias, A., Pipa, C., Boavida, J., Correia, L.: Programa nacional de prevenção e controlo da diabetes. Revista Portuguesa de Diabetes **2**, 5–13 (2007)
29. Goiana-da-Silva, F., et al.: Nutri-score: a public health tool to improve eating habits in Portugal. Acta Med. Port. **32**, 175–178 (2019)
30. Johnson, S.S., et al.: Transtheoretical model-based multiple behavior intervention for weight management: effectiveness on a population basis. Prev. Med. **46**, 238–246 (2008)
31. Lin, S.-P., Wang, M.-J.: Applying the transtheoretical model to investigate behavioural change in type 2 diabetic patients. Health Educ. J. **72**, 189–202 (2013)
32. Prochaska, J.O., DiClemente, C.C.: Transtheoretical therapy: Toward a more integrative model of change. Psychotherapy: Theory, Res. Practice **19**, 276 (1982)
33. WHO, O.: WHO guidelines on physical activity and sedentary behaviour. Geneva. World Health Organization (2020)

Machine Learning Based System for the Control and Evaluation of Programming Vulnerabilities

María Alonso-García[1,2(✉)], Rubén Fuente-Alonso[3], and Juan M. Corchado[1,2]

[1] AIR Institute. Edificio PCUVa, Módulo 117-118 Campus Miguel Delibes Paseo de Belén, 9A 47011 Valladolid, Spain
marialonsogar@air-institute.org, corchado@air-institute.com

[2] BISITE Research Group, Facultad de Ciencias, Universidad de Salamanca, Calle Espejo, sn, 37007 Salamanca, Spain

[3] Madison Experience Marketing., Calle de Enrique Cubero, 9, 47014 Valladolid, Spain
ruben.fuente@madisonmk.com

https://air-institute.com, https://bisite.usal.es

Abstract. With the increasing role of technologies in our daily lives, software security has become an important issue to provide protection from malicious attacks and ultimately avoid irreversible damage to systems. The collection of data reflecting the quality of programming allows the application and training of machine learning algorithms to predict if potential software vulnerabilities are in a script, knowing some of its features. We examine if these algorithms can be used to predict some score that reflects the severity of vulnerabilities in a script prior to release. To this end, we develop a crucial preprocessing stage to define a metric that reflects the programmer evolution over time and be aware of future flaws in his code. To provide a comprehensive performance assessment, we use a private dataset containing labelled vulnerabilities of a programmer over time recorded by code snippet. Ultimately, an effective early diagnosis system is obtained.

Keywords: Software security · Predictive models · Machine learning · Regression problem · Quantile regression

1 Introduction

Detecting software vulnerabilities post-release not only costs dee money, but also leads to reputational damage and damaging lawsuits. Therefore, techniques to prevent and detect vulnerabilities before release are valuable. We can define a vulnerability as a security flaw, weakness or glitch that could be exploited by a threat source [1] usually for malicious purposes. This must be differentiated from a bug or weakness in the system that may or not cause some vulnerability in the future.

The tools that perform analysis of the programs have limitations as they are based on predefined rules. The availability and accessibility of open-source scripts have enabled the development of more powerful vulnerability scanners [2, 3]. However, these analysis tools may not be effective in detecting vulnerabilities in certain situations [4].

V. Julián et al. (Eds.): ISAmI 2022, LNNS 603, pp. 179–186, 2023.
https://doi.org/10.1007/978-3-031-22356-3_17

Our aim is to add value and complement the analysis resulting from this type of tools. For this purpose, we developed a machine learning based system that estimates a metric that aims to quantify in some way the possible vulnerabilities and their severity based on the characteristics of a script and the user. This could give an estimate of what needs to be revised in a project over time or how to prioritize revisions.

The challenges in the development of this project have been the collection and compilation of useful data for the purpose outlined above. We have refined the selection and collection of metrics to those that best reflected the ultimate purpose. Lack of sufficient data to train a machine learning model makes it necessary to complete the initially collected dataset with additional synthetic data.

This system achieves the stated purpose and is able to estimate a vulnerability metric per script. The study concludes that the experience of a developer is a determining factor in the vulnerabilities committed. While the tendency to commit vulnerabilities and the severity of vulnerabilities decreases with the experience of the programmer, it does not exempt him from committing them. Future work includes collecting more data and completing the study with more user information.

The rest of the article is structured in such a way that the following section presents a review of existing works directly related to the present article, followed by a description of the proposed system. Section 4 presents the results obtained. Section 5 presents the conclusions obtained after carrying out the work and analyzing the results and, finally, the last section presents the lines of future work.

2 Background

Exploitable vulnerabilities in software can pose a potential threat to the safe operation of computer systems and affect millions of users every day.

One of the effective solutions to mitigate attacks is to identify software vulnerabilities before deployment. The basic code analysis techniques for vulnerability detection can be divided into static, dynamic and hybrid methods. Static techniques, such as B. rule/template-based analysis [5], code similarity detection [6] and symbolic execution [7], rely on the analysis of source code but they usually suffer from high false positives [8]. Dynamic analysis includes fuzzing [9] and taint analysis [10], and often suffers from low code coverage. Hybrid approaches [11] have drawbacks such as inefficiency. Applying traditional machine learning techniques to detect vulnerabilities still requires experts to define features, which mainly depend on human experience, domain expertise, and depth of domain knowledge [12]. Several studies examine whether software metrics obtained from source code and development history are discriminative and predict vulnerabilities in the code. In one of these studies [13], three types of software metrics (complexity, code changes, and developer activity) were examined for their suitability in building models to predict vulnerabilities. However, these can be too time-consuming.

Leveraging data-driven techniques to perform automated intelligent analysis directly on the source code requires addressing some challenges, such as B. Representing the source code in a suitable form for further analysis in ML algorithms and locating detected in the source code loopholes [3].

Other studies include attribute-based models that attempt to discover a relationship between code attributes and the number of vulnerabilities [14]. Automatic feature extraction from source code using a deep learning algorithm has also been studied for the detection of software vulnerabilities [2].

Recent vulnerability prediction techniques model vulnerabilities based on time, describe the relationship between the total number of vulnerabilities through calendar time, and use vulnerability detection models to treat calendar time as an independent variable [15].

In the literature, the applicability of many of these techniques is limited due to the inability to obtain the source code directly. In this paper, we present a system based on machine learning techniques and a useful vulnerability metric.

3 Proposed System

The system we propose is the combination of adding a metric reflecting vulnerabilities in a script and applying to the result a powerful ensemble regressor, a gradient boosted decision tree (GBDT).

For a machine learning algorithm to perform correctly on data that it has not seen before, an appropriate training data set is necessary. In this study, we have used a dataset containing information on scripts developed by the programmers of one company. As most of the registered scripts did not contain vulnerabilities, the set has been completed with some synthetic data generated from the information provided by the existing data. Initially, the variables collected are the ID of a user, his experience with a programming language, the time he spent developing the script (in minutes), the programming language used, the number of words a script contains, the number of lines a script contains, the number of vulnerabilities it contains (after analysis), the number of vulnerabilities it contains differentiated by severity (info, low, medium, high, critical), the date the script was created and the date it was updated. We decided not to differentiate the available data per user, but to use the information collected in a general way.

Before starting the modelling phase, it has been necessary to remove duplicate instances. Duplicate instances can be a problem later if they are not removed, as they can inflate metrics and lead to misinterpretations since training data could be repeated in the test set.

For the subsequent analysis and definition of the metric to assess the propensity of a script to contain vulnerabilities, we select the following characteristics from the ones above: the user experience with a language, the time per script, the programming language used the number of lines in a script and the number of vulnerabilities it contains differentiated by severity (info, low, medium, high, critical).

In order to define the metric used here, we start from the features describing vulnerabilities by type and assign to each type of vulnerability a penalty that reflects each severity as shown in the Table 1.

Then, the metric is defined as follows:

$$\text{score} = \frac{\sum_{i=0}^{5} \text{pi number\_of\_vulnerabilities}}{\text{script\_lines}} \quad (1)$$

where i is the type of severity ($i = \{info, low, medium, high, critical\}$), p_i is the normalized penalty associated with each of the above, *number_of_vulnerabilities* is the number of

Table 1. Penalties associated with each type of vulnerability by severity.

	Penalty	Normalized penalty
Info	0.0	0.0
Low	0.1	0.05
Medium	0.5	0.25
High	1.0	0.5
Critical	2.0	1.0

vulnerabilities of type *i* in the script and *script_lines* is the number of lines in a script. It is clear from the definition that the minimum of the score is 0 and the maximum 1.

The modelling phase starts by testing different machine learning algorithms to finally select the gradient boosted decision tree regressor. This technique is a generalization to arbitrary differentiable loss functions of boosting [16]. Gradient boosting models consists of a set of individual decision trees trained sequentially so that each new tree attempts to improve on the error of the previous one. Predictions for new observations are obtained by summing the predictions of all the individual trees that make up the model. Many forecasting methods produce global models in which a single equation applies to the entire sample space. When a use case involves multiple predictors, they interact in a complex and non-linear way, and it is difficult to find a single global model that reflects the relationship between the variables. Tree-based methods comprise a set of supervised non-parametric techniques that manage to partition the space of predictors into simple regions within which interactions are easier to manage [17]. Moreover, due to the variability that exists between different programmers beyond the characteristics of their scripts, we have considered it appropriate to fit the quantile version. Most supervised learning algorithms predict the expected value of the target variable as a function of some features or predictor variables (point estimation), but this approach ignores certain properties of the distribution of the target variable such as the associated uncertainty [18] necessary in this case.

We have fitted a gradient boosting model by training them with a quantile loss function for the 0.05, 0.5 and 0.95 quantiles, so that the confidence interval is 90% (0.95−0.05 = 90). We have subsequently tuned their hyperparameters to obtain an optimal model.

4 Results

In this section, we present the results obtained after the definition and implementation of the proposed system. We used 70% of the available data to train and tune the model and reserved 30% as a test set to evaluate the results. For the sake of comparison and to avoid potential unnecessary complications, simpler models are fitted and discarded first. These results are shown in the Table 2, where the metrics used are mean absolute error (MAE), mean squared error (MSE), root mean squared error (RMSE), R^2or coefficient of determination (R2) and mean absolute percentage error (MAPE).

Table 2. Evaluation metrics results for simpler regression models.

Model	MAE	MSE	RMSE	R2	MAPE
Lasso regression	0.1476	0.0341	0.1846	0.3060	2.0131
K neighbours regressor	0.1327	0.0291	0.1703	0.4087	1.5772
Ridge regression	0.0899	0.0124	0.1111	0.7478	1.3535
Linear regression	0.0899	0.0124	0.1111	0.7478	1.3537
Decision tree regressor	0.0641	0.0082	0.0907	0.8322	0.7238
AdaBoost regressor	0.0623	0.0067	0.0815	0.8644	0.7119
Gradient boosting regressor	0.0575	0.0057	0.0754	0.8840	0.6860

The table shows that the best of the models tested is the gradient boosting regressor, consequently it makes sense to keep adjusting it until it provides the best possible results. The hyperparameters of the final quantile gradient boosting regressor are a learning rate of 0.05, number of estimators of 100, maximum depth of the individual regression estimators of 5, minimum number of samples required to be at a leaf node of 9 and minimum number of samples to split an internal node of 9.

In the following analysis, the metrics used to measure and compare the performance of the models are the mean squared error and mean pinball loss (mpl). After fitting the quantile gradient boosting regressor, we obtain the results in the Table 3. It follows that the training set converges, as the minimum per column is obtained by training a model and evaluating it with the same metric. In addition, for the quantile 0.5, we obtain a MAE $= 0.053690$ and R2 $= 0.888763$ on the test set. Similar errors in both sets mean that models are correctly fitted to data.

Afterwards we evaluated if the quantile estimators are fine calibrated by computing the fraction of observations falling between the estimated intervals. The closer the coverage fraction is to the confidence value of the interval, the better the fit. We obtained coverage fractions for the train and test sets of 0.936 and 0.925, respectively, so we can conclude that the fit is correct.

In order to graphically complete the error analysis above, we add Figs. 1 and 2.

Finally, we can examine in Fig. 3 the feature importance given by the final model to the available predictors. According to this graph, the attribute that most influences the final prediction is user experience with a programming language.

5 Conclusion

We have shown the potential of applying machine learning algorithms to determine the possible vulnerability score attached to a code script based on some of its characteristics. For this purpose, we have used a dataset on the code characteristics of the programmers of a company. After data curation, we defined and add a metric that measures the severity and quantity of vulnerabilities in a script. Finally, a model selection process took place and a quantile gradient boosted trees regressor has been chosen between a variety of

Table 3. Evaluation metrics results for the fitted model.

Evaluation on train set				
Model	*mpl = 0.05*	*mpl = 0.5*	*mpl = 0.95*	*MSE*
Quantile 0.05	0.005326	0.051922	0.098518	0.021400
Quantile 0.5	0.022037	0.024512	0.026988	0.004628
Quantile 0.95	0.123050	0.064875	0.006700	0.023837
Evaluation on test set				
Model	*mpl = 0.05*	*mpl = 0.5*	*mpl = 0.95*	*MSE*
Quantile 0.05	0.005603	0.052163	0.098722	0.021496
Quantile 0.5	0.024441	0.026845	0.029249	0.005473
Quantile 0.95	0.125372	0.066370	0.007368	0.024520

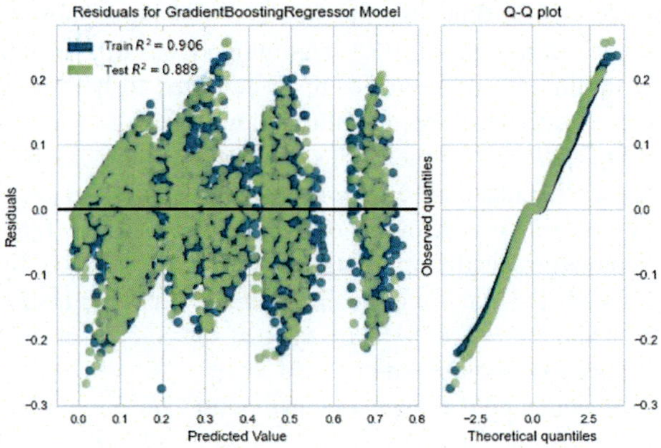

Fig. 1. Residuals and Q-Q plot of the final model.

machine learning algorithms. Subsequent to the fine-tuning of the chosen model, we achieved with that model the best results for the selected features with the available data.

This could be used to set revisions on developer's files of code and avoid future serious vulnerabilities.

6 Future Work

Future work could focus on collecting different features of a user code or of his personality, as his learning curve rate of a programming language. In addition, if sufficient data is available, a model can be differentiated and adapted for each developer. This would ensure the reduction of the overall variability, present in this case due to personal differences.

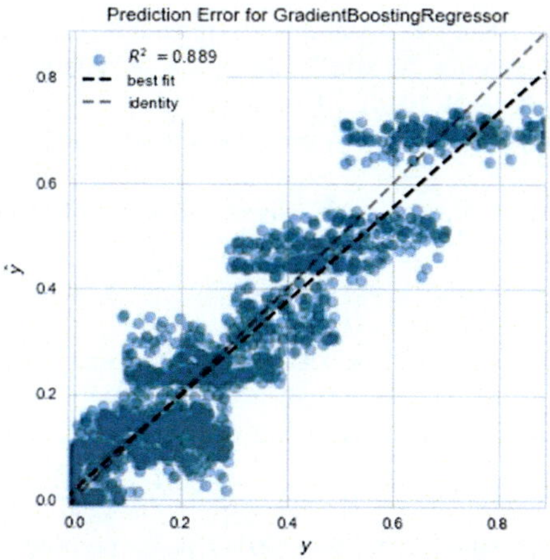

Fig. 2. Prediction error of the final model.

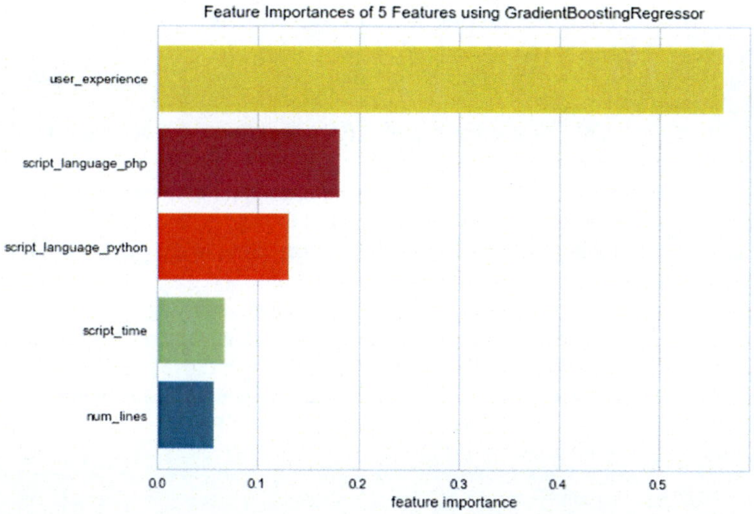

Fig. 3. Feature importance of predictors using the fitted gradient boosting.

Acknowledgements. This research work has been funded by the Spanish Ministry of Energy, Tourism and Digital Agenda, project "TrustedCoding: Plataforma de detección de vulnerabilidades y aprendizaje de soluciones para una codificación segura e inteligente" with ID: TSI- 100906-2019-6.

References

1. Dempsey, K., Paul, E., George, M., Takamura, E.: In: Automation Support for Security: Software Vulnerability Management. NIST (2020)

2. Rebecca, L., Russell, Kim, L.: Automated vulnerability detection in source code using deep representation learning. In: IEEE International Conference on Machine Learning and Applications (IEEE ICMLA 2018), Orlando, Florida, USA (2018)
3. Bilgin, Z., Ersoy, M.A.: Vulnerability prediction from source code using machine learning. pp. 8. (2020)
4. Delaitre, A.M., Stivalet, B.C.: Sate V report: ten years of static analysis tool expositions (2018)
5. Engler, D., Chen, D.Y.: Bugs as deviant behavior: a general approach to inferring errors in systems code. **35**(5) (2001)
6. Kim, S., Woo, S.: (s.f.). VUDDY: a scalable approach for vulnerable code clone discovery (2017)
7. Engler, D.A.: Under-constrained symbolic execution: correctness checking for real code (2015)
8. Li, Z., Zou, D., Xu, S., Ou, X., Jin, H., Wang, S., ... Zhong, Y. Vuldeepecker: a deep learning-based system for vulnerability detection (2018). arXiv preprint arXiv:1801.01681
9. M. Sutton, Greene, A.: Fuzzing: brute force vulnerability discovery (2007)
10. Song, J.N.: Dynamic taint analysis for automatic detection analysis and signature generation of exploits on commodity software (2005)
11. Li, X., Wang, L.: Automated software vulnerability detection based on hybrid neural network. Appl. Sci. (2021)
12. Jiang, J., Wen, S.: Identifying propagation sources in networks: state-of-the-art and comparative studies. **19**(1) (2017)
13. Shin, Y., Ersoy, A.M.: Evaluating complexity code churn and developer activity metrics as indicators of software vulnerabilities. **37**(6) (2011)
14. Rahimi, S., Zargham, M.: Vulnerability scrying method for software vulnerability discovery prediction without a vulnerability database. **2**(62) (2013)
15. Alhazmi, O.H., Malaiya, M.Y.: Quantitative vulnerability assessment of systems software. In: Reliability and Maintainability Symposium (2005)
16. Friedman, J.: Greedy function approximation: a gradient boosting machine. **5**(29) (2001)
17. Hastie, T.T.: The elements of statistical learning: data mining, inference and prediction (2009)
18. Johnson, M.K.: In: Applied predictive modeling. Springer (2013)
19. Ban, X., Liu, S., Chen, C., Chua, C.: A performance evaluation of deep-learnt features for software vulnerability detection. Concurrency and Comput.: Practice and Exper. **31**(19), e5103 (2019)

Quantum Optimization for IoT Security Detection

Vita Santa Barletta(✉), Danilo Caivano, Mirko De Vincentiis, Alessio Magrì, and Antonio Piccinno

Dipartimento di Informatica, Universitá di Bari Aldo Moro, Bari, Italy
vita.barletta@uniba.it, danilo.caivano@uniba.it, mirko.devincentiis@uniba.it, a.magri3@studenti.uniba.it, antonio.piccinno@uniba.it

Abstract. The digital transformation and innovation of today's environment requires a focus on the security of systems to preserve their confidentiality, integrity, and availability. Considering these needs, in this work an efficient and high-performance intrusion detection system based on Quantum Annealing is presented. The goal is to identify attacks in an IoT environment such as Smart Home. The goal is to reduce the processing time for identifying attacks and to respond promptly to implement measures appropriate to the threat. We use Qboost, an iterative training algorithm in which a subset of weak classifiers is selected by solving a hard optimization problem in each iteration. The models were tested on a IoT Fridge dataset, characterized by a large volume of traffic with a low number of features and highly imbalanced data distribution. The results show an optimization of computational time and an improvement in detection accuracy.

Keywords: Quantum optimization · Detection · Iot security

1 Introduction

Modern information technology has revolutionized work and daily activities by offering the possibility of carrying out such activities in an automated way. Take, for example the personal sphere, everyone has credentials to access social networks, mobile banking apps or services offered by the public administration. In the business sphere, on the other hand, the digital age has affected the activities of most companies around the world, which are constantly using IT systems to store data, manage business operations, manage personnel, provide services to customers and, most importantly, ensure the security of all these resources [1]. To confirm this, just think of Smart Cities where technology is a service for the people and a means for improving the quality of their economic and social life [3]. This requires to ensure the confidentiality, integrity and availability of data, services and systems [2].

V. Julián et al. (Eds.): ISAmI 2022, LNNS 603, pp. 187–196, 2023.
https://doi.org/10.1007/978-3-031-22356-3_18

Considering these needs and highly pervasive IoT systems contexts, in this research work the attention is focused on quantum computing in the cybersecurity field. Quantum computing is a new computing paradigm that has a great deal of direct applications, and many other potential ones, such as privacy and cryptography; supply chain and logistics; chemistry; economics, and financial services; energy and agriculture; medicine and health; defence and national security programs [16]. There are no works in the literature that have focused on the application of quantum computing to cybersecurity. In addition, the existing ones, only consider Denial of Service (DoS) detection. Reddy et al. [17] have worked on two different quantum approach QSE (Quantum Swarm Evolutionary) and QEA (Quantum-inspired Evolutionary Algorithm) for detecting DoS attacks on cloud computing. Instead in Payares et al. [15] have presented three quantum models to detect Distributed Denial of Service (DDoS) attacks, that is, QSVM (Quantum Support Vectors Machines), H-QNN (Hybrid Quantum-Classical Neural Networks) and two-circuit ensemble model running in parallel on two quantum processing units.

Therefore, in this research work, the goal is to optimize the detection time through the use of quantum computing. In particular, we compare classification of cyber attacks through SVC (Support Vector Classification) and Qboost [7], an iterative training algorithm in which a subset of weak classifiers is selected by solving a hard optimization problem in each iteration, through the platform made available by D-Wave, D-Wave Leap [6].

The rest of the paper is organised as follows. Section 2 presents the fundamental concepts of quantum computing. Section 3 introduces Quantum Annealing algorithm preparation for D-Wave QPU calculus. Section 4 introduces Qboost for IoT security detection and how it works. Section 5 shows the analysis of experiments. Finally, Sect. 6 presents the conclusions drawn after the analysis of the studies and introduces future research activities.

2 Quantum Computing

Computers currently are capable of performing computational processes through the use of classical mechanics, while quantum computers are capable of computation through the use of quantum mechanics [4,18]. This makes the two models completely distant in terms of comparison. A normal computer is able to recognize and perform algebraic computations through the use of values that are equivalent to either 0 or 1. Instead, a quantum computer, being subject to the laws of quantum mechanics, can exist in both states simultaneously (0 and 1). This property of quantum objects is called superposition [9,14].

The units of a quantum computer are called quantum bits or qubits. The qubit is very similar to the simple bit, it can be represented through the same two states of the bit (0 and 1) always taking into account that the qubit, however, is in a quantum state. The value of a qubit can be read and as such can be measured. Qubits can also be represented with only 0 and 1, taking into account that the states in which they collapse are allowed as superposition as much as the intermediate states. It is therefore useful from here on to denote the state of qubits with the Dirac notation [10]: $|0\rangle$ and $|1\rangle$. A qubit is often represented as an array or vector which within the Hilbert space is referred to as $|\psi\rangle$. The Hilbert space is a vector space of all possible real and complex numbers. The quantum state can be represented as mutually orthogonal two-base vectors:

$$|0\rangle = \begin{bmatrix} 1 \\ 0 \end{bmatrix} \tag{1}$$

$$|1\rangle = \begin{bmatrix} 0 \\ 1 \end{bmatrix} \tag{2}$$

Thus we can identify qubits with an even more formal notation:

$$|\psi\rangle = \alpha|0\rangle + \beta|1\rangle \tag{3}$$

Qubits differ from normal bits because it can be said that qubits are always in a complex linear state given by the combination of the states of the two bases. α and β show the probability with which the state will collapse into 1 or 0 without any certainty. To better understand the characterization of a qubit it is possible to represent it through a Bloch Sphere (Fig. 1). This is a three-dimensional sphere that is often used for the geometric representation of a qubit. The state of a qubit is represented as a point positioned on the surface of the sphere: "north pole" represents the $|0\rangle$ state, while the "south pole" represents the $|1\rangle$ state. Each point on the surface can be represented as the linear combination of a unit vector with origin in the center of the sphere. The vector can rotate on the

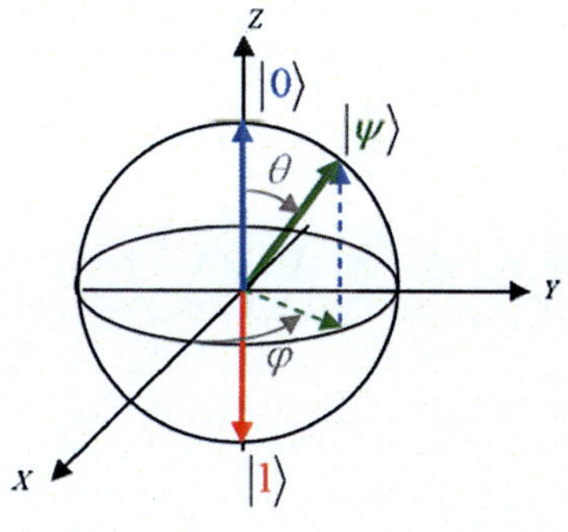

Fig. 1. Qubit bloch sphere model. [19].

surface of the bloch sphere using the following representation:

$$|\psi\rangle = \cos\frac{\theta}{2}|0\rangle + e^{i\phi}\sin\frac{\theta}{2}|1\rangle \tag{4}$$

3 Quantum Annealing

The Quantum Annealing algorithm is used to return the best solution with less energy, it has fields of application in optimization problems and probabilistic sampling problems. In optimization problems we look for the best possible combination also in terms of efficiency by minimization of energy. The use of quantum physics for solving sampling problems is very useful for machine learning problems where you want to implement a probabilistic model of reality. Qubits are the minimum energy state of a superconductor loop in a QPU (Quantum Processing Unit) D-Wave [5] and have a supply current and a magnetic field within which they circulate. At the end of a quantum annealing process they collapse into a state of 0 or 1 [8].

This physical process can be illustrated as an energy diagram. This diagram changes with time and assumes different states. It is illustrated below in the various steps (Fig. 2). In (a) it is possible to identify a single depression of the curve that tends to a superposition of the qubit with a single minimum. At the time of launching the quantum annealing algorithm, the depression of the curve splits into two parts given by the expression of the superposition of the qubit. In fact, at that time the qubit can assume in a given percentage probability both states. More specifically in (b) it has 50% chance to assume one of the two classical states. In (b) the left depression corresponds to 0 and the right one to 1. The qubit at the end of annealing ends up in one of the two. It is possible to analyze the probability of the qubit "falling out" by applying a magnetic field external to the qubit. The programmed amount of which controls the external magnetic field is called bias, in the presence of such bias the qubit minimizes its energy. The single use of the bias however is not sufficient for the measurement. It becomes important at the time of entanglement of two qubits.

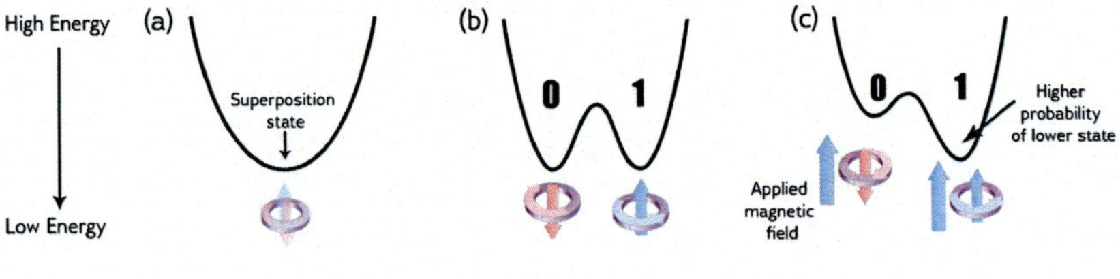

Fig. 2. Energy diagram over time [5].

4 Qboost for IoT Security Detection (Q4IoT)

Qboost is a classification boosting method that combines a set of predictions considered "weak" in order to produce a better performing one called "strong". This method selects weak classifications by minimizing the difference between observed and label predictions. The results of the optimizations are formulated as a binary quadratic model and solved by the D-Wave system. This approach is designed using the mathematical principles of "Training a Binary Classifier with Quantum Adiabatic Algorithm" by Hartmut Neven [13]. Given a set of N weak classifiers, Qboost uses problem optimization to select one of them. The goal of the optimization is to minimize the quadratic deviation between current and predicted labels. A penalty is also introduced for the complex model that includes many weak classifiers [12]. The architecture of the Qboost system is shown in Fig. 3: starting from the dataframe, it is inserted in a QUBO matrix where the values reported in a positional way are the labels of the dataframe and the weight is given by the calculation of the values assumed by the features (which will be useful to determine the position of the non-predicted rows in the testing phase).

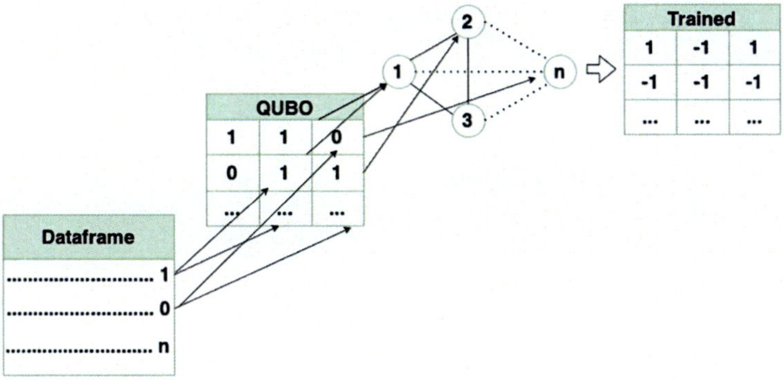

Fig. 3. Qboost operation diagram.

5 Empirical Evaluation

A comparison between traditional and quantum classifier was performed in order to analyze if the accuracy and speed execution improve. Support Vector Classifier (SVC) was used as the normal classifier. It is based on SVM (Support Vector Machine) that is a supervised learning model used for classification. The training algorithm for SVM forms a model that assigns the new examples to one of the allowed classes. It constitutes a linear classifier where uses distance from an ideal splitting line that is created on a plane. By calculating the Euclidean distance from it of each point, it is possible to define the class membership of the tested sample.

5.1 Dataset

We consider an IoT dataset that contains the analysis of Internet of Things (IoT) system behaviors, includes heterogeneous data cataloged from Telemetry datasets of IoT and IIoT sensors [11]. The datasets were collected from a realistic, large-scale network designed at the Cyber Range and IoT Labs, School of

Engineering and Information technology (SEIT), UNSW Canberra Australian Defence Force Academy (ADFA) . The dataset consists of 7 features and 587077 rows (Table 1): Message ID, Date, Time, Fridge Temperature, Temperature Condition, Attack or Normal Label, and Attack Type). The dataset IoT systems are characterized by a distributed and decentralized architecture that often runs into security issues. Not having the availability of a centralized server or "master" device that controls the behaviors of various devices. There are many ways to analyze IoT architecture security issues and many of them effectively. Transferring the analysis of device security into a broader discourse such as enterprise or public administration, it is possible to assert that the greater the security issues, the greater the Incident Response interventions to be implemented in case of attacks.

Table 1. IoT dataset overview "*IoT_Fridge*".

ID	Date	Time	Fridge temperature	Temp condition	Label	Type
1	31-Mar-19	12:36:52	13.1	High	0	Normal
...	...	...	...	...	...	...
413929	28-Apr-19	23:29:18	5.05	Low	1	Backdoor
...	...	...	...	...	...	...
436788	25-Apr-19	15:56:01	12.1	High	1	Ddos
...	...	...	...	...	...	...

5.2 Experimental Methodology

The methodology used to pursue the goal of this work consist of three main stages (Fig. 4): (i) *Data Selection*, the dataset is loaded from its source file into the system. The data transformation and preprocessing are common to the two classifiers; (ii) *Training*. Compared to the normal classification, Qboost requires a further step, that is the transformation of the data into a QUBO matrix. This transformation allows the software to insert examples in the training algorithm and then in the testing algorithm. The division of the examples for training, happens in random way selecting a prefixed number of examples of training and testing. After the testing phase, the outputs of the classifier Qboost are discretized since they are related to the processing of the quantum space of the qubits; (iii) *Data evaluation and interpretation*. It consists in two important phases such as the Plotting (example in Fig. 5) of the data and report that analyzed the confusion matrix (accuracy, recall, f1-score and support—Table 2 and relative time spent.

5.3 Results

Taking into analysis the data resulting from the various experiments (Table 3), it is possible to understand that despite some errors as in the case of 10000–10000 where the accuracy is greater in the SVC classifier, the two classifiers

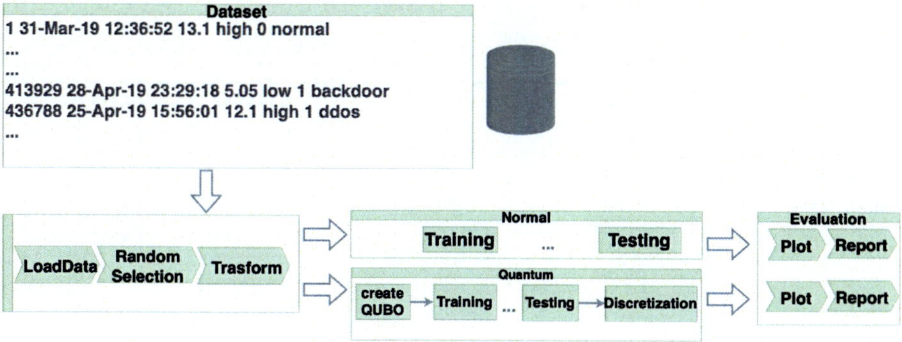

Fig. 4. Experimental process.

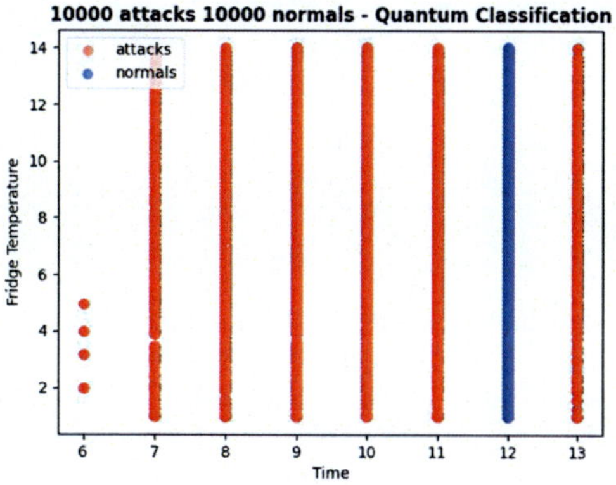

Fig. 5. Plot example.

Table 2. Example of confusion matrix.

	Precision	Recall	F1-score	Support
0	0.82	1.00	0.90	10,000
1	1.00	0.78	0.88	10,000
Accuracy			0.89	20,000
Macro avg	0.91	0.89	0.89	20,000
Weighted avg	0.91	0.89	0.89	20,000

are equivalent in terms of accuracy. In the last two experiments, where there are more examples than the others, even if the accuracy of both is clearly decreasing, the accuracy of the quantum classifier is higher.

We can see instead that the execution time decreases (in terms of seconds and milliseconds) in all cases. This is a good results in cybersecurity field because the reduction of the time to detect the attack is very fundamental in order to stop the

Table 3. Results of SVC and Qboost.

Attacks/normal classes	SVC accuracy	Qboost accuracy	SVC time (s)	Qboost time (s)
100, 100	1.00	1.00	0.00059	0.00043
200, 200	1.00	1.00	0.00066	0.00037
300, 300	1.00	1.00	0.00039	0.00082
400, 400	1.00	1.00	0.00099	0.00044
400, 500	1.00	1.00	0.00106	0.00039
500, 400	1.00	1.00	0.00125	0.00052
500, 500	1.00	1.00	0.00111	0.00042
1000, 1000	1.00	1.00	0.00181	0.00044
400, 5000	1.00	1.00	0.00425	0.00087
10,000, 300	1.00	0.80	0.00803	0.00125
300, 10,000	1.00	1.00	0.00993	0.00205
5000, 5000	1.00	1.00	0.00128	0.00074
10,000, 10,000	1.00	0.89	0.02348	0.00923
55,000, 55,000	0.25	0.29	0.05606	0.01944
60,000, 60,000	0.23	0.27	0.05947	0.01618

attack. Therefore, as shown in Fig. 6, at this early stage of analysis we can see, the quantum classifier using Qboost optimizes in terms of accuracy and timing the classification work; Qboost takes on average 0.00042 s per 1000 examples to classify, while the normal SVC classifier 0.00111 s; the curve describing the proportion of the increase in computational complexity of the normal classifier is similar to an exponential growth while the curve describing Qboost is similar to a logarithmic growth (thus slower to grow); Qboost improves the classification time thereby lowering it by 33.73%.

6 Conclusions and Future Developments

In this work it has been presented, through experimentation, the effectiveness of quantum classification. This method is not counted among the simplest because of difficulties in management, formatting of data input and interpretation of data output. The mathematical theory behind the training of a Quantum Machine Learning algorithm is quite complex and far from the usual computer science scenarios. Assuming a field of application such as Incident Response, it is advantageous to have an algorithm capable of classifying attacks as quickly as possible. Relying on quantum technology it is possible to implement, in SIEM and IDS, quantum classification algorithms in order to optimatize the detection. By expressing the relationship between computational complexity and time spent in the form of a logarithmic curve, the Qboost classifier demonstrates an excellent alternative to the simple classifier implemented with Sklearn's SVC. This opens up several prospects for future work such as improving the accuracy of the classifier; testing the proposed classifier with different datasets; extending it to the classification of different attack classes by using additional quantum technologies.

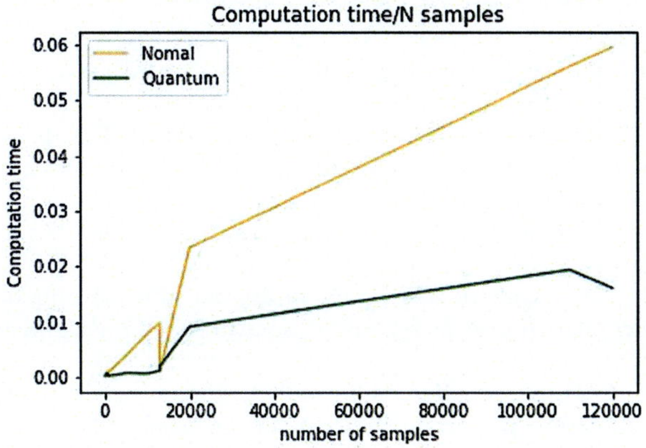

Fig. 6. Computation time/complexity proportion graph.

References

1. Baldassarre, M.T., Barletta, V.S., Caivano, D.: Smart program management in a smart city. In: 2018 AEIT International Annual Conference, pp. 1–6 (2018). https://doi.org/10.23919/AEIT.2018.8577379
2. Barletta, V.S., Buono, P., Caivano, D., Dimauro, G., Pontrelli, A.: Deriving smart city security from the analysis of their technological levels: a case study. In: 2021 IEEE International Conference on Omni-Layer Intelligent Systems (COINS), pp. 1–6 (2021). https://doi.org/10.1109/COINS51742.2021.9524268
3. Barletta, V.S., Caivano, D., Dimauro, G., Nannavecchia, A., Scalera, M.: Managing a smart city integrated model through smart program management. Appl. Sci. 10(2) (2020). https://doi.org/10.3390/app10020714, http://www.mdpi.com/2076-3417/10/2/714
4. Benioff, P.A.: The computer as a physical system: a microscopic quantum mechanical hamiltonian model of computers as represented by turing machines. Stat. Phys. 22 (1980). https://doi.org/10.1007/BF01011339
5. D-Wave: (Accessed 29 Apr 2022). https://docs.dwavesys.com/docs/latest/_images/change.png
6. D-Wave: D-wave leap (Accessed 29 Apr 2022). https://cloud.dwavesys.com/leap
7. D-Wave: Qboost (Accessed 29 Apr 2022). https://github.com/dwave-examples/qboost
8. Kadowaki, T., Nishimori, H.: Quantum annealing in the transverse ising model. Phys. Rev. E **58**, 5355–5363 (1998). https://doi.org/10.1103/PhysRevE.58.5355
9. Loredo, R.: Quantum computing with python and IBM quantum example (2020)

10. Maurice, D.P.A.: Theory of electrons and protons (1930). https://doi.org/10.1098/rspa.1930.0013
11. Moustafa, N.: A new distributed architecture for evaluating AI-based security systems at the edge: Network TON_IoT datasets. Sustain. Cities Soc. **72**, 102994 (2021). https://doi.org/10.1016/j.scs.2021.102994, https://www.sciencedirect.com/science/article/pii/S2210670721002808
12. Neven, H., Denchev, V., Rose, G., Macready, W.: Qboost: large scale classifier training with adiabatic quantum optimization. J. Mach. Learn. Res. **25**, 333–348 (2012)
13. Neven, H., Denchev, V.S., Rose, G., Macready, W.G.: Training a binary classifier with the quantum adiabatic algorithm. arXiv: Quantum Physics (2008). https://doi.org/10.48550/arXiv.0811.0416.
14. Pattanayak S.: Quantum Machine Learning with Python. Apress, Berkeley, CA (2021). https://doi.org/10.1007/978-1-4842-6522-2
15. Payares, E., Martinez-Santos, J.C.: Quantum machine learning for intrusion detection of distributed denial of service attacks: a comparative overview. In: OPTO (2021)
16. Piattini, M., Serrano, M., Perez-Castillo, R., Petersen, G., Hevia, J.L.: Toward a quantum software engineering. IT Prof. **23**(1), 62–66 (2021). https://doi.org/10.1109/MITP.2020.3019522
17. Reddy, P.R.K., Bouzefrane, S.: Analysis and detection of dos attacks in cloud computing by using GSE algorithm. In: 2014 IEEE International conference on high performance computing and communications. In: 2014 IEEE 6th International Symposium on Cyberspace Safety and Security. In: 2014 IEEE 11th International Conference on Embedded Software and System (HPCC, CSS, ICESS), pp. 1089–1096. https://doi.org/10.1109/HPCC.2014.183
18. Schrödinger, E.: An undulatory theory of the mechanics of atoms and molecules. Phys. Rev. **28**, 1049–1070 (1926). https://doi.org/10.1103/PhysRev.28.1049
19. Wikimedia: (Accessed 29 Apr 2022). https://commons.wikimedia.org/wiki/File:Sphere_bloch.jpg

IdeAir: IoT-Based System for Indoor Air Quality Control

Gleiston Guerrero-Ulloa[1,2], Alex Andrango-Catota[1], Martín Abad-Alay[1], Miguel J. Hornos[2](✉), and Carlos Rodríguez-Domínguez[2]

[1] Faculty of Engineering Science, State Technical University of Quevedo, 120501 Los Ríos, Ecuador
{gguerrero,alex.andrango2016,martin.abad2016}@uteq.edu.ec, gleiston@correo.ugr.es

[2] Higher Technical School of Computer and Telecommunications Engineering, University of Granada, 18071 Granada, Spain
{mhornos,carlosrodriguez}@ugr.es

Abstract. The burning of fossil fuels by cars and household appliances, which make people's lives easier, causes air quality to deteriorate. Fossil fuel-dependent (water and space) heaters, which help low-income people to live in better conditions, can cause their death. Many solutions to this social problem have been proposed; however, all of them suffer from some drawback that makes their application impossible for households with limited economic resources. This paper proposes the design of IdeAir, a low-cost Internet of Things (IoT)-based indoor air quality monitoring system that aims to reduce the disadvantages of existing systems. IdeAir captures data on harmful gases and determines their concentrations, issuing alarms and notifications according to their concentration levels. Its development has been carried out following the Test-Driven Development Methodology for IoT-Based Systems (TDDM4IoTS), which, together with the tool used (based on this methodology) for the automation of the system development process, has facilitated the work of its developers. Preliminary results on the functioning of IdeAir show a high level of acceptance by potential users.

Keywords: Internet of Things · Development methodology · Development tool · Microcontrollers · Air quality

1 Introduction

Air quality is measured by the concentration of toxic gases present in the environment. Gases, such as ozone (O_3) at the earth's surface, nitrogen dioxide (NO_2), carbon monoxide (CO), carbon dioxide (CO_2), ammonia (NH_3), sulphur dioxide (SO_2), benzene ($C_6 H_6$), ethanol (C_2H_5OH), and particulate matter (PM) are the main environmental pollutants. These gases are produced by the burning of fossil fuels in vehicles, refineries,

This publication is part of the R&D&i Project Ref. PID2019-109644RB-I00 funded by MCIN/AEI/10.13039/501100011033.

V. Julián et al. (Eds.): ISAmI 2022, LNNS 603, pp. 197–206, 2023.
https://doi.org/10.1007/978-3-031-22356-3_19

thermoelectric power generators, and industries [1], and even appliances used at home, such as heaters to heat the environment and water for daily activities. These gases, above certain concentration levels, cause potential harm to human health and even death, in the worst cases. This is especially problematic in populations located at high altitudes, where there is a low concentration of oxygen. The concern to prevent these pollution-related deaths worldwide has prompted the World Health Organisation (WHO) to update the indicators for air quality [2], in order to improve this universal right.

Water and space heaters are respectively used to condition the water and room temperature in cold climate homes. These appliances, being fuelled by fossil fuels, emit gases, such as CO, which are hazardous, due to their characteristics (odourless, invisible, tasteless and non-irritating to the eyes) [3]. In fact, CO is considered the *silent killer*. Hence, this problem and the consequences of the possible misuse of these heaters have concerned several governments, such as those of Argentina [4] and Ecuador [5].

Bommi et al. [6] propose a very important solution to monitor and try to solve the pollution problem. However, due to the investment required, it would not be feasible to implement it in households using space heaters. Another work on air quality monitoring is carried out by Zhou et al. [7], who present a system that issues messages to the user when pollution levels exceed thresholds. In addition, to prevent a serious event, the system can turn on a fan, which would clearly be insufficient in a closed space, such as inside a house.

As a contribution that aims to reduce the number of people suffering from poisoning by hazardous gases, such as CO, CO_2, C_2H_5OH, NO_2, and NH_3, this paper proposes an IoT-based system (IoTS) that is able to detect these gases indoors, and then make appropriate decisions. Therefore, it can be considered as a system to maintain the ideal air, hence its name: IdeAir. After its implementation, a technical demonstration of its execution was carried out in Quito, capital of Ecuador, which has a cold climate and is located at 2850 m above sea level, where there have been several deaths due to problems with heaters [8].

The main motivation for the development of this IoTS is to provide a reliable solution with a feasible and low-cost implementation to address the social problem described above. Another motivation is to perform an initial validation of the development methodology called TDDM4IoTS [9], since it is a new methodology and one of the few methodologies specifically designed for IoTS development.

The remainder of this paper is organised as follows: Sect. 2 presents the related work. Section 3 describes the proposed system, as well as the development methodology followed, and the results obtained by applying each of the phases of said methodology. Finally, Sect. 4 summarises the conclusions and future work.

2 Related Work

The world is concerned about air pollution and its fatal consequences. This is reflected in the fact that the WHO has issued new guidelines to control air pollution, considering particulate matter (PM) from 2.5 μm size upwards [2]. Moreover, the scientific community has carried out some works to help mitigate the consequences of air pollution as well as improve air quality, which we will present in this section.

Considering that vulnerable people usually spend a lot of time in indoor environments, where contamination is higher than in outdoor environments [10], solutions have been presented for these spaces. Bommi et al. [6] aim to control air pollution by reducing the toxicity of gases produced by combustion through a physical-chemical process that generates oxygen.

The work of Liu et al. [11] is related to pollution monitoring, since they monitor CO_2 and $PM_{2.5}$, among other parameters. In terms of pollutant monitoring, there is also AirSensEUR, which consists of a plug-and-play interoperable sensor node, designed as an open multisensor platform whose investment is around 1000 euros [12].

Taştan and Gökozan [13] propose a mobile real-time indoor air quality monitoring system, called *e-nose*, which is able to measure various air parameters, such as CO_2, CO, NO_2, PM_{10}, temperature and humidity. However, the only way to alert the user about dangerous pollution levels is through notifications sent to the mobile application. Another related work is that of Azma et al. [14], which provides air quality readings using low-cost sensors.

Although all the works reviewed provide solutions to help people prevent from environmental conditions harmful to their health, each of them has some characteristic that makes it impractical to be implemented in low-income households. In the case of the systems presented respectively by Bommi et al. [6] and Liu et al. [11], their main disadvantage is the high investment required. In addition, the laser used to obtain oxygen in the former must be activated manually when the CO value exceeds a threshold value, while the latter is intended for smart buildings. Moreover, the analysed systems, except that of Bommi et al. [6], do not have adequate mechanisms to try to improve air quality. Another disadvantage of all the reviewed works [6, 11–14] is the limitation of the means used to send notifications and alerts on air quality, since they only do so through the mobile application. As notifications are sent by email in the system proposed by Azma et al. [14], these are not very effective in emergency situations.

To counteract the disadvantages of the mentioned systems, we have developed a proof of concept for IdeAir, which is a low-cost IoTS. IdeAir consists of: (1) a device to be deployed inside the home to detect and control the pollution, in addition to emit notifications and alerts, (2) a mobile application to configure the device, view real-time data and receive notifications and alerts on pollution levels, and (3) a web application that, in addition to having the same functions as the mobile application, serves to view reports of the data captured by the system and additional information from IdeAir.

3 Proposed System

The system we propose, called IdeAir, in addition to providing information on air quality levels, alerts users in several ways, depending on the detected air pollution levels. For this work, four levels of CO have been considered to determine air quality, as shown in Table 1. In addition, for each level, the system displays the corresponding message on its LCD screen.

As shown in Table 1, the highest level (4), with a concentration above 800 ppm (parts per million), has been considered the most dangerous. This table also shows the notifications and actions that have been implemented in IdeAir for the four air quality levels considered.

Table 1. Air quality levels.

Level	Quality	Concentration (ppm)	Notifications (light)	Actions
1	Good	≤ 350	Green	No sound or action
2	Moderate	(350, 500]	Blue	Intermittent beeping sound
3	Low	(500, 800]	Orange	Medium and constant sound, open the window
4	Poor	> 800	Red	Loud and constant sound, open the window, turn on the fan, notification in the mobile app

IdeAir is an IoTS developed to support low-income families who may be at risk due to the use of (space and water) heaters or other similar equipment.

3.1 Development Methodology

IdeAir was developed as a proof of concept, to verify its functionality and effectiveness. The development methodology used was TDDM4IoTS, which specifies 4 roles to be played by the members of the project team [9]. In our case, the *project facilitator* role was played by an expert in IoTS development. The *development team* was made up of 2 developers with experience in developing both IoTSs and traditional information systems. The *counsellor* role was played alternatively by the two developers, depending on the mastery of each one on the topic to be applied (hardware, graphical user interfaces, web application, mobile application, web services, database, etc.). The aim of playing this role in this way is to try to balance the mastery of the topics needed in IoTS development. The role of *users/clients* was played by people living in the city of Quito. Therefore, real user requirements were taken into account.

Communication between project members was 90% telematic, as they were geographically dispersed. Meetings were held weekly by videoconference, and both consultations and advice as well as the delivery of software were provided by telematic means. GitHub was used for software version control. Hardware was provided through sporadic face-to-face meetings with the project facilitator or was directly ordered from suppliers via the corresponding online purchasing platform.

Project team members tried to carry out the activities involved in the 11 phases considered in the development methodology, in the order specified in TDDM4IoTS [9]. The execution of the activities and their results are shown in the following subsection.

3.2 Results of Applying TDDM4IoTS to This Case Study

This subsection details the work carried out in each of the development phases specified by TDDM4IoTS [9] in order to obtain the final product.

Preliminary analysis. Although the IoTS developed in this project is a system that can be considered relatively small, it is also necessary to perform the preliminary analysis phase in order to have a solid starting point. This phase, which allows knowing the initial conditions of the environment in which the system will be deployed as well as determining whether it is feasible to meet the requirements demanded by the user, involves the following activities:

Requirements Analysis. It is necessary to have a safe indoor environment at home. Poor installations, daily use or sudden breakdowns of (space and water) heaters can cause irregular amounts of toxic gases to be produced due to poor combustion. When it comes to the home environment, these problems can cause damage to the health of its inhabitants, and even their death. People need to stay at home resting, doing housework and/or working (as it happened in times of pandemic). To achieve peace of mind for users, the concentration of harmful gases should be determined from the concentration of PM2,5 and PM10, and users should be made aware as soon as possible of the danger to which they are exposed so that they can take appropriate precautions, if necessary.

Formal and informal interviews with a group of people living in Quito served to be aware of the existing problem in their households. After that, the requirements for IdeAir were established, among which are the following ones:

- Detecting harmful gases, in particular: CO, CO_2, C_2H_5OH, C_6H_6, NO_2, and NH_3.
- Alerting users who are inside the home.
- Notifying users who are away from home.
- Causing the detected gases to be dissipated into the environment and/or expelled to the outside.

Once the system requirements are known, it can be determined how to satisfy them through an appropriate solution.

Technology analysis. All the points raised by the methodology in this activity were answered positively. In other words, the necessary hardware resources were available. These resources are cheap and effective in fulfilling their functions. The software tools needed for the configuration of the IoTS hardware are available to everyone and free of charge, such as Arduino IDE and TDDT4IoTS (Test-Driven Development Tool for IoTS) [15].

In this methodology activity, the different existing options for both hardware and software components were compared. The main characteristics that determined which ones to use for this project were: popularity of use, integration between hardware and software, cost, functionalities, learning curve, and mastery by the development team.

The tools selected for the development of this case study were: Apache Netbeans IDE with Java, for the development of the web application; TDDT4IoTS, for the design and generation of models and code for both (web and mobile) applications, as well as for the hardware design and code generation for its configuration (it can record code on the Arduino board); Arduino IDE, to compile and complete the editing of the programming code for the hardware configuration; Android Studio, to complete, compile and generate the *apk* file of the application for the Android operating system; MySQL, as a database; and WebHost, as a web and database server.

Environment analysis. The device will be deployed in a low-income household. It will be installed in a strategic place near a source of electricity. In these times of pandemic, families were forced to contract internet services for students' homework and even for adults' teleworking, so they have this service via Wi-Fi. IdeAir will use this resource to send data to the cloud in real time, using web services. The user will be notified the instant the gas concentration changes. As the environment is silent, one way to alert users in the home will be through sounds, in addition to sending notifications to smartphones that have the mobile application installed.

Feasibility analysis. The study carried out determined the feasibility of the development of IdeAir in the three aspects considered:

- *Technical*: The skills and competencies of the development team were considered sufficient, given their previous experience. The technologies necessary for the development of the system exist in the market and are easy to acquire. And the estimated development time was considered adequate by the users.
- *Economical*: The hardware components to be used are low-cost and available on the local market.
- *Operational*: Once the system is deployed, the user does not need any technical knowledge to keep the system running.

Technology layer design. In this phase, the architecture of the IdeAir system was first obtained, which consists of 3 layers: (1) The top layer is the user interaction layer, where users interact with the web and/or mobile application. (2) The lower layer is the physical layer, which is made up of sensors and actuators, being responsible for controlling the environment. (3) The intermediate layer is the connectivity layer, which processes and stores the information, and uses the WebHost service to send data in real time to be stored in the MySQL database and be displayed on the (web and mobile) applications.

Figure 1 shows the device design (corresponding to the physical layer) to detect the level of air quality and act appropriately (according to its configuration) to alert the inhabitants of the house and safeguard their health when the air quality is not adequate. The diagram in Fig. 1, which was made using TDDT4IoTS [9], shows the device components, which are: (1) Arduino Uno board, configured to capture the data emitted by the MQ135 and MQ7 sensors, as well as to control (by sending commands) the actuators; (2) active buzzer, for audible notifications; (3) LEDs, for light notifications; (4) ESP3286 NodeMCU, development board to control the DHT11 sensor and for internet connectivity; (5) DHT11 sensor, to measure the temperature and humidity of the environment; (6) MQ135 sensor, to detect CO_2, C_2H_5OH, C_6H_6, NH_3 and NO_2; (7) MQ7 sensor, to detect CO; (8) servomotors, to open and close the window; (9) LCD display, to report on indoor air quality; and (10) fans, to expel polluted air.

Detailed requirements analysis. In this phase, extended use cases [16] were used, as recommended by TDDM4IoTS [9]. Use cases have proven to be a suitable tool for collecting the functional requirements of the system expressed by the users [17]. By doing a thorough analysis of each use case, and using TDDT4IoTS correctly, the development of an IoTS will be easier.

Generation of models, tests and software. In these 3 phases of TDDM4IoTS, the TDDT4IoTS development tool [15] was used. This tool, which is in its first version, can generate even the web application software, including its preliminary graphical user interfaces, from the use cases. It also allows the design of IoT devices and automatically

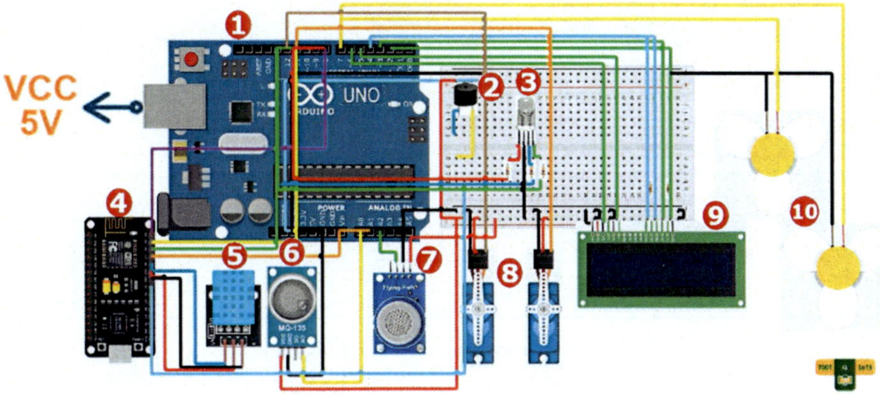

Fig. 1. IdeAir device design.

generates most of the software required for their configuration and operation (currently with support for Arduino). At the moment, TDDT4IoTS does not generate the mobile application. Therefore, the software for the graphical user interfaces of the IdeAir mobile app (front-end) was created entirely by the developers.

Regarding the web application, TDDTIoTS uses the model-view-controller pattern to generate source code from the class diagram. In addition, part of the business logic is transformed into classes to generate and publish the RESTful web services, which will be used to execute the CRUD (Create, Read, Update, and Delete) basic operations with the database.

Refinement and adaptation of models, tests and software. During model refinement, certain methods were added, such as constructors and other methods not specified in the use cases, as well as a class to specify air quality reference levels.

Figure 2 shows part of the software architecture (the class diagram), after being generated by TDDT4IoTS and refined for the generation of the IdeAir software and its tests. The automatically generated tests were refined and adapted to the needs of the end users. Integration tests were written by the developers in the detailed requirements analysis phase and executed in the deployment to verify their compliance. To test the correct functioning of the device before its deployment, the environment was contaminated by generating appropriate gases, basically to check the capture and sending of data in real time.

In addition, some methods specified in the class diagram had to be implemented by the developers, as their business logic could not be automatically generated in its entirety from the use cases.

Hardware and software deployment. The IdeAir device was deployed in a mock-up of a room that simulated a real room, in order to have a prototype of how the system would work. In it, low-power servomotors were used, which are not sufficient to drive a real window. With this proof of concept, the authors consider that there is sufficient evidence to determine that this system meets the requirements set by the client.

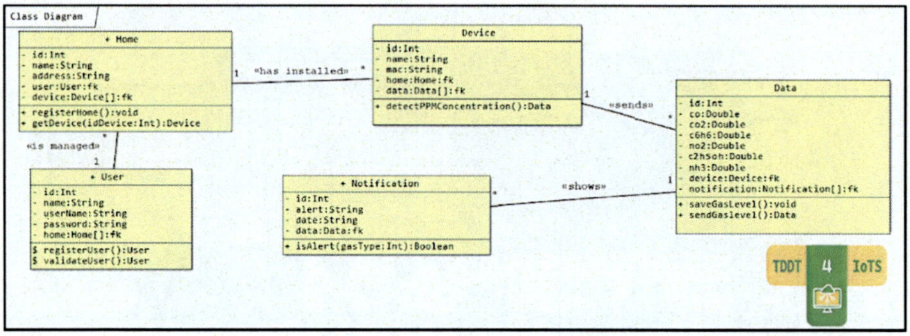

Fig. 2. IdeAir refined class diagram.

The mobile application allows registering the IdeAir device(s) installed to control it/them, and thus monitor data capture in real time and receive notifications when pollutant gas levels are harmful to health, as well as perform common operations, such as user registering, logging in and displaying information about the system. Figure 3 shows some screenshots of the mobile application. When the user is authenticated in the mobile application, s/he can access her/his information and options to view the installed devices and captured data, as shown in Fig. 3(a). To register a new device, it can be done through the Device option, which will display a screen like the one shown in Fig. 3(b). Once at least one IdeAir device is registered and the system is up and running, the user can monitor the indoor quality air of the environment in which the device is installed, since s/he will be able to view the data that the device is capturing and sending to the server in real time, as shown in Fig. 3(c).

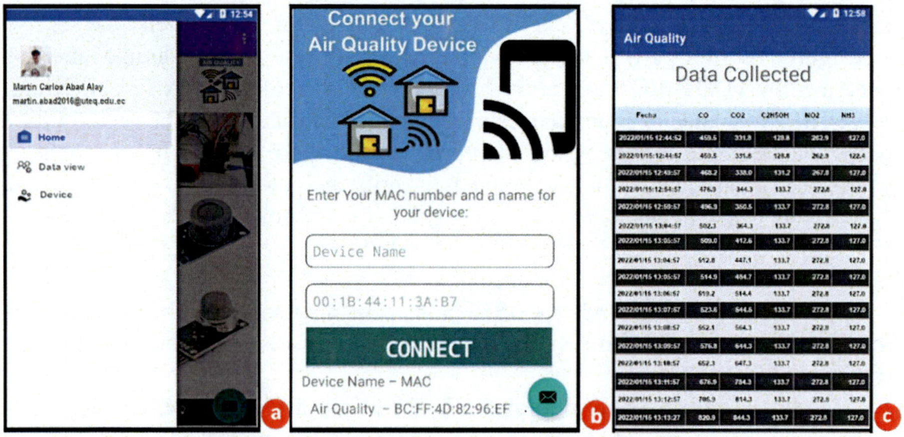

Fig. 3. Screenshots of the mobile application.

4 Conclusions and Future Work

We have presented the implementation of a proof of concept of a low-cost IoTS, called IdeAir, which makes it possible to determine the indoor air quality, and to act and alert people when its level of toxicity is dangerous. The use of web services and a WebHost server allows the environmental data captured by IdeAir to be known in real time. When testing IdeAir, users commented that they found it more useful to hear the alarm sound and see the light notifications in the room to alert them about air toxicity levels than the notifications sent to the mobile app.

Following the TDDM4IoTS methodology made developers more orderly in carrying out the activities to develop this IoTS. In addition, the use of the TDDT4IoTS tool helped them to be more productive. In fact, the code generated by the tool was fully used. Another advantage of using this tool is that it saves much of the time spent writing tests, which is a neuralgic part of the Test-Driven Development methodologies, being considered as "wasted time" by most developers.

As future work, after evaluating its functions as a proof of concept, a prototype of the IdeAir system will be developed and deployed to test its operation in a real home. Furthermore, we plan to further improve the TDDT4IoTS tool, especially regarding the generation of both more code and the user interfaces for the mobile application (for both Android and iOS).

References

1. Ezeonyejiaku, C., Okoye, C., Ezeonyejiaku, N., Obiakor, M.: Air quality in Nigerian urban environments: a comprehensive assessment of gaseous pollutants and particle concentrations. Curr. Appl. Sci. Technol. **22**(5) (2021)
2. World Health Organization: WHO Global Air Quality Guidelines. Particulate Matter (PM2.5 and PM10), Ozone, Nitrogen Dioxide, Sulfur Dioxide and Carbon Monoxide (2021). https://apps.who.int/iris/bitstream/handle/10665/345329/9789240034228-eng.pdf. Last accessed 2021/11/26
3. Gazis, A., Katsiri, E.: Smart home IoT sensors: principles and applications—a review of low-cost and low-power solutions. Int. J. Eng. Technol. Inform. **2**(1), 19–23 (2021)
4. Gobierno de Mendoza: Instant Heaters are the Main Cause of Carbon Monoxide Poisoning (in Spanish). https://www.mendoza.gov.ar/defensacivil/noticias/los-calefones-instantaneos-son-la-principal-causa-intoxicacion-por-monoxido-de-carbono/. Last accessed 2022/03/14
5. National Risk and Emergency Management Service (Servicio Nacional de Gestión de Riesgos y Emergencias): SGR Advises on the Use of Space Heaters (in Spanish). https://www.gestionderiesgos.gob.ec/sgr-aconseja-sobre-el-uso-de-calefones/. Last accessed 2021/11/15
6. Bommi, R.M., Monika, V., ArockiaKoncy, A.A., Patra, C.: A surveillance smart system for air pollution monitoring and management. In: Hemanth, J., Fernando, X., Lafata, P., Baig, Z. (eds.) Lecture Notes on Data Engineering and Communications Technologies, pp. 1407–1418. Springer, Cham (2019)
7. Zhou, M., Abdulghani, A.M., Imran, A.M., Abbasi, Q.H.: Internet of Things (IoT) enabled smart indoor air quality monitoring system. In: International Conference on Computing, Networks and Internet of Things, pp. 89–93. ACM, Sanya, China (2020)
8. Rodríguez, A.: A Heater in Bad Condition Caused a Tragedy in the Town Committee. Five Members of a Family Died (in Spanish). https://www.elcomercio.com/actualidad/quito/intoxicacion-monoxido-carbono-muerte-quito.html. Last accessed 2021/11/17

9. Guerrero-Ulloa, G., Hornos, M.J., Rodríguez-Domínguez, C.: TDDM4IoTS: a test-driven development methodology for Internet of Things (IoT)-based systems. Commun. Comput. Inform. Sci. **1193**, 41–55 (2020)
10. Shrestha, P.M., et al.: Impact of outdoor air pollution on indoor air quality in low-income homes during wildfire seasons. Int. J. Environ. Res. Public Health **16**(19), 3535–3555 (2019)
11. Liu, Z., Wang, G., Zhao, L., Yang, G.: Multi-points indoor air quality monitoring based on Internet of Things. IEEE Access **9**, 70479–70492 (2021)
12. Kotsev, A., Schade, S., Craglia, M., Gerboles, M., Spinelle, L., Signorini, M.: Next generation air quality platform: openness and interoperability for the Internet of Things. Sensors **16**(3), 403–418 (2016)
13. Taştan, M., Gökozan, H.: Real-time monitoring of indoor air quality with Internet of Things-based e-nose. Appl. Sci. **9**(16), 3435–3447 (2019)
14. Azma Zakaria, N., Zainal Abidin, Z., Harum, N., Chen Hau, L., Salih Ali, N., Azni Jafar, F.: Wireless Internet of Things-based air quality device for smart pollution monitoring. Int. J. Adv. Comput. Sci. Appl. **9**(11), 65–69 (2018)
15. Guerrero-Ulloa, G., Carvajal-Suárez, D., Brito-Casanova, G., Pachay-Espinoza, A., Hornos, J.M., Rodríguez-Domínguez, C.: Test-Driven Development Tool for IoT-based System. https://aplicaciones.uteq.edu.ec/tddt4iots/. Last accessed 2022/01/02
16. Usländer, T., Batz, T.: How to analyse user requirements for service-oriented environmental information systems. In: Hřebíček, J., Schimak, G., Denzer, R. (eds.) Environmental Software Systems. Frameworks of eEnvironment. ISESS 2011. IFIP Advances in Information and Communication Technology, vol. 359, pp. 161–168. Springer, Berlin, Heidelberg (2011)
17. Lee, S.C., Nadri, C., Sanghavi, H., Jeon, M.: Eliciting user needs and design requirements for user experience in fully automated vehicles. Int. J. Human-Comput. Interact. **38**(3), 227–239 (2022)

Automated Counting via Multicolumn Network and CytoSMART Exact FL Microscope

Sebastián López Flórez[1,2](✉), Alfonso González-Briones[1,2], Guillermo Hernández[1,2], and Fernando de la Prieta[1,2]

[1] Bisite Research Group, University of Salamanca, Edificio Multiusos I+D+i, Calle Espejo 2, 37007 Salamanca, Spain
alfonsogb@usal.es, guillehg@usal.es, fer@usal.es

[2] Universidad Tecnológica de Pereira Cra, 27 N 10-02 Pereira, Risaralda, Colombia
sebastianlopezflorez@usal.es,sebastianlopezflorez@utp.edu.co

Abstract. Neubauer chamber cell counting in microbiological culture plates is a laborious task that relies on technical expertise. As a result, efforts have been made to advance computer vision based approaches, increasing efficiency and reliability by quantitatively analyzing microorganisms and calculating their characteristics, biomass concentration and biological activity. However, the variability that still persists in these processes poses a challenge that is yet to be overcome. In this paper, a solution is proposed that adopts convolutional neural networks for automatic cell counting. The algorithm seeks to identify the characteristics of cells of va rious sizes using a multi-column network where there are general convolutional layers in the body of U-net. Furthermore, the solution has been implemented in the laboratory using CytoSMART Exact FL microscope images. The results show that the proposed method can handle different types of images with promising accuracy.

Keywords: Deep learning · Convolutional neural networks · Microscopic images · Image segmentation · Cell counting

1 Introduction

The assessment of cell viability in the study of bioengineering, tissue engineering, regenerative medicine, biomedical studies, food safety and drugs [31] plays an essential role in all forms of cell culture and this in turn depends on multiparametric analysis of characteristics depending on the culture process where new machine learning technologies through image processing have made great advances such as estimating the microbial content of cells [5], measuring cytotoxicity [23] discovering the role of particular genes in cell biology, microbiology and

V. Julián et al. (Eds.): ISAmI 2022, LNNS 603, pp. 207–218, 2023.
https://doi.org/10.1007/978-3-031-22356-3_20

immunology [36], although its major contributions are centered on cell counting, this consists of collecting a sample of cell suspension and pipetting a small volume, this is placed in a microcentrifuge tube and diluted in a suspension in blue trypan, living cells having intact cell membranes will exclude the dye, while dead cells will absorb the dye due to altered membrane integrity. The suspension is loaded into a hemocytometer mounted on a microscope. Finally, a technician measures the cell density of the sample and uses it to calculate the number of cells in the suspension, assuming that the cell density of the sample is equal to the suspension. This manual process is tedious and time-consuming, which increases the burden on technicians [19]. Therefore, researchers propose automatic models such as plate counting [51], real-time quantitative PCR [30], hemocytometers [47], automatic cell counting instruments [29] and flow cytometry counting in biological systems.

The task of counting cells in images can be approached in two different directions, first a cell detector and second a cell counter [32]. However, these systems use labels as supervised information, which consist of point annotations that locate the centroids of the cells [14] or bounding box annotations indicating cell borders [3]. Using this methodology in the real environment is a challenge as the variabilities in cell densities, shapes, sizes and colors can be extremely high. A different approach is based on applying semantic segmentation and predicting the spatial density of cells [46] or density maps [52]. So the system to be implemented is based on a convolutional neural network (CNN) semantic segmentation and spatial map predictions to capture cells or groups of cells with different scales, using different convolutional kernels to capture receptive fields with different sizes. To implement this system in a laboratory with images extracted by microbiology specialists with the help of a CytoSMART Exact FL(Fluorescence) microscope that captures the image of the image cells for counting. Cell counting application components are illustrated in Fig. 1.

This document is organized as follows: Sect. 1 is the introduction. Section 2 is the theoretical description of the manual method of how the counting and database is done. Section 3 is the fundamentals of the CNN. Section 4 presents the proposed method. Section 5 is the experimental study and analysis of the results.

1.1 Related Work

The proposed methodology for traceability in cell counting is based on image analysis, electronic and optical technologies [2,12]. Most of these methods are performed on a cell suspension sample. Image analysis techniques are mostly used in practical applications, which in turn encompasses two aspects image processing [1,15,16,37,38,40] and machine learning [6,9,18,24,26,34,53,54]. The advance in industrial computing technologies seeks to contribute and impact with new technologies of the industry 4.0 [4,10,11,13,17,20–22,25,45,48] following this paradigm some researches based on deep learning seek to apply cell counting to their possessions.

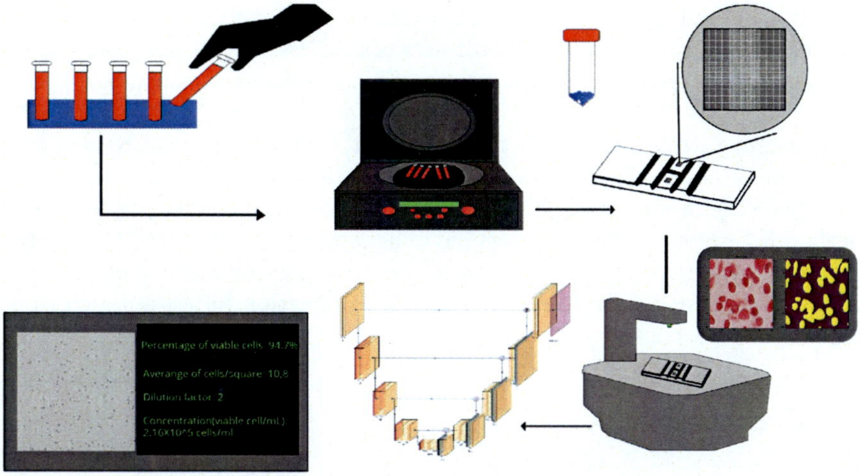

Fig. 1. Schematic diagram of cell counting application components.

1.2 Image Processing

Payasi and Patidar [39], proposed a methodology in which, tuberculosis bacilli diagnoses were processed and counted by converting the RGB image to an HSI color space image. Then the image was segmented according to the threshold. Then the noise was removed, to make it possible to label and count the bacilli in the image. Afterwards, the boundaries of the bacteria were detected and stored and the area and perimeter were calculated. Acharya and Kumar [2] proposed an image processing technique for red blood cell counting. The authors processed the blood smear image to count red blood cells along with the identification of normal and abnormal cells. They used the K-means algorithm to extract white blood cells from the image and granulometric analysis to separate red blood cells from white blood cells and then counted the number of cells using the labeling algorithm and a Circular Hough Transform (CHT).

Clarke et al. [15] proposed a low-cost, high-throughput colony counting system consisting of software that utilizes optical imaging techniques and uses extended minima function to find the center of colonies. Kaur et al. [28] proposed a method to count platelets automatically by applying a CHT on a microscopic image of blood cells. They used the platelet size and shape characteristics of the CHT in the counting process.

1.3 Deep Learning

Ferrari et al. [18] proposed a solution that adopts convolutional neural networks (CNN) to count the number of colonies contained in confluent agglomerates, applied data augmentation from horizontal spins, and applied three different artificial color distortions in RGB color space for optimal performance. Zhao et al. [54] proposed an automatic white blood cell identification and classification system using a convolutional neural network (CNN). First, they detected white blood cells from microscopic images, and then CNN was obtained to detect white blood cell types. Habibzadeh et al. [24] pretrained CNNs, ResNet and Inception Net,t o count white blood cells from segmented images. Images were segmented using color space analysis. Tamiev et al. [49] proposed a convolutional neural network (cCNN) for automatic bacterial classification and counting, and an efficient method for preprocessing microscopic images was presented. In [41] designed an imaging platform equipped with an on-board cell recognition function that has been obtained by developing a smartphone application for automatic cell counting with high accuracy.

2 Methods Network

The proposed method consists of two elements: image capture through a CytoSMART Exact FL microscope with open API for cells in the Neubauer plate. Since there are several types of Neubauer plates, it was necessary to search for an additional database in order to give robustness to the model, and then implement the images in laboratory conditions.

2.1 Handling of the Neubauer Cell Counting Chamber

Counting chamber is a precision system made of special optical glass. It is used to count cells or other particles in suspension, under the microscope, mainly in blood analysis, counting cells in liquor, bacteria, sperm and fungus spores. The counting grid consists of 9 large squares, each with an area of 1 mm^2. A series of steps are followed to load the mixture: (1) The pipette is removed from the stirrer and the first drops are ejected. (2) The outside of the pipette is cleaned and dried and held at an angle until a small drop has formed on the tip. This drop is placed at the point between the cover slip and the counting chamber. (3) By capillary effect the space between the cover slip and the bottom of the chamber is filled. Before the solution can overflow this part of the chamber, it must be removed. If air bubbles are present or if liquid has overflown into the slots, the chamber has to be cleaned and reloaded.

The different counting chamber systems differ in the pattern of the counting reticle and the depth of the chamber. The reticle is composed of a square division that can only be seen under the microscope the most commonly used counting chamber [8].

The microscope used to capture the information presented on CytoSMART Exact FL neubauer cameras. Using the 6.4 MP CMOS camera combined with 10× magnification, the CytoSMART Exact FL can view and count cells down to 4 μm in diameter [44].

2.2 Data Set

This research has used the 2018 Data Science Bowl Grand Challenges [7] dataset as the base learning, which contains 735 images in total. From the total image set, 650 images contain pixel-level annotations for training and the remaining 65 samples are unlabeled for use in testing. From the training set, 80% of the samples have been used for training and the remaining 20% have been used for validation. The numbers of training and validation samples were 536 and 134, respectively. In that 80% training set the matrices are divided into 90% training and 10% test random subsets to feed the model in generator mode.

In the case of the proprietary base, images have been captured with the cell microscope on the Neubauer plate in the laboratory in order to observe the performance of the system in a real environment.

2.3 The CNN Fundamentals

The architecture of the network is illustrated in Fig. 2. It consists of a semantic segmentation network based on an encoder (left side) and a decoder (right side). The encoder follows the typical architecture of a convolutional network. It consists of the repeated application of two 3×3 convolutions, each followed by a rectified linear unit (ReLU) and a maximum 2×2 grouping operation with step 2 for resolution reduction. Each step in the decoder network in an up-sampling of the function map followed by a 2×2 convolution that halves the number of function channels, a concatenation with the function map correspondingly trimmed from the shrinking path, and two 3×3 convolutions, each followed by a ReLU. The clipping is necessary due to the loss of edge pixels in each convolution. In the final layer, a 1×1 convolution is used to map each 64-component feature vector to the desired number of classes. In total the network has 23 convolutional layers [42].

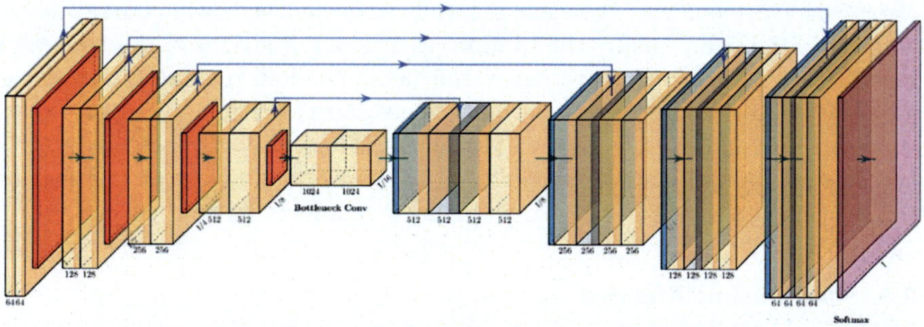

Fig. 2. The general vision of architecture.

Cross Entropy Loss This pixel-wise cross-entropy loss, which allows to change the unbalanced classes in the semantic segmentation, is described as follows:

$$L = \sum_{j=1}^{C} W(x) log \frac{\exp(\hat{y}_{y(x)}(x))}{\sum_{c=0}^{C} \exp(\hat{y}(x))} \tag{1}$$

where x is a pixel in the image domain, $\hat{y}$ is the predicted score for class C $\in \{0, \cdots, C\}$ is the number of classes y is the absolute truth segmentation map. Therefore, $\hat{y}_{y(x)}(x)$ the predicted score for the absolute $y(x)$ truth class at position x. The parameter is the weight loss map per pixel.

The weight of the pixel x is calculated by considering two important characteristics, the first part regions by using the weight map w_{bal} considers class imbalance. The second part, applies the separation of instances, but using the weight w_{sep} considers the labeling uncertainty, where weight for each pixel is used in the loss function according: $W = w_{bal} - \lambda w_{sep}$ where λ controls the importance of instance separation. Once the density map is generated, following the previous work [27], a normalized Gaussian function [33] G is taken to smooth the points and the cell labels become continuous, this is defined as:

$$L = \sum_{i=1}^{N} G(x_i; \alpha) \tag{2}$$

where x_i is the i-th cell point label, α controls how to smooth the points, $G(x_i; \alpha)$ [35] represents the normalized Gaussian function on x_i with N number of maps. The density map indicates how many cells per pixel it contains, which means that the integral over the entire image corresponds to the number of cells.

2.4 Data Augmentation

Data is augmented to enable the network to generalize cases that are novel and have few training samples. In the case of microscopic images, it is necessary to apply displacement and rotation invariance, as well as scale changes and gray value variations. Especially the random elastic deformations of the training samples. Subtle deformations are generated using random displacement vectors on a coarse grid of 3×3. The displacements are sampled from a Gaussian distribution with a standard deviation of 10 pixels. Then, the displacements per pixel are calculated by bicubic interpolation. The dropout layers at the end of the contracting path realize a higher implicit data augmentation [43].

2.5 Evaluation Metrics

Intersection over union IoU measures how well a predicted segmentation matches the corresponding true label annotation by dividing the intersection of two segments by their union. The performance evaluation metric is defined as the ratio of the overlap area and the area of the union. The overlap area is the total common area the intersection between the predicted bounding box Bp and the true label bounding box Bgt. The area of the union is the total area covered by Bp and Bgt according to equation.

$$IoU = \frac{area(B_{gt} \bigcap b_p)}{area(B_{gt} \bigcup b_p)} \tag{3}$$

3 Experiments and Results

The obtained results are summarized in Fig. 3, which represents the training curve according to the intersection over union (IoU) metric of the test with respect to the number of iterations. We demonstrate the application of U-net on a semantic segmentation task in cell counting. The first task is the segmentation of structures in electronic microscopic records obtained results. From the 2018 Data Science Bowl Grand Challenges database. λ controls the importance of the separation of instances in our case was 0.9 and α is 0.5. From Fig. 3, it can be seen that the model converged after 14 epochs. Training and evaluation continued to complete 24 epochs to ensure optimal convergence. From this experiment, a test accuracy of approximately 0.86% has been achieved in kernel segmentation with the U-Net model.

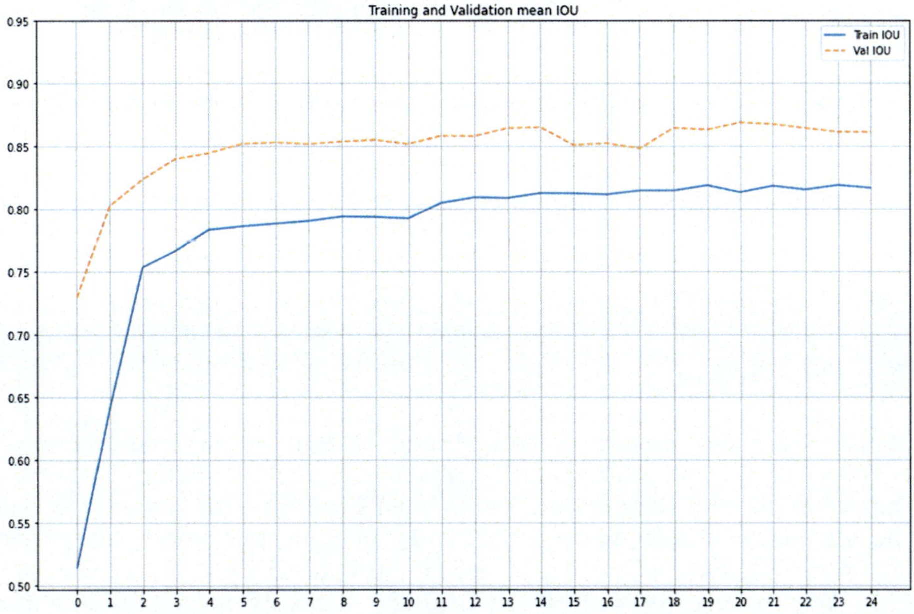

Fig. 3. Validation in terms of intersection over the junction (IoU)[50].

4 Analysis

Figure 4 shows some classification samples when using the U-Net model for kernel segmentation to represent the network outputs. According to these results, the proposed segmentation model displayed promising performance during the testing phase, since these are high novelty images, samples were extracted from the lab with nubar cameras different from the training database images. Furthermore, if we look at the input samples there is a strong separation between nuclei and similar objects of lower interest achieving segmentation of the samples of interest.

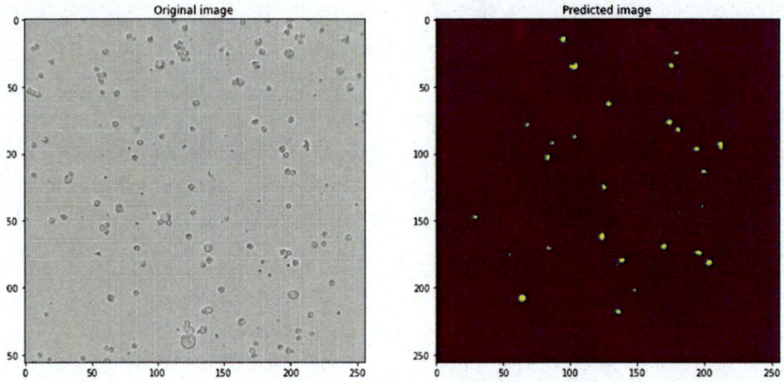

Fig. 4. Prediction result for the image originating from the laboratory.

This project seeks to implement a U-net architecture for CytoSMART Exact FL images in such a way that the model output and the density map is generated to perform the count, therefore, a test to evaluate the results of some images taken by the CytoSMART Exact FL is visualized according to Fig. 5.

5 Conclusion

In this paper, we demonstrate the application of deep learning model for semantic segmentation. We evaluated the multi-column, multi-resolution U-net-based model on the 2018 Data Science Bowl Grand Challenge 2018 dataset achieving an IoU metric of approximately 0.85%. In addition, the experimental results illustrate excellent qualitative performance, as well as adequate performance on real data implemented with the CytoSMART Exact FL microscope in semantic segmentation of some cells it is worth noting that these systems have been improved upon implementing and investigating different deep learning models for the classification and detection of nuclei in digital pathological images, which has enabled the laboratory to respond to the different needs that are structured

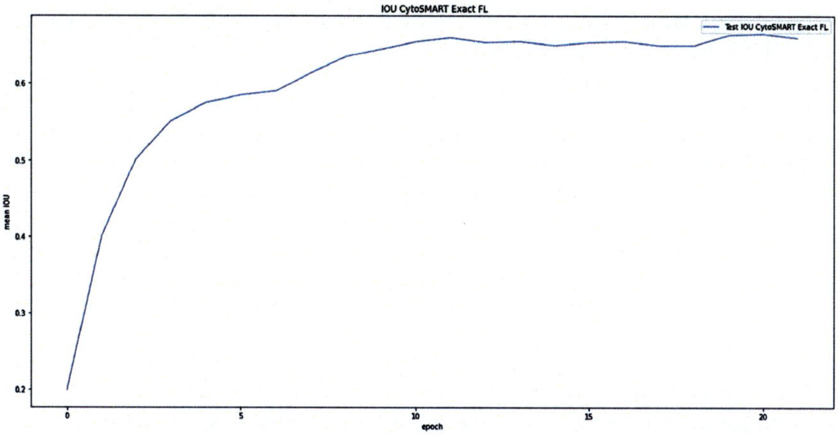

Fig. 5. Test in terms of intersection over the junction (IoU).

in the variability of the studied cells, where the model is able to perform optimally in the presence of new cells. Not only can we obtain the result of the estimated cell count, but also the positions.

As future work it is necessary to address the initial hypothesis relating the deep learning techniques applied in this problem so that a better generalization to novelty can be obtained, it is proposed to explore the Transformer models and evaluate their performance with different more complex scenarios.

Acknowledgements. This work has been supported by the project "XAI - XAI - Sistemas Inteligentes Auto Explicativos creados con Módulos de Mezcla de Expertos", ID SA082P20, financed by Junta Castilla y León, Consejería de Educación, and FEDER funds.

References

1. Acharya, V., Kumar, P.: Identification and red blood cell automated counting from blood smear images using computer-aided system. Med. Biol. Eng. Comput. **56**(3), 483–489 (2018)
2. Aijaz, A., Trawinski, D., McKirgan, S., Parekkadan, B.: Non-invasive cell counting of adherent, suspended and encapsulated mammalian cells using optical density. BioTechniques **68**(1), 35–40 (2020)
3. Akram, S.U., Kannala, J., Eklund, L., Heikkilä, J.: Cell segmentation proposal network for microscopy image analysis. In: Deep Learning and Data Labeling for Medical Applications, pp. 21–29. Springer, Heidelberg (2016)
4. Alonso, R.S.: Deep symbolic learning and semantics for an explainable and ethical artificial intelligence. In: International Symposium on Ambient Intelligence, pp. 272–278. Springer, Heidelberg (2020)
5. Anderson, M., Hinds, P., Hurditt, S., Miller, P., McGrowder, D., Alexander-Lindo, R.: The microbial content of unexpired pasteurized milk from selected supermarkets in a developing country. Asian Pacific J. Tropical Biomed. **1**(3), 205–211 (2011)

6. Basarslan, M.S., Kayaalp, F., et al.: Sentiment Analysis with Machine Learning Methods on Social Media (2020)
7. Caicedo, J.C., Goodman, A., Karhohs, K.W., Cimini, B.A., Ackerman, J., Haghighi, M., Heng, C., Becker, T., Doan, M., McQuin, C., et al.: Nucleus segmentation across imaging experiments: the 2018 data science bowl. Nat. Methods **16**(12), 1247–1253 (2019)
8. Cardona-Maya, W., Berdugo, J., Cadavid, A.: Comparación de la concentración espermática usando la cámara de makler y la cámara de neubauer. Actas urológicas españolas **32**(4), 443–445 (2008)
9. Carneiro, J., Meira, J., Novais, P., Marreiros, G.: Using machine learning to predict the users ratings on tripadvisor based on their reviews. In: Practical Applications of Agents and Multi-Agent Systems, pp. 127–138. Springer, Heidelberg (2021)
10. Carolis, B.D., D'Errico, F., Paciello, M., Palestra, G.: Cognitive emotions recognition in e-learning: exploring the role of age differences and personality traits. In: International Conference in Methodologies and intelligent Systems for Technology Enhanced Learning, pp. 97–104. Springer, Heidelberg (2019)
11. Casado-Vara, R., Novais, P., Gil, A.B., Prieto, J., Corchado, J.M.: Distributed continuous-time fault estimation control for multiple devices in IoT networks. IEEE Access **7**, 11972–11984 (2019)
12. Chakraborty, S., Das, C., Ghoshal, K., Bhattacharyya, M., Karmakar, A., Chattopadhyay, S.: Low frequency impedimetric cell counting: analytical modeling and measurements. IRBM **41**(1), 23–30 (2020)
13. Chmielarz, B., Herpson, C., Bredimas, A., Guessoum, Z.: Auction mechanisms for management of steam generated by high temperature reactors. In: International Conference on Practical Applications of Agents and Multi-Agent Systems, pp. 64–76. Springer, Heidelberg (2021)
14. Cireşan, D.C., Giusti, A., Gambardella, L.M., Schmidhuber, J.: Mitosis detection in breast cancer histology images with deep neural networks. In: International Conference on Medical Image Computing and Computer-Assisted Intervention, pp. 411–418. Springer, Heidelberg (2013)
15. Clarke, M.L., Burton, R.L., Hill, A.N., Litorja, M., Nahm, M.H., Hwang, J.: Low-cost, high-throughput, automated counting of bacterial colonies. Cytometry Part A **77**(8), 790–797 (2010)
16. Costa, A.C., et al.: Elements for the Agent-Based Modeling of Slavery Systems (2020)
17. Duque-Méndez, N.D., Tabares-Morales, V., Ovalle, D.A.: Intelligent agents system for adaptive assessment. In: International Conference in Methodologies and intelligent Systems for Technology Enhanced Learning, pp. 164–172. Springer, Heidelberg (2019)
18. Ferrari, A., Lombardi, S., Signoroni, A.: Bacterial colony counting by convolutional neural networks. In: 2015 37th Annual International Conference of the IEEE Engineering in Medicine and Biology Society (EMBC), pp. 7458–7461. IEEE (2015)
19. Gallego Albiach, V., Pérez Igualada, L.M.: Estimación de la densidad celular mediante el uso de cámaras de recuento (2021)
20. García-Retuerta, D.: Clifford algebras: a proposal towards improved image recognition in machine learning. In: International Symposium on Ambient Intelligence, pp. 258–261. Springer, Heidelberg (2020)
21. Gasperis, G.D., Mascio, T.D., Tarantino, L.: Toward a cognitive robotics laboratory based on a technology enhanced learning approach. In: International Conference in Methodologies and Intelligent Systems for Techhnology Enhanced Learning, pp. 101–109. Springer, Heidelberg (2017)

22. Gonçalves, C., Rebelo, D., Silva, F., Analide, C.: Sleep performance and physical activity estimation from multisensor time series sleep environment data. In: International Symposium on Ambient Intelligence, pp. 166–176. Springer, Heidelberg (2020)
23. Gray, T.E., Thomassen, D.G., Mass, M.J., Barrett, J.C.: Quantitation of cell proliferation, colony formation, and carcinogen induced cytotoxicity of rat tracheal epithelial cells grown in culture on 3t3 feeder layers. In Vitro **19**(7), 559–570 (1983)
24. Habibzadeh, M., Jannesari, M., Rezaei, Z., Baharvand, H., Totonchi, M.: Automatic white blood cell classification using pre-trained deep learning models: Resnet and inception. In: Tenth International Conference on Machine Vision (ICMV 2017), vol. 10696, pp. 274–281. SPIE (2018)
25. Hernández, G., Rodríguez, S., González, A., Corchado, J.M., Prieto, J.: Video analysis system using deep learning algorithms. In: International Symposium on Ambient Intelligence, pp. 186–199. Springer, Heidelberg (2020)
26. Jasim, Y.A., et al.: High-performance Deep Learning to Detection and Tracking Tomato Plant Leaf Predict Disease and Expert Systems (2021)
27. Jiang, N., Yu, F.: Multi-column network for cell counting. OSA Continuum **3**(7), 1834–1846 (2020)
28. Kaur, P., Sharma, V., Garg, N.: Platelet count using image processing. In: 2016 3rd International Conference on Computing for Sustainable Global Development (INDIACom), pp. 2574–2577. IEEE (2016)
29. Kim, S.I., Kim, H.J., Lee, H.J., Lee, K., Hong, D., Lim, H., Cho, K., Jung, N., Yi, Y.W.: Application of a non-hazardous vital dye for cell counting with automated cell counters. Anal. Biochem. **492**, 8–12 (2016)
30. Knight, G.M., Dyakova, E., Mookerjee, S., Davies, F., Brannigan, E.T., Otter, J.A., Holmes, A.H.: Fast and expensive (pcr) or cheap and slow (culture)? a mathematical modelling study to explore screening for carbapenem resistance in UK hospitals. BMC Med. **16**(1), 1–11 (2018)
31. Langer, R.S., Vacanti, J.P.: Tissue engineering: the challenges ahead. Sci. Am. **280**(4), 86–89 (1999)
32. Lavitt, F., Rijlaarsdam, D.J., van der Linden, D., Weglarz-Tomczak, E., Tomczak, J.M.: Deep learning and transfer learning for automatic cell counting in microscope images of human cancer cell lines. Appl. Sci. **11**(11), 4912 (2021)
33. Li, T.c., Su, J.y., Liu, W., Corchado, J.M.: Approximate gaussian conjugacy: parametric recursive filtering under nonlinearity, multimodality, uncertainty, and constraint, and beyond. Front. Inf. Technol. Electronic Eng. **18**(12), 1913–1939 (2017)
34. Li, T., Chen, H., Sun, S., Corchado, J.M.: Joint smoothing and tracking based on continuous-time target trajectory function fitting. IEEE Trans. Autom. Sci. Eng. **16**(3), 1476–1483 (2018)
35. Li, T., Corchado, J.M., Sun, S.: Partial consensus and conservative fusion of gaussian mixtures for distributed phd fusion. IEEE Trans. Aerosp. Electronic Syst. **55**(5), 2150–2163 (2018)
36. Li, Y., Hetet, G., Maurer, A.M., Chait, Y., Dhermy, D., Briere, J.: Spontaneous megakaryocyte colony formation in myeloproliferative disorders is not neutralizable by antibodies against il3, il6 and gm-csf. Br. J. Haematol. **87**(3), 471–476 (1994)
37. Márquez-Sáncheza, S., Mora-Simonb, S., Herrera-Santosa, J., Roncerod, A.O., Corchadoa, J.M.: Intelligent dolls and robots for the treatment of elderly people with dementia. Adv. Distributed Comput. Artif. Intell. J. **9**(1), 99–112 (2020)
38. Márquez-Sáncheza, S., Mora-Simonb, S., Herrera-Santosa, J., Roncerod, A.O., Corchadoa, J.M.: Intelligent dolls and robots for the treatment of elderly people with dementia. Adv. Distributed Comput. Artif. Intell. J. **9**(1), 99–112 (2020)

39. Payasi, Y., Patidar, S.: Diagnosis and counting of tuberculosis bacilli using digital image processing. In: 2017 International Conference on Information, Communication, Instrumentation and Control (ICICIC), pp. 1–5. IEEE (2017)
40. Pimpalkar, A.P., Raj, R.J.R.: Influence of pre-processing strategies on the performance of ml classifiers exploiting tf-idf and bow features. ADCAIJ: Adv. Distributed Comput. Artif. Intell. J. **9**(2), 49 (2020)
41. Rabha, D., Biswas, S., Hatiboruah, D., Das, P., Rather, M.A., Mandal, M., Nath, P.: An affordable, handheld multimodal microscopic system with onboard cell morphology and counting features on a mobile device. In: Analyst (2022)
42. Ronneberger, O., Fischer, P., Brox, T.: U-net: Convolutional networks for biomedical image segmentation. In: International Conference on Medical Image Computing and Computer-Assisted Intervention, pp. 234–241. Springer, Heidelberg (2015)
43. Ronneberger, O., Fischer, P., Brox, T.: U-net: convolutional networks for biomedical image segmentation. In: International Conference on Medical Image Computing and Computer-Assisted Intervention, pp. 234–241. Springer, Heidelberg (2015)
44. Sanderson, M.J., Smith, I., Parker, I., Bootman, M.D.: Fluorescence microscopy. Cold Spring Harbor Protocols 2014(10), pdb–top071795 (2014)
45. Sati, V., Sánchez, S.M., Shoeibi, N., Arora, A., Corchado, J.M.: Face detection and recognition, face emotion recognition through nvidia jetson nano. In: International Symposium on Ambient Intelligence, pp. 177–185. Springer, Heidelberg (2020)
46. Shelhamer, E., Rakelly, K., Hoffman, J., Darrell, T.: Clockwork convnets for video semantic segmentation. In: European Conference on Computer Vision, pp. 852–868. Springer, Heidelberg (2016)
47. Song, B., Zhuge, B., Fang, H., Zhuge, J.: Analysis of the chromosome ploidy of candida glycerinogenes. Wei Sheng wu xue bao= Acta Microbiologica Sinica **51**(3), 326–331 (2011)
48. Stavropoulos, T.G., Meditskos, G., Papagiannopoulos, S., Kompatsiaris, I.: ehealth4ms: Problem detection from wearable activity trackers to support the care of multiple sclerosis. In: International Symposium on Ambient Intelligence, pp. 3–12. Springer, Heidelberg (2020)
49. Tamiev, D., Furman, P.E., Reuel, N.F.: Automated classification of bacterial cell sub-populations with convolutional neural networks. PloS One **15**(10), e0241200 (2020)
50. Wang, W., Fu, Y., Pan, Z., Li, X., Zhuang, Y.: Real-time driving scene semantic segmentation. IEEE Access **8**, 36776–36788 (2020)
51. Wilson, C., Lukowicz, R., Merchant, S., Valquier-Flynn, H., Caballero, J., Sandoval, J., Okuom, M., Huber, C., Brooks, T.D., Wilson, E., et al.: Quantitative and qualitative assessment methods for biofilm growth: a mini-review. Research & reviews. J. Eng. Technol. **6**(4) (2017)
52. Xie, W., Noble, J.A., Zisserman, A.: Microscopy cell counting and detection with fully convolutional regression networks. Comput. Methods Biomech. Biomed. Eng.: Imaging Visualization **6**(3), 283–292 (2018)
53. Yigitcanlar, T., Kankanamge, N., Regona, M., Ruiz Maldonado, A., Rowan, B., Ryu, A., Desouza, K.C., Corchado, J.M., Mehmood, R., Li, R.Y.M.: Artificial intelligence technologies and related urban planning and development concepts: how are they perceived and utilized in Australia? J. Open Innov.: Technol. Market Complexity **6**(4), 187 (2020)
54. Zhao, J., Zhang, M., Zhou, Z., Chu, J., Cao, F.: Automatic detection and classification of leukocytes using convolutional neural networks. Med. Biol. Eng. Comput. **55**(8), 1287–1301 (2017)

Using Deep Learning for Implementing Paraphrasing in a Social Robot

Laura Martín Galván(✉), Enrique Fernández-Rodicio, Javier Sevilla Salcedo, Álvaro Castro-González, and Miguel A. Salichs

Robotics Lab, Carlos III University of Madrid, Madrid, Spain
100444738@alumnos.uc3m.es, enrifern@ing.uc3m.es, jasevill@pa.uc3m.es, acgonzal@ing.uc3m.es, salichs@ing.uc3m.es

Abstract. Social robotics is intended for interacting and communicating with people. This communication needs to be natural and fluid, similar to a human conversation. For instance, people do not always use exactly the same words when transmitting the same message in different situations. One way to imitate this is through paraphrase generation, allowing the robot to use different sentences while keeping the same meaning. In this paper we propose the application of deep learning models to generate these paraphrases in a social robot speaking in Spanish. This application has been integrated into the Mini robot both in English and Spanish, and can be run locally or from an external server. We integrated and evaluated 3 different models. The T5 and Pegasus Transformers have been considered as a base, and we have chosen three different models finetuned for paraphrase generation, one based on Pegasus, and the other two on T5 (Finetuned T5 and Parrot). Compared to the Parrot and Pegasus models, that provide similar results, the Finetuned T5 model can generate paraphrases that do not vary as much from the original sentence, but keep more of the meaning. However, the main advantage of the Finetuned T5 model is its average execution time, around one second faster than the other considered models.

Keywords: Deep Learning · NLP · Paraphrase generation · Social robotics

1 Introduction

The main purpose of social robotics is to communicate with people and its surroundings, providing an easy way to access information or be entertained. This can be accomplished by using different forms of communication simultaneously, like voice, gestures or images. It is also essential to make sure that the user feels comfortable when interacting with the robot. A friendly appearance can make it more approachable and easier to talk to, and allows the user to get used to

V. Julián et al. (Eds.): ISAmI 2022, LNNS 603, pp. 219–228, 2023.
https://doi.org/10.1007/978-3-031-22356-3_21

it more quickly. Moreover, the interaction needs to be natural and fluid, so it becomes easier for the user to keep the conversation without losing interest.

Nevertheless, reproducing a natural human conversation can be complex. One of the issues we found is that, in many occasions, every sentence the robot says has to be explicitly programmed in the code. This means that the robot is always going to reproduce the same phrases. Specially in some interactions with the user, these sentences can be repetitive and can lead to monotonous interactions that make the user lose interest.

A possible solution to this problem is paraphrasing. Each sentence the robot has to say can be transformed into a paraphrase, this is, a sentence with the same meaning as the original, but constructed differently. This allows the interaction with the robot to seem more diverse and natural. But these generated paraphrases have to meet some requirements. Firstly, for them to be considered acceptable, they have to keep as much meaning as possible from the original sentence, while providing diversity in its form. It is also important to keep the processing times limited for the conversation to keep its coherence and naturalness. Some studies have proposed that the maximum acceptable response time for the robot is two seconds [10]. In this work, we have also considered this threshold.

Lastly, it is important to note the language in which this application is going to be implemented. Most of the available resources are in English, but a solution has to be found in order to do this in other languages, such as Spanish.

The rest of the paper is structured as follows. In Sect. 2, we present a brief overview of the state of the art on paraphrase generation. In Sect. 3, there is a description of the approach followed in this work, explaining the methods and models used. Then, in Sect. 4, we described the integration of this application in a social robot, that we then evaluated in Sect. 5. Lastly, we present an example of this application in use in Sect. 6, and the conclusions of this work in Sect. 7.

2 State of the Art

Paraphrase generation has been a topic of interest for decades, mostly as a means of solving other problems, since it can be an important part of other NLP (Natural Language Processing) tasks, like summarization, question answering or machine translation. Different approaches have been developed over time to solve this task.

In 2005, some early studies have focused on bilingual pivoting [1]. This method consists on translating the sentence to a pivot language, and then translating it back to the target language, generating a paraphrase. This method has also been implemented with the help of neural networks [4].

More recently, significant progress has been made in monolingual paraphrasing using neural networks, where the sequence-to-sequence framework (seq2seq) [11] has been widely used. The models based on this method contain an encoder and a decoder, which are, in most cases, based on convolutional or recurrent neural networks. Here, the encoder takes the input and generates an internal state, that is then used by the decoder to produce the output. In [6], they use this framework to generate paraphrases using stacked residual LSTM networks.

To improve the quality of the results while reducing the complexity of the framework, it has become more common to use the Transformer architecture, that relies entirely on attention mechanisms, substituting the complex convolutional or recurrent neural networks of previous seq2seq models. It was first introduced for translation, where it proved to be able to outperform previous architectures, with better results and allowing training to be considerably faster [12]. This method has been used for paraphrase generation in various languages [2,3].

Additionally, Transformer models can benefit from transfer learning, that intends to further speed up the process of training. This method consists of pretraining the model on more general tasks, where plenty of data is available, to generate an extensive base model. Then it can be finetuned on more specific downstream tasks, making the training easier and faster for paraphrase generation [13].

In this work, we have considered a combination of these approaches. Firstly, different Transformer models finetuned on paraphrase generation have been considered for this task. Since these models are in English, and considering that our system is in a different language, we translate the sentences before feeding them to the model, and then translate back the generated paraphrases to the original language. What differentiates our model from bilingual pivoting is that, in our case, we include the Transformer model between the two translations. This makes the generated paraphrases more diverse every time. Lastly, the innovation of this work relays on its application, since it has been integrated into a social robot, which has not been done before in the state of the art.

3 Our Approach Towards Paraphrases in Spanish

We have contemplated two different approaches based on the language of the models. Firstly, a Spanish model has been considered. To do this, the mT5 model [14], a multilingual transformer pretrained on NLP tasks, has been used as a base. It has been then finetuned using a Spanish paraphrase dataset, Paws-x [15]. Given the difficulty to find other resources in Spanish, such as different pretrained models or paraphrase datasets, this first approach has not been developed further.

The second approach consists on using an English model, as there are multiple available transformers that are already pretrained and finetuned on paraphrasing tasks. Three of these models have been chosen and compared based on the average paraphrasing time and on two scales widely used in other NLP tasks: BLEU and BERT. In order to integrate them in our system, the original sentence has to be translated from Spanish to English, then passed through the model to generate a paraphrase, and lastly, translated back to Spanish, as shown in the diagram in Fig. 1.

In this work, the English models used as a base for the paraphrase generation task are T5 and Pegasus. Both of these models are based on the Transformer

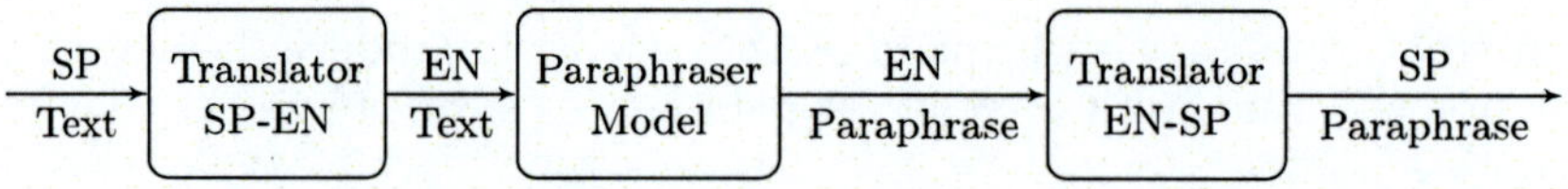

Fig. 1. Diagram of the exploitation process.

architecture and rely on transfer learning, since they have been pre-trained and then finetuned for paraphrase generation, as shown in Fig. 2. These models have been retrieved from the Huggingface Transformers library, that provides plenty of NLP state-of-the-art resources.

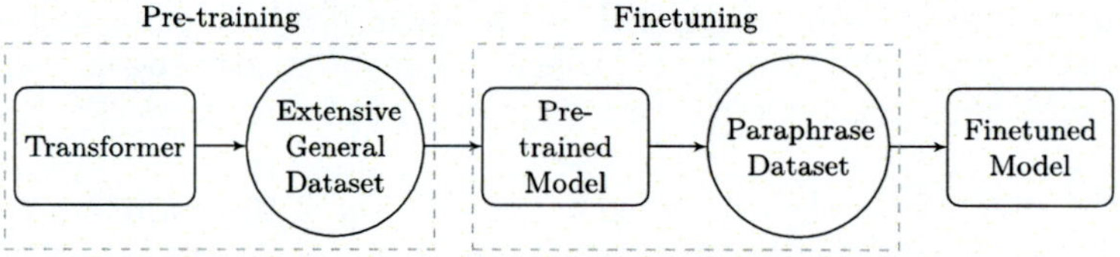

Fig. 2. Diagram of the training process.

T5 The T5 (Text-To-Text Transfer Transformer) [7] intends to solve all the most relevant text-based tasks, by turning them into a text-to-text problem. This framework shows the advantages of scaling the model and corpus size, by using up to 11 billion parameters in the training model, and the C4 (Colossal Clean Crawled Corpus), a dataset that contains hundreds of gigabytes of English text. This pre-trained model can be then finetuned on many different specific NLP tasks. In this case, it needs to be finetuned for paraphrase generation.

For that, we have chosen two different available models based on the same pre-trained T5 Transformer, but finetuned with different datasets and configuration. Here they will be referred to as Finetuned T5,[1] and Parrot.[2]

PEGASUS The PEGASUS Transformer (Pre-training with Extracted Gap-sentences for Abstractive SUmmatization Sequence-to-sequence models) [17] is also based on transfer learning. In this case, pre-training is focused on abstractive summarization, which tries to rewrite the input text to generate a summary that can be constructed with different words, contrary to extractive summarization, that copies sentences from the input directly to the output.

It is evaluated on different abstractive summarization tasks, where the pre-trained model can be finetuned with scarce data, and the results have proven to

[1] https://huggingface.co/ceshine/t5-paraphrase-paws-msrp-opinosis.
[2] https://huggingface.co/prithivida/parrot_paraphraser_on_T5.

be similar or better than many state-of-the-art's. The PEGASUS model used in this work has been finetuned on paraphrasing.[3]

4 Integration in the Robot's Architecture

This application has been integrated in the HRI (Human-Robot Interaction) architecture of a social robot. This architecture is in charge of managing all the interactions between the robot and the user, as it is connected to the input and output modules of the robot, that are used to communicate with its environment. Inside it, there is the Expression Manager block, shown in Fig. 3.

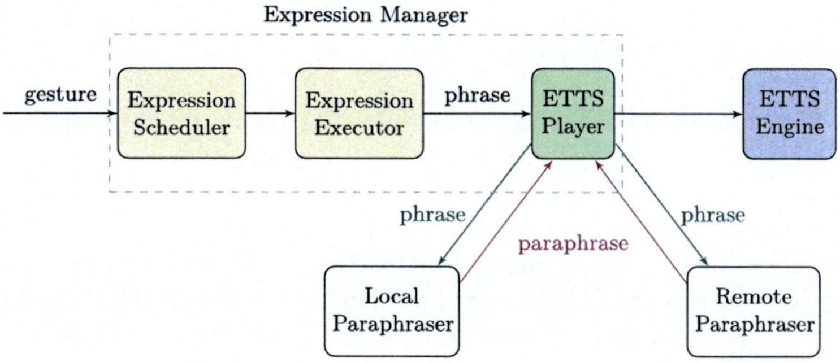

Fig. 3. Expression manager architecture.

Inside the Expression Manager, the Expression Scheduler receives the gestures or *Expressions*, this is, the combination of different outputs the robot uses to communicate the information. This block schedules these *Expressions*, and checks that there are no conflicts between them. Then, the Expression Executor loads the necessary *Expression* and executes it. Lastly, the Players communicate directly with the output interfaces.

In this case, the ETTS (Emotional Text-To-Speech) Player receives the phrase the robot has to say, and sends it directly to the hardware, that reproduces that sentence. This is where the paraphrase function is integrated, as the phrase that is received by the Player is converted into a paraphrase, and that is what is finally sent to the ETTS engine.

An external server was also introduced to speed up the application. This way, the ETTS Player can be configured to send the phrase to either the local paraphraser, or the one on the remote server.

Inside the robot, the communications are performed using ROS messages [8]. This way, the text is sent to the local paraphraser and then the paraphrase is sent back to the ETTS Player. On the other hand, communication between the server and the robot is done using sockets. This connection has to be established before running the paraphrasing application, and allows to transmit the text both ways.

[3] https://huggingface.co/tuner007/pegasus_paraphrase.

5 Evaluation

As mentioned before, for a paraphrase to be considered acceptable, it needs to keep as much meaning as possible from the original sentence, while providing diversity in its form. To evaluate these factors, we have used two different complementary scores that are widely used for other NLP tasks. These are BLEU Score and BERT Score.

BLEU Score (BiLingual Evaluation Understudy) [5] allows to measure the quality of an automatic translation, where quality refers to the similarity to a human translation. If the evaluated sentences are identical, the BLEU Score is 1. To evaluate the quality of a paraphrase, we are actually looking for the opposite: we need the paraphrase to be syntactically diverse from the original phrase. We have chosen two different BLEU scores to evaluate our models: BLEU-2 and BLEU-3.

On the other hand, BERT Score [16] lets us evaluate how semantically similar both phrases are, using BERT (Bidirectional Encoder Representations from Transformers), a language representation model. This way, this method compares the phrase and paraphrase using pretrained BERT contextual embeddings, instead of looking for exact matches between the two sentences.

Since we want sentences to be syntactically diverse but semantically similar, for paraphrase generation we are looking for a low value of BLEU, and a high value of BERT.

To evaluate the different models using these scores, we have created a list of more than 600 sentences that were sent to the different models we evaluated. These sentences are processed by each of the models individually to generate a paraphrase. Then the mentioned scores are calculated for each sentence, comparing the paraphrase to the original, and then averaged. The different considered models have been tested using this list, and the results are shown in Table 1.

Firstly, the models have been evaluated by comparing the original sentences and paraphrases in Spanish (SP). But also, considering that some of the quality of the paraphrase can be lost on the translation of the sentences back and forth, the paraphrase models on their own have also been compared directly in English (EN). The best values for each row are highlighted in bold in the table.

Table 1. Scores for considered paraphrase models.

	Finetuned T5		Parrot		Pegasus	
	EN	SP	EN	SP	EN	SP
BERT	**0.88**	0.77	0.69	0.50	0.66	0.51
BLEU-2	0.76	0.56	0.44	**0.31**	0.44	0.32
BLEU-3	0.70	0.46	0.38	**0.25**	0.37	**0.25**
Local time (s)	**0.85**	2.04	2.76	3.16	2.57	3.46
Remote time (s)	**0.61**	1.61	1.2	2.27	1.12	2.05

Firstly, in Spanish, the Parrot and Pegasus models provide very similar results, with the BERT Score higher than the BLEU Scores. For the results obtained with the Finetuned T5 model, all the scores are higher. This means that, in the generated paraphrases, the meaning is more similar to the original sentences', because they are also more syntactically similar, which is not desirable.

For the paraphrase generation in English, although the results continue in the same trend, all of the studied scores are increased. This difference shows that the translation brings more diversity into the sentences.
In both cases, the Parrot and Pegasus models provide smaller BERT and BLEU Scores than the Finetuned T5 model. This means that they can generate more diverse paraphrases, but some of the original sentence's meaning can be lost. Most of the times the generated paraphrases are suitable and fit in the context of the sentence, but sometimes a paraphrase can be generated that does not make sense. This is even more likely to happen in our case, since the translations also introduce more diversity. The Finetuned T5 model is more reliable in that aspect, since the form of the phrase does not change that much, guaranteeing a wider part of the meaning to be kept.

Lastly, the average execution times have been compared for the different models running locally and on the remote server. In the case of the server, these times only include the paraphrase generation, not the communication delays.

Here, the Finetuned T5 model proves to be around one second faster than the rest of the considered models, on average. Even though the translations add up about one second too, the fastest average time in Spanish, 1.61 s, can be achieved using the Finetuned T5 model on the remote server.

6 Use Case

This application has been integrated in the Mini robot [9]. Mini is a small desktop social robot whose main function is to communicate with elders, providing a variety of social skills that allow to entertain and assist the user, while expressing emotion. To do this, the robot has different interfaces that it uses to communicate, such as the movement of the body, lights or voice. It can speak thanks to an ETTS system, where the paraphrasing application has been integrated.

One of the many skills available in the robot to entertain the user is telling tales. We can see videos of the robot performing this skill with a short story[4] and its paraphrase.[5]

In this case, we have used an external server to generate the paraphrases. This server has 20 Intel i9 CPU cores and two GPUs that allow a faster execution than the robot, with just 7 Intel i7 CPU cores.

The results are shown in English in Table 2. Here, the sentences have been processed by the models one at a time, generating the paraphrases. The sentences that change with respect to the original have been highlighted in bold.

[4] https://youtube.com/shorts/rERpBROzhtw?feature=share.
[5] https://youtube.com/shorts/WGWZ4NN6fz8?feature=share.

Table 2. Paraphrased short story in English. The sentences that change with respect to the original are highlighted in bold.

Text	Finetuned T5	Parrot	Pegasus
The first time Julia went to the beach, she liked it so much that she ran to get three jars to take home.	Julia liked the beach the first time she went there, so much so that she ran to get three jars to take home.	julia went to the beach the first time she liked it so much she ran to get three jars for her trip home.	Julia ran to get three jars when she went to the beach the first time.
She filled one with sand, another with seawater and the third with rays of sunlight.	She filled one with sand, another with seawater, and the third with rays of sunlight.	She filled one with sand, another with seawater and the third with rays of sunlight.	She put sand, seawater, and rays of sunlight in them.
One day, after some time, his father threw the sand away on the garden and the water down the sink and put the three cans in the storage cupboard, thinking that they were all empty.	One day, after some time, his father threw the sand away on the garden and the water down the sink, and put the three cans in the storage cupboard, thinking that they were all empty.	he threw sand and water under the sink one day and kept the three cans in a storage cupboard thinking it was empty.	After some time, his father threw the sand away on the garden and the water down the sink and put the three cans in the storage cupboard, thinking they were empty.
In a corner of the garden a fragile sand castle appeared that vanished in the wind and a wave loaded with tiny fish slid through the pipes of the house.	A fragile sand castle appeared in a corner of the garden, which vanished in the wind and a wave loaded with tiny fish slid through the pipes of the house.	in a corner of the garden appeared a fragile sand castle that disappeared in the wind and a wave stuffed with tiny fish slid through the pipes of the house.	In a corner of the garden there was a sand castle that vanished in the wind and a wave with tiny fish slid through the pipes of the house.
But Julia's dad didn't see it, because by then he had already gone back to his chores.	But Julia's dad didn't see it because by then he had already gone back to his chores.	but julia's father didn't see her because he was already back to his work.	By the time Julia's dad saw it, he had already gone back to his chores.
The sun is still kept in the closet.	The sun is still kept in the closet.	the sun is still stored in the closet.	There is a closet with the sun in it.
But don't worry, it doesn't expire and it will light up on the day when Julia or her father need a ray of hope.	But don't worry, it doesn't expire and will light up on the day when Julia or her father need a ray of hope.	but don't worry it's not over and it'll show up on the day when julia or her father need a ray of hope.	It will light up on the day when Julia or her father need a ray of hope.

This Table shows that the Finetuned T5 model is the one that changes the least, while the rest of them present changes in almost all sentences of the story, substituting some words and altering the order of others.

We have also compared the paraphrased text in Spanish, and, in that case, it is clear how the translations bring more diversity into all the models, specially for the Finetuned T5, making more differences appear between the two texts. In fact, all the sentences change in the paraphrased text in Spanish for all the models except for the second sentence, that remains the same for the Finetuned T5 and Parrot models.

7 Conclusions

In this work, we have implemented a paraphrasing application in the architecture of a social robot. For this, different models have been considered both in Spanish and English. The tested Spanish model, the mT5 finetuned on paraphrasing, did not provide good results, as the generated paraphrases were too similar to the original sentences. On the other hand, the models in English are proven to be more promising.

We have used a Pegasus model finetuned on paraphrasing, and the Parrot and Finetuned T5 models, both finetuned on paraphrasing using T5 as a base. To implement these models in our application, the sentences in Spanish have to be translated to English, then passed through the models, and then the generated paraphrase has to be translated back to Spanish. These translations add more diversity to the text. While the Finetuned T5 model on its own does not modify the sentences as much as the other considered models, the translations help make the results more diverse, while also being the fastest model.

When running paraphrasing on the robot, the amount of time required for processing was excessive. The interaction between the robot and the user must be as fluid as possible, so these silences should be avoided. To do this, a remote server was also introduced to make this process faster, allowing to generate the paraphrases, on average, within the two second threshold.

Lastly, the paraphrasing application has been tested in one of the skills of the robot, telling tales. In practice, the paraphrase generation should not be used for all the applications on the robot. For some specific tasks, we want the text to be explicitly read by the robot, like telling jokes, which rely mostly on puns that do not make sense if the words are changed.

Acknowledgement. This work was partially supported by project Robots sociales para estimulación física, cognitiva y afectiva de mayores (ROSES) RTI2018-096338-B-I00 funded by Agencia Estatal de Investigación (AEI), Ministerio de Ciencia, Innovación y Universidades. This publication is part of the R&D&I project PLEC2021-007819 funded by MCIN/AEI/10.13039/501100011033 and by the European Union NextGenerationEU/PRTR. This work has been supported by the Madrid Government (Comunidad de Madrid-Spain) under the Multiannual Agreement with UC3M ("Fostering Young Doctors Research", SMM4HRI-CM-UC3M), and in the context of the V PRICIT (Research and Technological Innovation Regional Programme).

References

1. Bannard, C., Callison-Burch, C.: Paraphrasing with bilingual parallel corpora. In: ACL'05, pp. 597–604. ACL, Michigan, USA (2005)
2. Egonmwan, E., Chali, Y.: Transformer and seq2seq model for paraphrase generation. In: WNGT, pp. 249–255. ACL, Hong Kong (2019)
3. Fenogenova, A.: Russian paraphrasers: paraphrase with transformers. In: BSNLP, pp. 11–19. ACL, Kiyv, Ukraine (2021)
4. Mallinson, J., Sennrich, R., Lapata, M.: Paraphrasing revisited with neural machine translation. In: EACL, vol. 1, Long Papers, pp. 881–893. ACL, Valencia, Spain (2017)
5. Papineni, K., Roukos, S., Ward, T., Zhu, W.: BLEU: a method for automatic evaluation of machine translation. In: ACL'02, pp. 311–318. ACL, Philadelphia, USA (2002)
6. Prakash, A., Hasan, S.A., Lee, K., Datla, V., Qadir, A., Liu, J., Farri, O.: Neural paraphrase generation with stacked residual LSTM networks. In: COLING 2016: Technical Papers, pp. 2923–2934. COLING, Osaka, Japan (2016)
7. Raffel, C., Shazeer, N., Roberts, A., Lee, K., Narang, S., Matena, M., Zhou, Y., Li, W., Liu, P.J.: Exploring the limits of transfer learning with a unified text-to-text transformer. JMLR **21**(140), 1–67 (2020)
8. ROS Messages. http://wiki.ros.org/Messages. Last accessed 19 Apr 2022
9. Salichs, M.A., Castro González, Á., Salichs, E., Fernández Rodicio, E., Maroto Gómez, M., Gamboa Montero, J.J., Marqués Villarroya, S., Castillo, J.C., Alonso Martín, F., Malfaz, M.: Mini: a new social robot for the elderly. Int. J. Soc. Robot. (12), 1231–1249 (2020)
10. Shiwa, T., Kanda, T., Imai, M., Ishiguro, H., Hagita, N., Anzai, Y.: How quickly should communication robots respond? In: ACM/IEEE International Conference on HRI, pp. 153–160 (2008)
11. Sutskever, I., Vinyals, O., Le, Q. V.: Sequence to sequence learning with neural networks. In: NIPS, vol. 2, pp. 3104–3112. Montreal, Canada (2014)
12. Vaswani, A., Shazeer, N., Parmar, N., Uszkoreit, J., Jones, L., Gomez, A., Kaiser, Ł., Polosukhin, I.: Attention is all you need. In: NIPS, pp. 6000–6010. California, USA (2017)
13. Witteveen, S., Andrews, M.: Paraphrasing with large language models. In: WNGT, pp. 215–220. ACL, Hong Kong (2019)
14. Xue, L., Constant, N., Roberts, A., Kale, M., Al-Rfou, R., Siddhant, A., Barua, A., Raffel, C.: mT5: a massively multilingual pre-trained text-to-text transformer. In: NAACL: Human Language Technologies, pp. 483–498. ACL (2021)
15. Yang, Y., Zhang, Y., Tar, C., Baldridge, J.: PAWS-X: a cross-lingual adversarial dataset for paraphrase identification. In: IJCNLP, pp. 3685–3690. ACL, Hong Kong, China (2019)
16. Zhang, T., Kishore, V., Wu, F., Weinberger, K.Q., Artzi, Y.: BERTScore: evaluating text generation with BERT. In: ICLR (2020)
17. Zhang, J., Zhao, Y., Saleh, M., Liu, P.: PEGASUS: pre-training with extracted gap-sentences for abstractive summarization. In: ICML, pp. 11328–11339 (2020)

Epilepsy Seizure Detection Using Low-Cost IoT Devices and a Federated Machine Learning Algorithm

Marcos Lupión[1(✉)], Juan F. Sanjuan[1], Javier Medina-Quero[2], and Pilar Martínez Ortigosa[1]

[1] Department of Informatics, University of Almería, CeiA3, Almería, Spain
marcoslupion@ual.es, jsanjuan@ual.es, ortigosa@ual.es
[2] Department of Computer Science, University of Jaen, Jaen, Spain
jmquero@ujaen.es
https://hpca.ual.es

Abstract. Epilepsy is one of the most common chronic brain diseases, but with the proper medication, patients enjoy everyday life. New trends in Internet of Things approaches are now introduced to detect seizures in users with epilepsy to adjust treatments, reducing mortality rates due to the fatal consequences they develop. Wearable devices on the market can detect motor seizures thanks to their built-in acceleration sensors. However, the limitations of this type of device are the short battery life, a non-ergonomic design, and the high price. In this work, we propose a low-cost system that allows the monitoring of users in residential centers using a wearable device equipped with an acceleration sensor. The epileptic seizure detection algorithm has been built following federated machine learning, creating a general model for all possible users from the models learned in each residential center. Based on the first results, the system has more than 60 h of autonomy, obtaining favorable detection rates in the first simulations. In this case, the information of each user is kept in the local environment, maintaining privacy.

Keywords: Epilepsy seizure detection · Federated machine learning · Fog computing · Low cost devices

1 Introduction

Epilepsy is a disease that affects the brain and causes the brain to behave incorrectly. This disease causes convulsions or loss of consciousness and, therefore, conditions the life of the person who suffers from it, paying particular attention to avoiding fatal consequences. Nowadays, around 50 million people suffer from this disease [1], and these people are more likely to die prematurely [7]. Although medication is available to control episodes of epileptic seizures, it is necessary to know how many episodes the person suffers during the day in order to be able to

V. Julián et al. (Eds.): ISAmI 2022, LNNS 603, pp. 229–238, 2023.
https://doi.org/10.1007/978-3-031-22356-3_22

adjust the treatment correctly [16]. For this reason, research is conducted into the use of devices that can detect the instant when an epileptic seizure occurs.

The more common seizures are classified in those [15] that cause jerky muscle movements (clonic and myoclonic); the ones that cause a loss of muscle control (tonic and atonic), which can lead to falls; and the most severe, general tonic-clonic seizures (GTCS), which can even lead to unconsciousness. We note that other kinds of epileptic seizures with different characteristics make a single device unable to detect them.

These clonic and myoclonic seizures can now be detected using an Electroencephalogram (EEG). This device has a series of electrodes placed on the person's skull and can measure electrical activity in the brain. Currently, machine learning solutions have been created that make it possible to detect the different types of seizures in the diagrams provided by these sensors with a high degree of certainty [13]. However, this device cannot be applied in a person's daily life, and it is necessary to create devices that are as non-invasive as possible in the person's life.

The seizures mentioned above involve an alteration in the movement of different parts of the person's body, so devices incorporating Accelerometry (ACM) sensors are being used to detect such seizures, with reasonable detection rates and low false-positive rates [3].

For the development and validation of this type of device, in [5] several phases are (from phase 0 to phase 4) depending on factors like the number of users participating in the study, the number of epileptic seizures recorded, and the number of centers in which the devices are deployed.

Among the devices at the most advanced stage is the Empatica Wristband [14], the first commercially available device to measure the physiological hallmarks of ongoing GTCS. It incorporates acceleration sensors and electrodermal activity (EDA) measurement. EpiCare [4] detects tonic-clonic seizures through the use of an accelerometer. Nightwatch [2] incorporates an acceleration sensor in addition to an optical heart rate sensor (PPG), allowing detection of up to 86% of generalized tonic-clonic and myoclonic-tonic seizures. However, the device is not designed to be worn all day by the user [6]. The devices mentioned above are priced at over £1000, with an autonomy of no more than 24 h, making them unaffordable for democratic implantation and health centers with a large volume of users. Thus, there are limitations at the hardware level that make it challenging to create a device that can monitor continuously for several days and in a non-invasive way [17].

On the other hand, Federated Learning (FL) has emerged as a distributed learning proposal [18] where devices are located in an edge layer collecting and processing data locally. The upper elements maintain data processing, using the relevant information from lower layers, increasing computational performance, and distributing the computation intelligently in the new [9] application models.

In this work, we propose an Internet of Things (IoT) system that allows the detection of epileptic seizures associated with movement through low-cost devices. We propose using a wearable device with an acceleration sensor, which sends information via Bluetooth Low Energy (BLE) to devices placed in a residential environment. This information is processed in a central device that executes an epileptic seizure detection algorithm created from data from other users and residential centers. It maintains their privacy through a federated machine learning approach. This solution allows for greater autonomy of the wearable device, with the processing falling solely on the central device. The main contributions are incorporating a federated learning process in the epilepsy seizure model generation, preserving user data, and a new device having nearly 60 h of continuous streaming of accelerometer data.

The rest of the manuscript is organized as follows: Sect. 2 describes the proposed system, Sect. 3 evaluates the system, and Sect. 4 provides the conclusions of the work and future improvements.

2 Methodology

The current section details the methods for deploying and training an epileptic seizure detection system using low-cost devices. This section firstly develops the IoT system in charge of collecting data from the environment and centralizing it in a local device. The embedded learning process is also developed to generalize epileptic seizure detection in different environments.

2.1 Fog Computing System

The detection of most types of epilepsy, such as tonic-clonic, clonic and myoclonic, is computed using an acceleration sensor. Currently, many devices on the market have these sensors, so algorithms can be built to process such data and detect patterns corresponding to epileptic seizures.

The developed system is shown in Fig. 1. In our approach, data are firstly collected on wearable devices. Then, they transmit the data to a central device (Raspberry), which stores the data, executes an epilepsy detection algorithm, and allows users to tag the data to improve the epilepsy detection algorithm.

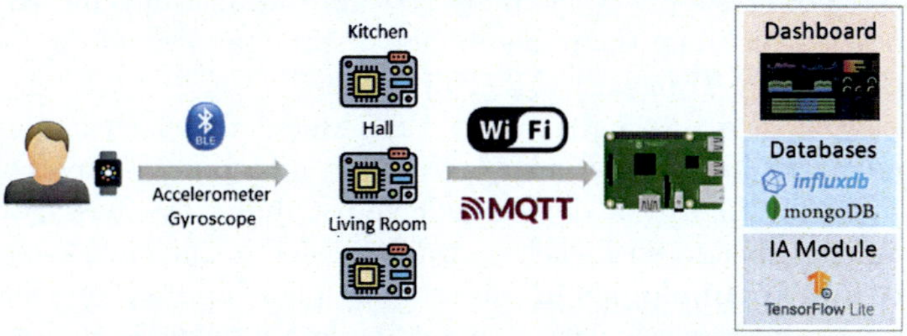

Fig. 1. Fog computing system deployed in each residence.

IoT Devices and communications In the proposed system, users are to be monitored wear a wearable device that collects and transmits acceleration data in real-time. This device is a smartwatch [11] running WearOS as an operating system, which allows the programming of native applications with access to these sensors. However, in the previously mentioned work, battery consumption is very high, with a maximum battery life of 5 h, according to the realized experiments. Moreover, the price of these devices is high.

In this case, the Arduino Nano 33 BLE Sense device was selected, with a 1000 mA battery. According to the experiments, this device allows acceleration data to be continuously obtained and transmitted via Bluetooth Low Energy (BLE) for 66 h. The charger allows a 500 mAh charge for battery charging, so the charging time is approximately 2 h.

Once the wearable device collects data from the sensors, this information must be distributed to the central device to build an algorithm to detect a hypothetical epileptic seizure. The transmission is developed via the BLE protocol, which improves battery-saving power compared to other types of communication such as Wi-Fi. However, the system must be deployed in a large environment, and the BLE connection has a theoretical maximum distance of 20 m.

ESP32, a low-cost, energy-efficient IoT device that enables communication with other devices via BLE and Wi-Fi, has been selected to use BLE communications. Therefore, in the current system, different ESP32 devices are deployed in different locations in the home to cover all possible locations where users can locate.

On the one hand, the communication between the wearable device and the ESP32 uses the BLE protocol. In this case, the wearable device is configured as a client, and the ESP32 generates the server connection. Therefore, the wearable device selects among the possible ESP32 devices the one whose signal is more intense, meaning that it is the closest one to the wearable device. Thus, the wearable device sends the information using the *WRITE* operation in the ESP32. This operation lets different devices be connected to the same ESP32 device. Otherwise, if the wearable device acts as a server, only one ESP32 can establish a connection with it, not being available to another device.

On the other hand, a Wi-Fi connection is used to forward the information from the wearable device to the central device because these devices are connected to a power supply without battery limitations. Specifically, we provide communication-based on the pub-sub model through messages by a fog Message Queuing Telemetry Transport (MQTT) broker.

The central device is a Raspberry Pi 4 Model B, which has up to 8GB of RAM, and allows communications via Wi-Fi. It incorporates an MQTT broker to obtain data from the ESP32 devices, which distributes the messages in real-time with a topic associated with each wearable device in the environment.

Software and technologies In this section, a machine learning algorithm classifies the wearable device's data to detect epileptic seizures. However, this algorithm has to be previously trained on real data sequences. In addition to data classification, the training of these algorithms in real-time is allowed in the proposed system. For this purpose, the central device incorporates different software tools.

First, when data is received from the ESP32, a consumer process running on the Raspberry device is in charge of collecting the data from the broker, pre-processing, eliminating erroneous data, and storing it in a database.

Second, the system uses two databases: InfluxDB and MongoDB. The first one, InfluxDB, is a database optimized for efficiently storing time-series data with the timestamp of each received data as the primary key of each object in the database. This storing mechanism allows faster querying of this time-series data. In this case, InfluxDB stores sensor data streams and displays them graphically on a dashboard. The second database, MongoDB, is responsible for storing the records with the information used by the machine learning algorithm to perform the training.

Third, the system also incorporates a dashboard that displays the acceleration data from the wearable devices. This dashboard has been built using the Grafana tool and InfluxDB database. It shows a real-time visualization of the data incorporated in it, allowing different types of graphs and time aggregations.

The data are labeled using the *notation* functionality of the Grafana tool, where time regions can be defined in addition to the labels. Therefore, a script is incorporated that makes a query on the InfluxDB database, obtaining the acceleration data in that interval and creating the corresponding records with the tag "epileptic seizure" in the MongoDB database.

The deployment and configuration of the software involved in data reception, storage, and visualization have been containerized using Docker for its automation and deployment in different environments.

2.2 Federated Machine Learning Process

Typically, algorithms are trained using a large amount of data in machine learning-based solutions. These data come from different devices and applications and are treated with very high computing power in data processing centres. This process is computationally expensive, and, in some cases, the handled information is confidential or sensible.

The concept of Federated Machine Learning was born As a solution to these problems of computational burden and privacy of the solutions. It consists of creating and executing local machine learning models in devices and, from these models, creating a general model that generalizes the information learned by these particular ones.

In this case, this concept is well suited to the proposed system (Fig. 2), which deals with epileptic seizure detection in residential environments. Each residential environment trains and executes a machine learning model per user suffering epileptic seizures. The system periodically updates a general model that involves the set of residences and tries to generalize epileptic seizure detection.

Learning and inference process The training of federated learning models lets create a general model without sharing specific data of users. However, the training is difficult, needing a large quantity of data to create a non-over-trained model. In this work, the main steps of the learning process are the following:

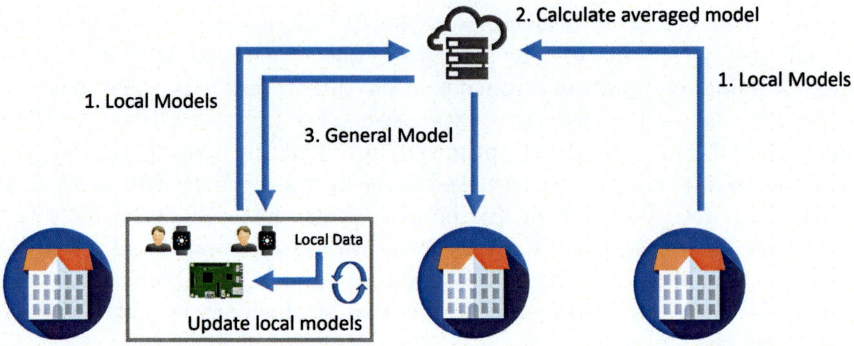

Fig. 2. Proposed training and inference process using federated machine learning.

- **Phase 1**: Build a general epileptic seizure model. This model is loaded in the different IoT devices.
- **Phase 2**: Each IoT device fine-tunes the general model with the user's data.
- **Phase 3**: Aggregation of the models to a general model in the cloud. At the end of the day, all trained models are uploaded to a server using MQTT. The server makes aggregation of the different models.
- **Phase 4**: Transfer the general model to the IoT devices. Continues with phase 2.

The key factor, in this case, is the fine-tuning of the general model in each IoT device. It lets the local model be adapted to the specific user.

As for the inference, the model runs on the Raspberry device and produces an output with the given user-defined frequency that has been set to 0.5 s. In general, epilepsy seizures last more than 18 s [8]; therefore, a window of 10 s is established to detect these seizures and avoid false positives. Smaller windows with regular movements similar to those produced by seizures could be detected, producing a false alarm.

Then, when an epileptic seizure is detected for more than 10 s, an alert message is sent to the staff who have installed the application via a Telegram bot. Once the epileptic seizure ceases, another message is sent with the seizure duration. At this point, the caregiver or researcher is validating the record. If it is correct, an affirmative message is sent to the bot, and the records corresponding to that interval are labeled as "epileptic seizure" and become part of the database.

Therefore, at the end of each day, the controlling devices transfer their models to a server hosted in the private cloud of the University of Almeria through a Virtual Private Network (VPN) using MQTT. In this server, the different models are pooled, creating a general model and sending this new model to the central devices of the residences using MQTT. Since only the models are transferred to the server, the acceleration data of the users do not cross the border of the local router, protecting their privacy.

Proposed neural network architecture Detecting an epileptic seizure requires acceleration data (having 3 coordinates). These data are obtained at a frequency 100 Hz to cover the relevant data to detect a possible epileptic seizure from the signal. A higher frequency implies a higher query of the sensor data by the Arduino and a more frequent data transmission through BLE, thus increasing the battery consumption.

In addition, a given time window is defined to encompass data and perform a classification of the sequence as a whole. In this case, a window size of 50 data is defined, corresponding to 0.5 s.

The neural network developed in this case is configured with convolutional layers, temporal learning, a recurrent layer, and several fully connected layers, based on the structure proposed in [10]. First, the convolutional layers process the data to extract morphological features from the signals as if they were images. This structure allows the selection of the essential signal features and reduces the input dataset dimensional. Then, a recurrent Long-Short-Therm-Memory layer (LSTM) is incorporated, which acts as a memory and allows the data processing, taking into account the temporal component of the data. Subsequently, three fully connected layers of 100, 50, and 10 neurons and Rectified Linear Unit (ReLU) activation function are incorporated. As an output layer, there is a layer with one neuron and the Sigmoid activation function since, in this case, the output is binary.

3 Preliminary Results

The system has been deployed in a real environment (Smart Home of the University of Almeria), configuring the communication of the different devices and the tools to annotate the dataset.

The IoT devices allow an average of no more than 5 s between generating the record and obtaining the classification by the neural network, so the system allows knowing the user's status in real-time.

Regarding the machine learning system training, we first conducted a literature search to use a public dataset to create an initial model. However, no dataset that collects acceleration data in patients with epileptic seizures has been found. Therefore, we had to create a simulated dataset with four healthy users for two hours [12], simulating several epileptic seizures.

The data of each user has been split into five days. At the end of the day, the local IoT device trains a local model adapted to each user and sent this model to the server. There, an aggregation of the local models is made, returning the newly generated general model to each of the IoT devices. Once received the general model, each IoT device fine-tunes the general model with the dataset of each user during only a small number of epochs.

In Table 1, the federated approach is compared with the traditional machine learning approach: train a local model each day from scratch with the whole dataset without sharing data with other users. The metrics Precision, Recall, and F-Score are obtained from each user in the "Epilepsy Seizure" registers, along with the training time of each model on the last day. It is important to remark that both models have similar values in terms of Precision, Recall, and F-score. However, training the local model in the federated approach requires less time.

Table 1. Federated approach vs traditional approach.

Approach	Time train(s)	Recall	Precision	F-Score
Federated	278	0.88	0.88	0.88
Traditional	389	0.9	0.87	0.88

In Fig. 3 a confusion matrix per user on the last day is displayed. In three users, the model tends to produce more false negatives than false positives. It may produce that an epileptic seizure is not detected, being critical to detect it. The small and imbalanced dataset causes this. In order to obtain a better model with federated learning, a huge quantity of data has to be recorded, as the main problem of the traditional approach is that it tends to be over-trained with the user's data and federated learning overcomes this, providing a robust general model.

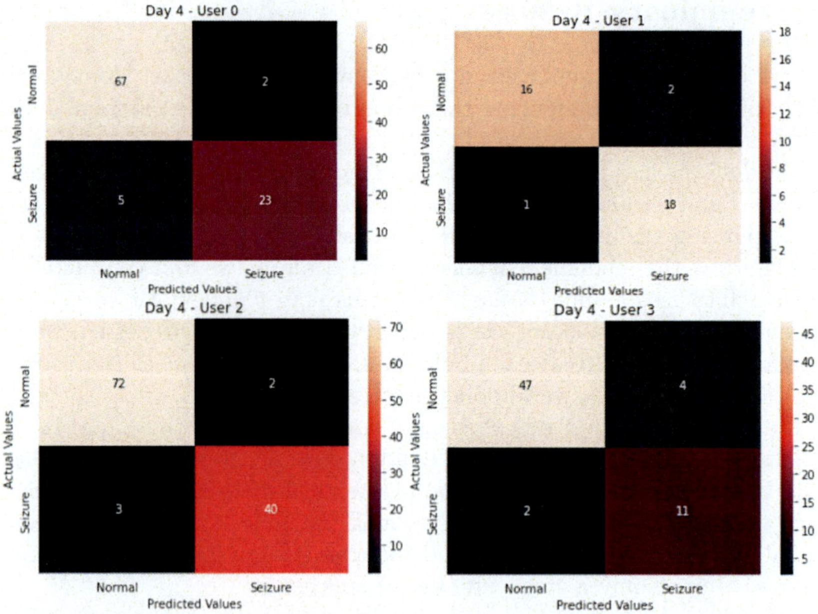

Fig. 3. Confusion matrix per user in the last day.

As discussed in Sect. 1, in [5] a guideline is defined to measure the maturity status of epileptic seizure detection systems. Following the considerations defined in this document, the system is in phase 0. However, even though the system is in phase zero, some incorporated features correspond to higher phases. It is the case of real-time data collection and classification, required in phases 3 and 4. Thus, other features that are optional in phases 0, 1, and 2 but compulsory in phases 3 and 4, such as continuous data recording using a proprietary device, are fulfilled in our system.

4 Conclusions

An epileptic seizure detection system based on the Internet of Things has been presented in this work. In this system, a wearable device worn by the user communicates through a network of low-cost devices via BLE to send acceleration information to a central device. The proposed device processes this information using a machine learning algorithm to detect possible epileptic seizures. The embedded devices have a lower overall price than other solutions and a higher autonomy.

The model is executed locally, allowing the data not to leave the environment and maintaining the users' privacy. In addition, it requires less time to train each local model. Thus, this model can be periodically shared with other models learned in other residential centers, allowing for learning a more robust and general model adapted to different conditions of people and enhancing the joining of new people on the platform.

After simulations with test data, the autonomy and epileptic seizure detection results are encouraging. The system is currently in phase zero. In future work, the system will be deployed imminently in four residential centers in order to validate the system in real environments and to be able to reach phase 2.

Acknowledgements. This research has been funded by the R+D+i project RTI2018-095993-B-I00, financed by MCIN/AEI/10.13039/501100011033/ and ERDF "A way to make Europe"; by the Junta de Andalucá with reference P18-RT-1193; by the University of Almería with reference UAL18-TIC-A020-B and by the Department of Informatics of the University of Almería. M. Lupión is supported by FPU program of the Spanish Ministry of Education (FPU19/02756).

References

1. Epilepsy. www.who.int/news-room/fact-sheets/detail/epilepsy. Accessed: 19 Apr 2010
2. Arends, J., Thijs, R.D., Gutter, T., Ungureanu, C., Cluitmans, P., Van Dijk, J., van Andel, J., Tan, F., de Weerd, A., Vledder, B., et al.: Multimodal nocturnal seizure detection in a residential care setting: a long-term prospective trial. Neurology **91**(21), e2010–e2019 (2018)
3. Beniczky, S., Jeppesen, J.: Non-electroencephalography-based seizure detection. Curr. Opin. Neurol. **32**(2), 198–204 (2019)

4. Beniczky, S., Polster, T., Kjaer, T.W., Hjalgrim, H.: Detection of generalized tonic-clonic seizures by a wireless wrist accelerometer: a prospective, multicenter study. Epilepsia **54**(4), e58–e61 (2013)
5. Beniczky, S., Ryvlin, P.: Standards for testing and clinical validation of seizure detection devices. Epilepsia **59**(S1), 9–13 (2018)
6. Bruno, E., Viana, P.F., Sperling, M.R., Richardson, M.P.: Seizure detection at home: do devices on the market match the needs of people living with epilepsy and their caregivers? Epilepsia 61(S1) (2020)
7. Christensen, J., Pedersen, C.B., Sidenius, P., Olsen, J., Vestergaard, M.: Long-term mortality in children and young adults with epilepsy—a population-based cohort study. Epilepsy Res. **114**, 81–88 (2015)
8. Jenssen, S., Gracely, E.J., Sperling, M.R.: How long do most seizures last? A systematic comparison of seizures recorded in the epilepsy monitoring unit. Epilepsia **47**(9), 1499–1503 (2006)
9. Li, T., Sahu, A.K., Talwalkar, A., Smith, V.: Federated learning: challenges, methods, and future directions. IEEE Sign. Proces. Mag. **37**(3), 50–60 (2020)
10. Li, Z., Liu, Y., Guo, X., Zhang, J.: Multi-convLSTM neural network for sensor-based human activity recognition. J. Phys. Conf. Ser. **1682**(1), 012062 (2020)
11. Lupión, M., Medina-Quero, J., Sanjuan, J.F., Ortigosa, P.M.: Dolars, a distributed on-line activity recognition system by means of heterogeneous sensors in real-life deployments—a case study in the smart lab of the university of almería. Sensors 21(2) (2021). https://doi.org/10.3390/s21020405
12. Lupión Lorente, M.: Simulated epileptic seizures in UAL smart home 1 (2022). https://doi.org/10.17632/37f8w8f7tm.1, https://data.mendeley.com/datasets/37f8w8f7tm/1
13. Park, Y., Luo, L., Parhi, K.K., Netoff, T.: Seizure prediction with spectral power of EEG using cost-sensitive support vector machines. Epilepsia **52**(10), 1761–1770 (2011)
14. Regalia, G., Onorati, F., Lai, M., Caborni, C., Picard, R.W.: Multimodal wrist-worn devices for seizure detection and advancing research: focus on the Empatica wristbands. Epilepsy Res. **153**, 79–82 (2019)
15. Sarmast, S.T., Abdullahi, A.M., Jahan, N.: Current classification of seizures and epilepsies: scope, limitations and recommendations for future action. Cureus **12**(9), e10549
16. Ulate-Campos, A., Coughlin, F., Gaínza-Lein, M., Fernández, I.S., Pearl, P., Loddenkemper, T.: Automated seizure detection systems and their effectiveness for each type of seizure. Seizure **40**, 88–101 (2016)
17. Verdru, J., Van Paesschen, W.: Wearable seizure detection devices in refractory epilepsy. Acta Neurol. Belg. **120**(6), 1271–1281 (2020). https://doi.org/10.1007/s13760-020-01417-z
18. Yang, Q., Liu, Y., Cheng, Y., Kang, Y., Chen, T., Yu, H.: Federated learning. Synth. Lect. Artif. Intell. Mach. Learn. **13**(3), 1–207 (2019)

Electroencephalographic Signal Processing from Brain-Computer-Interface Following Image-Based Emotion Induction

Eloy García-Pérez[1], Daniel Sánchez-Reolid[1], Roberto Sánchez-Reolid[1,2], Antonio Fernández-Caballero[1,2,3(✉)], José M. Latorre[4], and Alejandro L. Borja[5]

[1] Unidad Multidisciplinar de la Investigación en Neurocognición y Emoción en Entornos Virtuales y Reales, Instituto de Investigación en Informática, Albacete, Spain
Daniel.Sanchez@uclm.es, Roberto.Sanchez@uclm.es, Antonio.Fdez@uclm.es
[2] Departamento de Sistemas Informáticos, Universidad de Castilla-La Mancha, Albacete, Spain
[3] CIBERSAM-ISCIII (Biomedical Research Networking Centre in Mental Health, Instituto de Salud Carlos III), Madrid, Spain
[4] Departamento de Psicología, Universidad de Castilla-La Mancha, Albacete, Spain
Jose.Latorre@uclm.es
[5] Departamento de Ingeniería Eléctrica, Electrónica, Automática y Comunicaciones, Universidad de Castilla-La Mancha, Albacete, Spain
Alejandro.Lucas@uclm.es

Abstract. In this paper, the identification of arousal and valence levels in people monitored by means of electroencephalograms is presented. More specifically, different images were exposed to several volunteers in order to induce different basic emotions i.e. happiness, sadness, fear, and anger, where brain activity was recorded employing a BCI Emotiv EPOC X helmet. The measured signals were processed and modified accordingly to determine the main areas of the brain activated during the experiments through statistical analyses.

Keywords: Electroencephalography · Brain-computer interface · Signal processing · Emotion induction

1 Introduction

Understanding and recognizing human emotions has become one of the areas of interest of intelligent systems. For this purpose, the main thing is to perform a

V. Julián et al. (Eds.): ISAmI 2022, LNNS 603, pp. 239–248, 2023.
https://doi.org/10.1007/978-3-031-22356-3_23

classification of these emotions. Russell's circumplex model relies on two independent variables to perform this classification, valence and arousal [15]. Valence describes how pleasant this emotion is for the subject, thus differentiating pairs of emotions such as joy-sadness or calm-boredom. On the other hand, arousal defines the level of brain activation of the subject with respect to that emotion, thus contrasting emotions such as excitement-calm or stimulated-relaxed. Arousal levels are variable and play an important role in day-to-day activities.

Brain-Computer Interface (BCI) systems are commonly used to measure these arousal levels [2]. BCI is defined as a communication system capable of monitoring brain activity and translating certain characteristics of brain activity, such as the user's intentions to control a device [3–5]. These systems can be very useful for dependent people or those with a certain degree of disability, providing a new communication channel compared to conventional ones. Emotional state detection can be performed using electroencephalography (EEG) alone [1,8,10,16] or in conjunction with other physiological signals such as electrodermal activity (EDA) [17,18], blood volume pressure (BVP) [22], electrocardiogram (ECG) and electromyography (EMG), among others. Electroencephalography is defined as a technique by which electrical brain activity is obtained in real time. This measurement is based on the electrical signals from the surface of the skull from electrodes that capture the voltage difference between large groups of neurons. In this case, an electrode helmet has been used, which provides a simple and non-invasive technique, in addition to the portability of the device.

Many studies on BCI systems try to understand the emotional state to provide a viable connection between humans and machines. Methods based on neurological responses like electroencephalography make it possible to accurately recognize several human emotions, based on brain wave parameters such as entropy [14]. Thus, there is an area of study called affective computing in charge of recognizing, interpreting, processing, and simulating human feelings. This field is rapidly growing thanks to BCI systems, novel emotion induction procedures and classification methods [20].

This paper describes a process in which subjects were stimulated by exposure to a sequence of images. The subjects of the experiment have been a series of volunteers from Castilla-La Mancha, Spain. In this case, each of the participants was shown a series of blocks with different images, each of them related to a type of emotion according to its arousal and valence. For this study, a segmentation has been carried out in order to obtain, from each of the subjects, the signals corresponding to each of the blocks for comparison. A filtering and independent component analysis is also carried out to eliminate noise and possible artifacts.

2 Materials and Methods

This section describes the methodology and the material necessary to carry out the proposed study. To begin with, an introduction about electroencephalography and brain waves is given. Subsequently, the material used to obtain the signals from the subjects is described and the participants are introduced. This is followed by a detailed explanation of the experiment performed. Finally, the process of segmentation and filtering of the signals as well as the extraction of the results are explained.

2.1 Electroencephalography

The electroencephalogram or electroencephalography is defined as a technique for functional exploration of the central nervous system through which the recording of brain electrical activity is obtained in real time. The origin of these waves or electrical impulses occurs in the pyramidal cells of the cerebral cortex, being these a specific type of multi-polar neuron of the brain that constitutes the primary source of excitation of the prefrontal cortex.in order to obtain and analyze the electrical signals of the different regions that exist from the surface of the skull, electrodes that capture the differences in voltage or potential between them are needed. These electrodes capture the potential differences of large groups of neurons due to the impossibility of measuring them one by one.

Electroencephalography, therefore, deals with the study and analysis of the electric fields generated in the brain based on the amplification of the voltage that occurs between the electrodes receiving the signal. In this case, a non-invasive EEG technique is used using electrodes placed on the scalp. Electroencephalography is a technique with various uses such as monitoring a seizure-prone patient or determining brain death in an individual. Brain waves can be defined as the patterns of activation of certain groups of neurons by electrical impulses, i.e. brain activity. A number of characteristic parameters such as frequency, topographical distribution, shape, amplitude or duration are necessary to define these waves.

2.2 Data Acquisition and Emotiv EPOC X Device

The EEG signal extraction capability of the Emotiv EPOC X device has been used to carry out this experiment [9,19]. It is a portable electrode helmet designed for human developmental research that provides easy, advanced-level access to brain data. It is considered a Brain-Computer Interface (BCI) system. The Emotiv EPOC X helmet, as well as the software distributed by Emotiv, has much more functionality than just recording the user's brain activity data. In this case, the Emotiv PRO software will be used to record the brain activity of the person wearing the helmet.

Something fundamental for the signals taken from the helmet to be correct is the proper placement of the helmet on the subject's scalp. The helmet has arms to place the electrodes in these positions, this electrode is completed by placing in each of the positions some felt pads that must be soaked in a saline solution that the model comes with. One of the most important things when taking data is to wet these electrodes well between each data taking so that the conductivity is correct.

2.3 Image-Based Emotion Induction and Participants

As previously mentioned, images will be used to induce emotions in subjects, in a similar way to previous work [12,13,21]. These images belong to the IAPS (International Affective Picture System) [11], database of images designed and tested by the Center for Emotions and Mindfulness at the National Institute of Mental Health, University of Florida, used primarily in psychological research.

The images are divided according to valence and arousal, combining them and giving rise to 4 main types of emotions. Each of these types of emotions is related to a block of images, selected with the purpose of generating in the subject the emotion they represent. The images of high valence and high arousal (HH) will be related to emotions such as joy or excitement, showing images of strong emotions such as a roller coaster or skydiving. The emotions of high valence and low arousal (HL), correspond to calm or tranquility, showing familiar or peaceful rural environments. The emotions of low valence and high arousal (LH), represent disgust or fear, corresponding to bloody and unpleasant images. Finally, low valence and low arousal (LL) emotions, which show sadness or boredom, are more related to people in poverty or abandoned places. A fifth type of emotions is also added, the neutral (N), designed not to generate stimuli in the subject. Each block is composed of 8 images of 2 s each.

On the other hand, the participants are 24 college-age subjects who volunteered to participate. All were in good health at the time of the experiment. All were from Albacete, Spain.

2.4 Experimental Design

The E-Prime software was used to create the basis of the experiment. This software is the most widely used in the field of psychology for conducting experiments. E-Prime allows to randomize the blocks of images, synchronize them, add questionnaires between them and many other functions.

The order in which each block is shown to the subjects is completely randomized so that the data are not influenced by the order of appearance, except for the neutral block which always appears first. In addition, between each block of images, a questionnaire is given to the subject in which they have to rate from 1 to 9 how pleasant the images were, i.e. their valence, and how activated they felt when viewing them, i.e. arousal. After this questionnaire, a distraction task will appear. This allows the subject to forget the previous emotion so that this emotion does not influence the next one, thus providing erroneous data. The process followed in the experiment can be seen in Fig. 1.

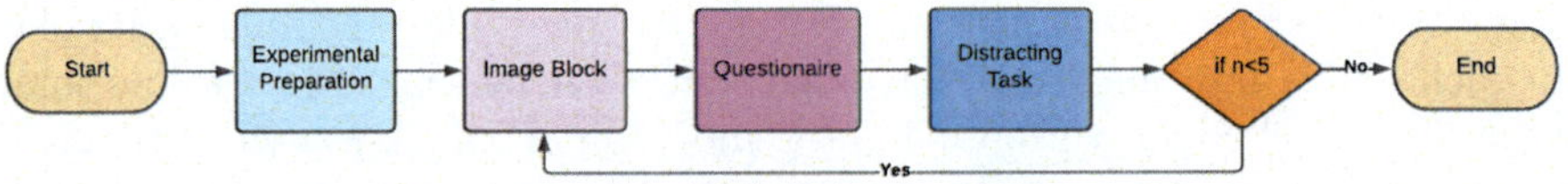

Fig. 1. Flow diagram of the experiment.

As you can see, the main thing is to put on the helmet and prepare the subject for the experiment. Once the experiment starts, the blocks of images will be shown, followed by a questionnaire and a distraction task. Once the 5 blocks of images have been shown, the experiment comes to an end. To obtain the brain activity data it will be necessary to use the Emotiv PRO software at the same time as the experiment is being carried out, thus recording a complete signal of similar duration to that of the experiment.

2.5 Brain Activity Preprocessing

The signal obtained from each subject is raw and corresponds to its brain activity during the whole duration of the experiment. As only the part corresponding to the image blocks is needed, a segmentation of the signal is performed. Using Matlab software and thanks to the fact that E-Prime records the time in which the events occurred, we can synchronize the signal obtained with the beginning and the end of each block of images. Thus, 5 signals are obtained from each subject, one for each type of emotion being studied.

With the raw signals already segmented thanks to Matlab, they are subjected to processing in order to eliminate artifacts and parts of the signal originated as a result of noise. In this case, using EEGLAB and FASTER, two Matlab toolboxes, the signal is filtered. A notch filter 50 Hz is used to eliminate the noise coming from the electrical network, a low pass filter with a cutoff frequency 43 Hz is used to eliminate components higher than this frequency that are not relevant in the analysis of EEG signals and, finally, a high pass filter 4 Hz is used to eliminate the very low frequency components that are also not important in this analysis since they are more related to states of sleep or deep meditation than to the appearance of emotions.

In addition to filtering, the signal is subjected to an independent component analysis (ICA). This is a computational method of signal processing whose purpose is to separate a signal into sub-components that show the value of the signal that really corresponds to what is to be measured and not to external distortions. It is, therefore, a process of elimination of noise and artifacts chosen mainly by the interferences produced by the other electrodes on the sensor that records each signal due to its proximity. In addition, when comparing signals from different volunteers, it must be taken into account that due to multiple personal factors, such as gender, age or origin, the brain activity signals of each individual are different in intensity from those of the rest.

Finally, using Matlab, the characteristics to be analyzed will be extracted from the signals. These will be the mean of the signal values over time and the logarithmic and Shannon entropies [6,7].

The average is calculated

$$\bar{x} = \frac{1}{n} \sum_{i=1}^{n} x_i \tag{1}$$

Shannon entropy corresponds to

$$(X) = -\sum_{i=1}^{n} p(x_i) log_2(p(x_i)) \tag{2}$$

Finally, the logarithmic entropy will correspond to

$$(X) = \sum_{i=1}^{n} log_2(p(x_i)^2) \tag{3}$$

Defining in the last two cases $p(x_i)$ as the probability of the term x_i.

These three values for each of the signals obtained are used for the final analysis. Each volunteer is identified with a number from 1 to 24 and another number from 1 to 5 is assigned to each emotion. That is, "HH" will become emotion 1, "HL" will become emotion 2, "LH" will become emotion 3, "LL" will become emotion 4 and finally the neutral block "NN" will become emotion number 5. The data are imported into the SPSS statistical software. Thus, a database is obtained which indicates the mean and the two entropies of each electrode of each emotion of each patient.

2.6 Analysis

To analyze the data obtained in the previous section, the analysis of variance or ANOVA method will be used. This method is used to decompose the variation of an experiment into independent components to assign these to different causes, which makes it the most commonly used statistical test when the comparison of the means of two or more groups is required. Thus, the ANOVA test with a single factor is usually used when the data are not paired to study whether there are significant differences between the means of a variable (continuous and random) at different levels of another qualitative variable, which is known as a factor. When an analysis of variance is significant, it implies that at least two of all the means that have been compared are significantly different from each other, but it is not stated which ones. To find these different means, the means of all the groups that have been entered in the study must be compared two by two. This analysis is called post-hoc. In this case, the Bonferroni test will be used because it is ideal when there are not many comparisons to be made and the groups are homogeneous in variances. This test is a conservative analysis.

A series of comparisons are made between the different types of emotions in order to see if they are representative. This is done for each of the three previously obtained characteristics, mean, Shannon entropy and logarithmic entropy, measured at each of the helmet electrodes. The 5 emotions are compared with each other and, subsequently, the high arousal emotions with the low arousal emotions and the high valence emotions with the low arousal emotions.

3 Results and Discussion

This section compiles the results obtained during the experiment, divided into three different comparisons. In the first case, high valence emotions are compared with low valence and neutral emotions. In the second, high arousal emotions are compared with low arousal and neutral emotions. Finally, all types of emotions are compared with each other. IBM SPSS Statistics version 23 is used for this analysis. Only cases in which the p-value is less than 0.05 will be considered statistically significant. An example of the case of comparisons between all the emotions is shown in Table 1.

Table 1. Example of multiple comparison.

Dependent variable: AF3Log

(I) Emotion	(J) Emotion	Mean difference (I-J)	Std. error	Sig.	95% Confidence interval	
					Lower bound	Upper bound
1	2	1161.97676	318.0301761	0.004	251.7096918	2072.243820
	3	1190.75466	318.0301761	0.003	280.4875942	2101.021723
	4	1105.21931	318.0301761	0.007	194.9522469	2015.486376
	5	1287.50120	318.0301761	0.001	377.2341403	2197.768269

3.1 High-Low Valence Comparison

In the comparison between the high and low valence block, significant differences in the Log signal are obtained with the electrodes of the left rear part of the brain and the frontal lobe, especially the right part. A comparison will also be made between the signals coming from the high valence imaging block with those coming from the neutral block. In this case, for Log type signals, it is again observed that the outermost electrodes show the most significant differences while the central brain area does not show any. The frontal lobe area predominates above all. No significant difference was found in Sha and medial signals. On the other hand, for the comparisons between low valence feelings and the neutral block no significant differences were found in any of the three signals to be analyzed.

In Fig. 2a the comparison of the Log signal between the high and low valence signals for all the electrodes of the hull is observed, showing the p-value for each electrode.

3.2 High-Low Arousal Comparison

In this case, the comparison between high and low arousal emotions shows significant differences in the left and right extremes of the brain, mainly on the right side of the frontal lobe. This difference is only observed in the Log type signal, while in the Sha and medium signals there is nothing remarkable. On the other hand, when a comparison is made between the high arousal block and the neutral block, significant differences are obtained in the electrodes of the left hemisphere of the brain, especially in the frontal lobe. Finally, it should be noted that no significant difference was observed for any of the three types of signal in the comparison of low arousal signals with neutral emotions.

In Fig. 2b the electrode diagram of the comparison between high and low arousal is observed, showing how the difference in the frontal lobe stands out.

3.3 Multiple Comparison

In this case, all types of emotions are compared with each other, i.e., each of them with the other 4. It is found that most of the relevant results come from the comparison of high arousal and high valence emotions with the rest. First, it will be compared with the NN block. It is observed that in the central area of the brain no very significant differences are found, while in the more distant areas differences appear in almost all the electrodes located in these areas. It should be noted that differences are only found in the Log signals. For the HH-HL comparison, very similar results are obtained to those of the NN comparison, finding even fewer differences in the central electrodes. Subsequently, the HH-LH comparison is performed where, as in the previous cases, the central zone does not show significant differences. It is worth noting that this comparison shows the least differences with the HH block. Finally, when comparing HH and LL we observed somewhat greater differences in the outer areas of the brain than in the central areas.

As for the comparison of the rest of the signals between them, no significant differences are found, so the main source of differences are found in the emotions generated by the HH images. In Fig. 2c the electrode diagram of the HH-LL comparison can be seen.

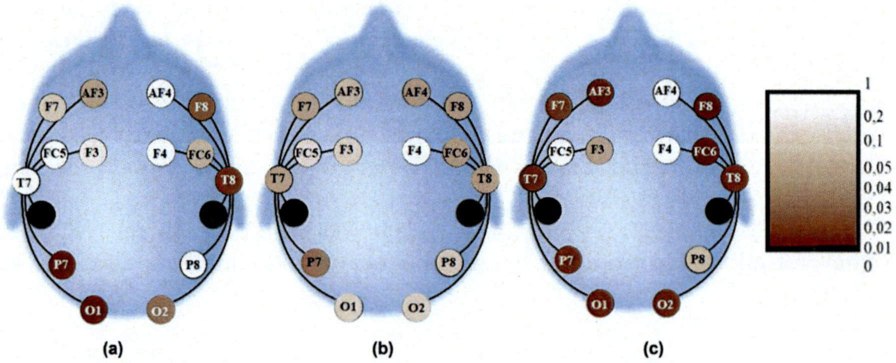

Fig. 2. (a) High-low valence comparison, Log type signal. (b) High-low arousal comparison, Log type signal. (c) HH-LL comparison, Log type signal.

4 Conclusions

This paper describes an experiment on the differences in brain activity of a series of subjects exposed to different types of images that generate an emotion in them. To carry out the experiment, 24 people from Albacete, Spain, volunteered to participate. While the experiment was shown to each of the subjects, their brain activity was monitored and recorded thanks to the Emotiv EPOC X electrode helmet. A prior differentiation is made in 5 types of images according to the arousal and the valence of the emotion they generate. The aim is to analyze the differences in the brain activity of the volunteers depending on the emotion they feel. The analysis consists of comparing the brain activity signals of the different types of emotion, in the search for any significant difference between them. In this case it is observed that there is a difference between emotions that produce a high arousal response in the patient and those that generate neutrality or a low arousal emotion. In the case of the comparisons of high and low valence something similar occurs, there being differences between the high emotion and the low and neutral emotion, but not between the latter two. It is observed that the differences are more significant in the frontal lobe, part of the brain in charge of emotion processing, in the occipital lobe, in charge of the interpretation of complex images, and in the right temporal lobe, where high intensity emotions such as anger or anxiety are also regulated. Within these differences, the most significant are those of the right frontal lobe, which would fit with the theory that it is the frontal lobe and, specifically, the right hemisphere, which is responsible for emotional processing.

Acknowledgments. .5 Grants PID2020-115220RB-C21 and EQC2019-006063-P funded by MCIN/AEI/10.13039/501100011033 and by "ERDF A way to make Europe". Grants BES-2017-081958 and PTA2019-016876-I funded by MCIN/AEI/10.13039/501100011 033 and by "ESF Investing in your future". This work was also partially supported by CIBERSAM of the Instituto de Salud Carlos III.

References

1. Alhagry, S., Fahmy, A.A., El-Khoribi, R.A.: Emotion recognition based on EEG using LSTM recurrent neural network. Emotion **8**(10), 355–358 (2017)
2. Atkinson, J., Campos, D.: Improving BCI-based emotion recognition by combining EEG feature selection and kernel classifiers. Expert Syst. Appl. **47**, 35–41 (2016)
3. Chowdhury, P., Shakim, S.K., Karim, M.R., Rhaman, M.K.: Cognitive efficiency in robot control by emotiv epoc. In: 2014 International Conference on Informatics, Electronics & Vision, pp. 1–6. IEEE (2014)
4. Edla, D.R., Dodia, S., Bablani, A., Kuppili, V.: An efficient deep learning paradigm for deceit identification test on EEG signals. ACM Trans. Manag. Inf. Syst. **12**(3), 1–20 (2021)
5. Fakhruzzaman, M.N., Riksakomara, E., Suryotrisongko, H.: EEG wave identification in human brain with Emotiv EPOC for motor imagery. Procedia Comput. Sci. **72**, 269–276 (2015)

6. García-Martínez, B., Martinez-Rodrigo, A., Alcaraz, R., Fernández-Caballero, A.: A review on nonlinear methods using electroencephalographic recordings for emotion recognition. IEEE Trans. Affect. Comput. **12**(3), 801–820 (2021)
7. García-Martínez, B., Fernández-Caballero, A., Alcaraz, R., Martínez-Rodrigo, A.: Application of dispersion entropy for the detection of emotions with electroencephalographic signals. IEEE Trans. Cogn. Dev. Syst. (2022). https://doi.org/10.1109/TCDS.2021.3099344
8. Hajek, P., Barushka, A., Munk, M.: Neural networks with emotion associations, topic modeling and supervised term weighting for sentiment analysis. Int. J. Neural Syst. **31**(10), 2150013 (2021)
9. Holewa, K., Nawrocka, A.: Emotiv EPOC neuroheadset in brain-computer interface. In: 15th International Carpathian Control Conference, pp. 149–152. IEEE (2014)
10. Kimmatkar, N.V., Babu, B.V.: Novel approach for emotion detection and stabilizing mental state by using machine learning techniques. Computers **10**(3), 37 (2021)
11. Lang, P.J., Bradley, M.M., Cuthbert, B.N.: International affective picture system (IAPS): technical manual and affective ratings. NIMH Center Study Emot. Attention **1**(3), 39–58 (1997)
12. Martínez-Rodrigo, A., García-Martínez, B., Alcaraz, R., González, P., Fernández-Caballero, A.: Multiscale entropy analysis for recognition of visually elicited negative stress from EEG recordings. Int. J. Neural Syst. **29**(2), 1850038 (2019)
13. Martínez-Rodrigo, A., García-Martínez, B., Zunino, L., Alcaraz, R., Fernández-Caballero, A.: Multi-lag analysis of symbolic entropies on EEG recordings for distress recognition. Front. Neuroinf. **13** (2019). https://doi.org/10.3389/fninf.2019.00040
14. Patel, P., Raghunandan, R., Annavarapu, R.N.: EEG-based human emotion recognition using entropy as a feature extraction measure. Brain Inf. **8**(1), 1–13 (2021). https://doi.org/10.1186/s40708-021-00141-5
15. Russell, J.A.: A circumplex model of affect. J. Pers. Soc. Psychol. **39**(6), 1161 (1980)
16. Sánchez-Reolid, R., García, A.S., Vicente-Querol, M.A., Fernández-Aguilar, L., López, M.T., Fernández-Caballero, A., González, P.: Artificial neural networks to assess emotional states from brain-computer interface. Electronics **7**(12), 384 (2018)
17. Sánchez-Reolid, R., Martínez-Rodrigo, A., López, M.T., Fernández-Caballero, A.: Deep support vector machines for the identification of stress condition from electrodermal activity. Int. J. Neural Syst. **30**(7), 2050031 (2020)
18. Sánchez-Reolid, R., de la Rosa, F.L., López, M.T., Fernández-Caballero, A.: One-dimensional convolutional neural networks for low/high arousal classification from electrodermal activity. Biomed. Sign. Proces. Control **71**, 103203 (2022)
19. Stytsenko, K., Jablonskis, E., Prahm, C.: Evaluation of consumer EEG device Emotiv EPOC. In: MEi:CogSci Conference, p. 99. University of Vienna (2011)
20. Torres, E.P., Torres, E.A., Hernández-Álvarez, M., Yoo, S.G.: EEG-based BCI emotion recognition: a survey. Sensors **20**(18), 5083 (2020)
21. Yeşilkaya, B., Güren, O., Bahar, M.T., Turhal, L.N., Akan, A.: Estimation of emotion status using IAPS image data set. In: 28th Signal Processing and Communications Applications Conference, pp. 1–4. IEEE (2020)
22. Zangróniz, R., Martínez-Rodrigo, A., López, M.T., Pastor, J.M., Fernández-Caballero, A.: Estimation of mental distress from photoplethysmography. Appl. Sci. **8**(1), 69 (2018)

Real-Time Algorithm Recommendation Using Meta-Learning

Guilherme Palumbo[1,2], Miguel Guimarães[1], Davide Carneiro[1,2], Paulo Novais[2](✉), and Victor Alves[2]

[1] CIICESI, Escola Superior de Tecnologia e Gestão, Instituto Politécnico do Porto, Porto, Portugal
pg42832@alunos.uminho.pt, 8150520@estg.ipp.pt, dcarneiro@estg.ipp.pt
[2] Centro ALGORITMI, Universidade do Minho, Braga, Portugal
pjon@di.uminho.pt, valves@di.uminho.pt

Abstract. As the field of Machine Learning evolves, the number of available learning algorithms and their parameters continues to grow. On the one hand, this is positive as it allows for the finding of potentially more accurate models. On the other hand, however, it also makes the process of finding the right model more complex, given the number of possible configurations. Traditionally, data scientists rely on trial-and-error or brute force procedures, which are costly, or on their own intuition or expertise, which is hard to acquire. In this paper we propose an approach for algorithm recommendation based on meta-learning. The approach can be used in real-time to predict the best n algorithms (based on a selected performance metric) and their configuration, for a given ML problem. We evaluate it through cross-validation, and by comparing it against an Auto ML approach, in terms of accuracy and time. Results show that the proposed approach recommends algorithms that are similar to those of traditional approaches, in terms of performance, in just a fraction of the time.

Keywords: Machine learning · Meta-learning · Algorithm recommendation

1 Introduction

The process of Machine Learning (ML) has always been very dependent on Human intuition and expertise, especially in certain tasks such as feature engineering, model configuration, or hyper-parameter tuning. An experienced Data Scientist will be able to more efficiently train a satisfactory ML model, based on her/his observation of the input data (e.g. number of rows/columns, type of problem, type of features) and on her/his knowledge of ML algorithms, their strengths and their weaknesses.

V. Julián et al. (Eds.): ISAmI 2022, LNNS 603, pp. 249–258, 2023.
https://doi.org/10.1007/978-3-031-22356-3_24

This valuable intuition is one of the reasons why the Human expert still remains a valuable element in the ML and Data Science (DS) process, in a moment in which automated ML tools are growing and automating the ML process end-to-end [2].

However, there is currently a general shortage of ML and DS experts in face of the tremendous growth of the use of data by organizations. Moreover, the growth of these automated tools may favor the notion (an erroneous one in our opinion) that human expertise is no longer necessary in the ML process, and contribute to the shortage of human experts by demotivating prospective experts to follow such a career.

At the same time, these automated tools, which work mostly through extensive search or brute force, consume considerable resources to test significant number of combinations [4]. This happens for instance in feature engineering (in which automated tools try numerous random combinations of features to find potentially relevant ones) or in model training (in which hundreds of models are trained to find the best one). These approaches are thus very resource and time-intensive.

There are thus two clear constraints: limited experienced human resources, and limited computational resources and time.

In this paper we propose an approach that is inspired in the way human ML experts build their intuition and sensitivity. Specifically, we use meta-learning to gather knowledge about many different ML experiments, and then use it to predict the potentially best models for a new problem. The main goal is to avoid training a large number of models to find the single best one. By using the proposed recommendation system, the user could just train a few of the best predicted models, and choose the best from that group of candidates. This would, on the one hand, decrease the dependence on human expertise and, on the other hand, make the ML process a more efficient one in terms of time and resources.

This paper describes the methodology followed for developing this algorithm recommendation system and the results obtained in validating it and in testing it in three ML problems.

2 Meta-Learning

In recent years, the adoption of Machine Learning (ML) in all types of industries around the world has been increasing exponentially, as it accelerates business processes, reduce costs and serves customers better [5].

However, there are still many open challenges waiting to be solved [3], and one of them can be deciphering which of the numerous algorithms available is the best to solve a given problem.

Meta-learning can be seen as a potential solution to this challenge, and can be used to automate the application and maintenance of a ML system in a company or, at least, help with the adoption of ML in companies that can't afford to have dedicated ML experts. Also, it can be useful to help inexperienced and experienced data scientists alike. Another challenge for data scientists is the vast and continuous increase of data and the need to re-train or update models [1], or suggest a different algorithm to tackle the new data. Meta-learning can help with both problems.

Meta-learning can be generally described as the ability of *learning to learn* [8]. It does this by using meta-data with the aim to learn more about the data itself. It relies on meta-features, that are like hyperparameters (parameters whose values are used to control the learning process in a ML model [6,9]), that in the context of meta-learning characterize the original data source. These meta-features can be mainly grouped into three categories: (i) features that describe the attributes of the original dataset, (ii) relationships between the attributes, and (iii) relationships between the attributes and the target column.

These features can range from very simple measures, such as the number and distribution of classes, to more complex ones, such as statistical information (e.g. mean kurtosis of attributes, mean skewness, etc.), information-theoretic characteristics (e.g. noise signal ratio, class entropy, etc.), and others. This is a growing field, and new meta-features are still being developed and added to these systems that use meta-learning [7].

A dataset that includes meta-features is often called a *meta-dataset*, as in a dataset that describes the features of other data. Conversely, a model trained with such data is deemed a *meta-model*. While meta-learning can be used for very different purposes, in the case of this work it is used to predict the performance metrics of specific algorithms/configurations for a given ML problem, as a ranking problem.

Specifically, the meta-dataset was built with data from many ML problems, in which different algorithms/configurations were used. From these ML problems, two groups of features were extracted: the former containing the meta-features of each ML problem, and the latter containing the performance metrics of the corresponding model. Based on this, a meta-model is trained that is able to predict the performance of a given algorithm/configuration when applied to a specific ML problem. This meta-model is then used in a ranking application, that suggests to the user the best predicted models for their ML problem.

3 Methodology

In order to present a new way for data scientists to find the best ML algorithm, instead of focusing on trial-and-error or brute force procedures, it was developed an approach for algorithm recommendation based on meta-learning. This approach is capable of predicting the best n algorithms and its configurations for a given ML problem, and its ranking is based on a selected performance metric.

The first step of this approach, described in Fig. 1, consists in doing some data preparation, namely removing the outliers from some datasets, since it was concluded after some experiments that the outliers were not relevant and would adulterate the final results. To develop the meta-model currently in use, a total of 40 datasets from different ML problems were used, together with 49 different configurations of 4 algorithms.

For each dataset, the meta-features are extracted. Then, 49 different models are trained (for each dataset), containing all the considered configurations of the 4 algorithms used. Next, the meta-features are combined with the performance metrics of each model trained and the training time. Thus, each line in the meta-dataset represents a ML experiment, performed on a dataset with given characteristics (meta-features), using a specific algorithm/configuration, which resulted on a model with the given performance metrics.

The meta-dataset is then used to build the meta-model, a model that learns from the given meta-features. The purpose of the meta-dataset is to predict a performance metric (e.g. RMSE) given the characteristics of a ML problem. Thus, the developed meta-dataset can be used to train different meta-models: one for predicting each error metric considered, and another for predicting training time.

In order to decide on the algorithm for the meta-model and its best configuration, an extensive grid-search was performed encompassing several state of the art algorithms. The performance of the meta-model was evaluated using 5-fold cross-validation.

The best model found by this process was a *H2OGradientBoostingEstimator* with a $score\_tree\_interval = 5$, $ntrees = 99$, $max\_depth = 17$, $min\_rows = 5$, $nbins = 20$ and $nbins\_top\_level = 1024$. The RMSE of the resulting meta-model was 0.045 and the R^2 was 0.953.

Each of the meta-models trained can thus be used to predict a specific performance metric, given the characteristics of a ML problem and an algorithm/configuration. A service was then developed that, given a ML problem, returns to the user the ranking of the best n algorithms/configurations. To this end, this service uses the meta-model to make predictions for each algorithm/configuration considered, and it then sorts the results by decreasing error metric, and returns them to the user.

The rationale is that, instead of using brute force or extensive search approaches, which take a considerable time and are computationally expensive, the user can only train a small set of models: those suggested by the ranking service as being more promising.

4 Results

This section describes the results obtained so far. While multiple meta-models were developed, one for each performance metric, in this paper we analyze the results based on the RMSE. The meta-dataset used is composed of 1960 rows, resulting from the combination of 40 datasets and 49 algorithm configurations.

As previously mentioned, the meta-model was validated during training using 5-fold cross validation. The RMSE was 0.045 and the R^2 was 0.953.

Then, the meta-model was used on 3 ML problems whose datasets were not used for the training of the meta-model. In this case, the datasets for this ML problem can be seen as the hold-out set. These datasets went through the same process previously depicted in Fig. 1 for meta-feature extraction. The quality of

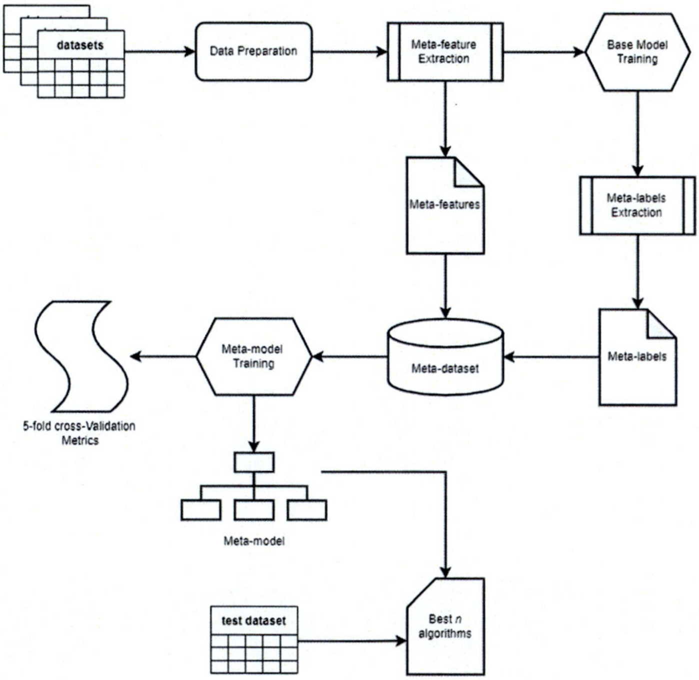

Fig. 1. Methodology followed to validate the propose approach for algorithm recommendation.

the meta-model was then assessed through the difference between the RMSE predicted by the meta-model, and the RMSE observed when the actual model is trained. If the RMSE predicted is relatively close to the observed one it means the meta-model is capable of accurately predicting its RMSE meaning it's a good meta-model and can be used for the algorithm recommendation system.

Given that the RMSE depends on the order of magnitude of the dependent variable, it may vary significantly between applications. I.e., it is problem-dependent. For this reason, the inputs are normalized before being fed to the meta-model, and de-normalized again after the prediction, to be compared with the actual values.

The results of this analysis are summarized in Fig. 2, in which each dot represents the RMSE predicted by the meta-model for a given algorithm/configuration versus the observed RMSE. The Figure shows that different algorithms are best suited for different ML problems: for the *2016.csv* dataset the best algorithm is the *H2ODeepLearningEstimator* but for the *auto_ mpg_ dataset.csv* dataset both *H2ODeepLearningEstimator* and *H2ORandomForestEstimator* present similar predictions compared with the true observed values of RMSE.

Once the meta-model was validated and its performance was deemed adequate, the next step was to implement a recommendation system capable of predicting the best n algorithms for a given ML problem, as previously described.

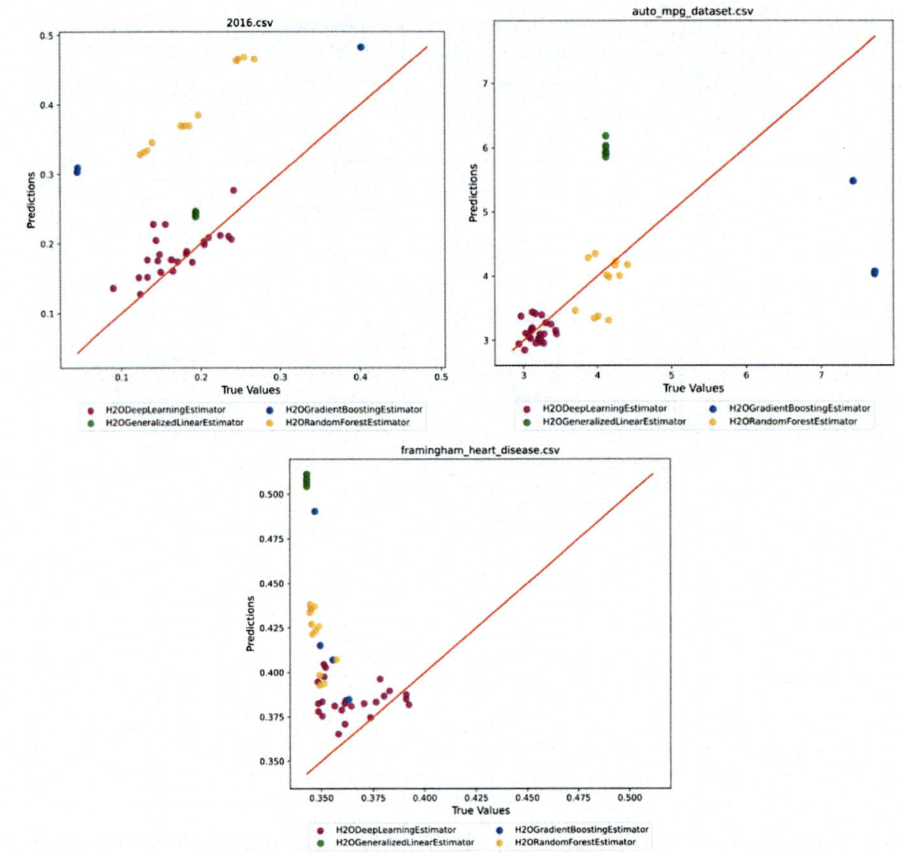

Fig. 2. Comparison between predicted RMSE versus observed RMSE for the three different ML test problems.

The Recommendation System was validated by comparing it with an exhaustive approach (H_2O's AutoML) and comparing both Top-N results in terms of the algorithms used, their configurations and the corresponding performance metrics. It should be noted that the use of the Recommendation System does not imply the training of any model, it simply suggests the potentially better algorithms to use. The user can then decided whether to train one or several models. In that sense, it is a significantly faster approach that using brute-force approaches such as AutoML, which imply the training of dozens or hundreds of models.

The two criteria used for comparison were the RMSE and the elapsed time of the process. Given that stopping criteria had to be defined for the AutoML process, the decision was to stop it once 49 models had been trained. While other criteria could have been used, we did not resort to performance or time-based criteria, since this could eventually be used to turn results in our favor (e.g. by giving it a short amount of time to train). Allowing it to train the same number of models that we consider seemed to be the fairest approach. The AutoML process also included a stopping metric for each model trained: the training process stopped as soon as the deviance (error metric) between iterations didn't change significantly (with a tolerance of 0.001). This means that if the model is not improving any further, the AutoML will stop the training process early.

Tables 1, 2 and 3 detail, for the 3 test ML problems, the predicted RMSE of the Top-5 given by the Recommendation System, the RMSE of these models when actually trained (observed) and the RMSE of the 5 best models found by the AutoML approach.

Table 1. Comparison between the RMSE predicted by the recommendation system (for a Top-5 scenario), the actual RMSE when training the proposed models (observed), and the 5 best models obtained by the AutoML for the *2016.csv* dataset.

2016.csv					
	Recommended		Observed	AutoML	
Top 5	Algorithm	RMSE	RMSE	Algorithm	RMSE
1	H2ODeepLearningEstimator	0.138	0.124	H2OGeneralizedLinearEstimator	0.009
2	H2ODeepLearningEstimator	0.155	0.090	H2OGeneralizedLinearEstimator	0.027
3	H2ODeepLearningEstimator	0.158	0.122	H2OGradientBoostingEstimator	0.040
4	H2ODeepLearningEstimator	0.163	0.160	H2ODeepLearningEstimator	0.050
5	H2ODeepLearningEstimator	0.168	0.150	H2ODeepLearningEstimator	0.051

For the *2016.csv* dataset (Table 1), it can be noted that the values of RMSE predicted by the Recommendation System are very close to the observed ones, with an average distance of 0.027. Additionally, comparing both approaches, the recommended results have two algorithms in common with the AutoML. The algorithms found by the AutoML process however achieve a lower RMSE.

More specifically, the top algorithm recommended by the proposed system was a *H2ODeepLearningEstimator* with $epochs = 100$, $rectifier = activation\ function$, and $hiddenlayersize = 200, 200$. The best model found by the AutoML was a *H2OGeneralizedLinearEstimator* with $max\_iterations = 300$ and $max\_active\_\ preditors = 5000$.

In the case of the *auto_ mpg_ dataset.csv* ML problem (Table 2), the AutoML presented a better performance for the first two models but the last three where beat by the recommended models. The Recommendation System proposed 4 algorithms in common with AutoML (out of five). The average distance between the predicted and observed RMSE is 0.117.

The best algorithm recommended by the proposed solution was a *H2ODeep LearningEstimator* with $epochs = 100$, $rectifier = activationfunction$, $hidden\ layersize = 200, 200$. The best algorithm found by the AutoML process was the *H2ORandomForestEstimator* with $ntrees = 50$.

Table 2. Comparison between the RMSE predicted by the recommendation system (for a Top-5 scenario), the actual RMSE when training the proposed models (observed), and the 5 best models obtained by the AutoML for the *auto_mpg_dataset.csv* dataset.

auto_mpg_dataset.csv					
	Recommended		Observed	AutoML	
Top 5	Algorithm	RMSE	RMSE	Algorithm	RMSE
1	H2ODeepLearningEstimator	3.049	3.013	H2ORandomForestEstimator	2.932
2	H2ODeepLearningEstimator	3.067	3.167	H2ODeepLearningEstimator	3.043
3	H2ODeepLearningEstimator	3.079	2.935	H2ODeepLearningEstimator	3.100
4	H2ODeepLearningEstimator	3.101	3.274	H2ODeepLearningEstimator	3.141
5	H2ODeepLearningEstimator	3.101	3.232	H2ODeepLearningEstimator	3.208

Table 3. Comparison between the RMSE predicted by the recommendation system (for a Top-5 scenario), the actual RMSE when training the proposed models (observed), and the 5 best models obtained by the AutoML for the *world_food_scrubbed.csv* dataset.

world_food_scrubbed.csv					
	Recommended		Observed	AutoML	
Top 5	Algorithm	RMSE	RMSE	Algorithm	RMSE
1	H2ODeepLearningEstimator	0.161	0.035	H2OGradientBoostingEstimator	0.001
2	H2ODeepLearningEstimator	0.163	0.036	H2OGradientBoostingEstimator	0.009
3	H2ODeepLearningEstimator	0.165	0.029	H2OGradientBoostingEstimator	0.010
4	H2ODeepLearningEstimator	0.167	0.054	H2ODeepLearningEstimator	0.130
5	H2ODeepLearningEstimator	0.167	0.037	H2ODeepLearningEstimator	0.170

Lastly, in the *world_food_scrubbed.csv* ML problem, shown in Table 3, the recommender system presents a slightly higher RMSE than the AutoML, but recommending 4 out of 5 of the same algorithms as the AutoML. However, only the first three models found by AutoML beat those proposed by our approach. This happens despite the fact that the predicted and observed results are a bit apart for the recommended algorithms.

The best algorithm recommended by the proposed solution was a *H2ODeep LearningEstimator* with $epochs = 200$, $rectifier = activationfunction$, and $hidden\_layer\_size = 100, 100, 100, 100$, while the best algorithm found by AutoML was the *H2OGradientBoostingEstimator* with $ntrees = 157$, $maxdepth = 13$, and $minrows = 5$.

The performance of the proposed system was also evaluated time-wise. Table 4 compares the elapsed time of the Recommendation System and of the AutoML process. The elapsed time of the Recommendation System includes: 1) the time for extracting the meta-features of the test dataset (which are the input for the meta-model), which may take a significant amount of time, depending on the number of features and size of the dataset; and 2) the actual time of predicting the RMSE for all the configurations considered.

Time-wise, there is a significant difference between both approaches. Moreover, further analysing shows that, despite the fact that the RMSE results of each ML problem are not perfect as previously discussed, looking resource wise, the proposed solution presented excellent results due to the fact the elapsed time of each Top 5 recommended by the Recommendation System was significantly lower than the AutoML approach. In average the Recommendation System is 155x faster to return a recommendation given a Top-N scenario compared to the AutoML. The user would then still have to train the desired number of models, but the elapsed time would always be inferior.

Table 4. Elapsed time of the recommendation system for a Top-5 scenario (meta-feature extraction + prediction) and the AutoML process.

Dataset	Recommendation time (s)	AutoML time (s)
2016.csv	5	849
auto_mpg_dataset.csv	27	1308
world_food_scrubbed.csv	5	1240

Overall, these results show that although the proposed solution isn't able to find the best algorithms in terms of performance, the RMSE of the proposed algorithms isn't far away from the models found by the AutoML process. Thus, there is a trade-off that may be interesting, especially in scenarios of BigData: a slightly smaller performance is traded for a much smaller running time (and needed computational resources). Specifically, the proposed solution is in average 155x faster. This means that the proposed approach can successfully recommend algorithms that are similar to those of traditional approaches, in terms of performance, but in just a fraction of the time of said traditional approaches.

5 Discussion and Conclusions

Despite the advances in the last years in the field of Machine Learning, especially in what concerns algorithms and computational resources, there are still significant challenges to overcome. Nowadays, these stem mostly from the volume of the data, and the velocity at which it changes. Simultaneously, there is an increasing shortage of human experts who are usually able to, based on their experience and intuition, minimize some of these challenges.

While new automated ML tools are being quickly adopted by organizations, these usually require much more computational resources than more traditional human-centered approaches, as they tend to rely on brute force or extensive search strategies.

In this paper we proposed an approach to address these challenges, based on meta-learning. We took inspiration from the way human ML experts gain their intuition and sensibility: by training a large number of models, observing the characteristics of the dataset and understanding how these interact with the algorithm configuration to produce better or worse models.

We realized this in the form of a recommender service that, given a new ML problem, will suggest the best n algorithms. The user can then direct her/his search towards this group of candidate algorithms instead of searching the whole search space. We show that it is possible to do this very satisfactorily in what concerns the error metric of the models, and in a much faster time than automated approaches such as AutoML: 155x faster according to the criteria adopted in our methodology.

While the meta-model will be further improved with more models and datasets, these results show that this is a valid and promising path to follow towards the development of more efficient ML pipelines.

Acknowledgments. This work has been supported by national funds through FCT - Fundação para a Ciência e Tecnologia through projects UIDB/04728/2020 and EXPL/CCI-COM/0706/2021.

References

1. Carneiro, D., Guimaraes, M., Carvalho, M., Novais, P.: Using meta-learning to predict performance metrics in machine learning problems. Expert Syst. (2021)
2. Duros, V., Grizou, J., Sharma, A., Mehr, S.H.M., Bubliauskas, A., Frei, P., Miras, H.N., Cronin, L.: Intuition-enabled machine learning beats the competition when joint human-robot teams perform inorganic chemical experiments. J. Chem. Inf. Model. **59**(6), 2664–2671 (2019)
3. Elshawi, R., Maher, M., Sakr, S.: Automated machine learning: state-of-the-art and open challenges. arXiv preprint arXiv:1906.02287 (2019)
4. He, X., Zhao, K., Chu, X.: Automl: a survey of the state-of-the-art. Knowl.-Based Syst. **212**, 106622 (2021)
5. Lee, I., Shin, Y.J.: Machine learning for enterprises: applications, algorithm selection, and challenges. Bus. Horiz. **63**(2), 157–170 (2020)
6. Probst, P., Boulesteix, A.L., Bischl, B.: Tunability: importance of hyperparameters of machine learning algorithms. J. Mach. Learn. Res. **20**(1), 1934–1965 (2019)
7. Rivolli, A., Garcia, L.P., Soares, C., Vanschoren, J., de Carvalho, A.C.: Meta-features for meta-learning. Knowl.-Based Syst. 108101 (2022)
8. Vanschoren, J.: Meta-learning: a survey. arXiv preprint arXiv:1810.03548 (2018)
9. Weerts, H.J., Mueller, A.C., Vanschoren, J.: Importance of tuning hyperparameters of machine learning algorithms. arXiv preprint arXiv:2007.07588 (2020)

Author Index

V. Julián et al. (Eds.): ISAmI 2022, LNNS 603, pp. 259–260, 2023.
https://doi.org/10.1007/978-3-031-22356-3